Chemistry of Biomolecules

By the Same Author

- ❒ Practical Organic Chemistry: Qualitative Analysis
- ❒ Organic Reactions with Mechanisms

Ane's Active Chemistry Series:

- ❒ Organic Polymers
- ❒ Amino Acids, Peptides and Proteins
- ❒ Carbohydrates

Chemistry of Biomolecules

Dr. S. P. Bhutani

Formerly Associate Professor
Department of Chemistry, Rajdhani College
University of Delhi

Ane Books Pvt. Ltd.

New Delhi ♦ Chennai

Chemistry of Biomolecules

S.P. Bhutani

First Edition : 2010

Reprint : 2015, 2016, 2017, 2018, 2019, **2020, 2023**

Published by

Ane Books Pvt. Ltd.

4821, Parwana Bhawan, 1st Floor, 24 Ansari Road, Darya Ganj,
New Delhi - 110 002, Tel.: +91(011) 23276843-44, Fax: +91(011) 23276863
e-mail: kapoor@anebooks.com, Website: www.anebooks.com

Branch Office:

- Avantika Niwas, 1st Floor, 19 Doraiswamy Road, T. Nagar,
  Chennai - 600 017, Tel.: +91(044) 28141554, 28141209
  e-mail: anebookschennai@gmail.com, rathinam@anebooks.com

ISBN : 978-81-9084-065-1

Printed at : Repro Knowledgecast Limited, Thane

This book is dedicated to my wife Komal and son Nikhil for their love, understanding and emotional support.

Preface

We know that organic chemistry is the study of compounds of carbon and hydrogen and their derivatives. It deals with a huge variety of compounds both synthetic and natural that influence our lives directly or indirectly. Organic chemistry also deals with the molecules of life called 'Biomolecules' which are very important for all of us.

Carbon, hydrogen and oxygen combine to make up many different kinds of biomolecules, such as carbohydrates, oils and fats and fatty acids. The addition of nitrogen as well as sulphur makes the amino acids that combine to form polypeptides and proteins. Further, addition of phosphorus provides the ingredients for making DNA, RNA and complex lipids. Carbohydrates, proteins and lipids are important constituents of our daily diet. The enzymes which regulate the various metabolic reactions in living system are all proteinous in nature. Thus, all known forms of life are comprised of biomolecules.

Biomolecules are those molecules that are involved in the maintenance and metabolic processes of living organisms. They can also be called organic compounds that naturally occur in living organisms. The study of these biomolecules becomes the subject-matter of organic chemistry. The reactions of biomolecules can be described by the methods of organic chemistry, which is based on the reactions of their functional groups. Important classes of biomolecules have characteristics functional groups that determine their reactions. Therefore, it is very important that students should have an understanding of the basic reactions and fundamentals of organic chemistry before starting the study of biomolecules.

The subject matter of the present book has been divided into six chapters. Chapter 1 deals with the chemistry of carbohydrates. It gives a detailed account of the structures of glucose, fructose, disaccharides-maltose, lactose, cellobiose and sucrose, and polysaccharides-cellulose and starch. Chapter 2 provides a detailed study of amino acids, peptides and proteins. It describes chemical reactions, synthesis of amino acids, various methods of synthesizing peptides besides giving methods for determining the primary structure of peptides and proteins. Chapter 3 is on enzymes, alcoholic fermentation, glycolysis and Krebs cycle. Chapter 4 deals with the study of pyrimidines, purines and nucleic acids-DNA and RNA. Chapter 5 covers the study of some important monoterpenoids like citral, dipentene, α-terpineol, α-pinene and camphor. Chapter 6 is on lipids-their classifcation and biological functions. The detailed study of oils and fats is also described in this chapter.

The topics of the present book are included in the syllabi of graduate students of most Indian universities. But all these topics are not available at one place in the present textbooks. For example, topics like enzymes, lipids, nucleic acids etc are not included in most organic chemistry books. Therefore, students have to consult biochemistry books for these topics. My experience of teaching these topics to graduate students at Delhi University for more than 37 years has helped me to write the present book in a logical manner. I have tried my best to fulfill the requirements of the students. The book will not only be useful to 3rd year 'Chemistry Honours' students of Delhi University but it will also help graduate students of most Indian Universities.

I express my sincere thanks to my students and teachers of various colleges of Delhi University for encouraging me to write this book.

I also express my thanks to Mr. Sunil Saxena and Mr. Jai Raj Kapoor of Ane Books Pvt. Ltd. for the help and co-operation extended to me for the preparation of this book.

I invite constructive comments and valued suggestions from the students and learned teachers for improving the book in subsequent editions.

S.P. Bhutani

Contents

CHAPTER 1

CARBOHYDRATES

Learning Objectives

In this chapter, we will study:

- Carbohydrates—their occurrence in nature and definitions
- Formation of carbohydrates
- Classification of carbohydrates-monosaccharides, oligosaccharides and polysaccharides
- Configurational relationship of monosaccharides -aldotetroses, aldopentoses and aldohexoses
- Structure and properties of D-glucose-open-chain structure, cyclic structure and conformations of D-glucose, configuration of D-glucose and Haworth projections
- Mutarotation and its mechanism
- Lengthening and shortening of carbon-chain of aldoses
- Methods for determining the ring size of glucose
- Osazone formation and its mechanism
- Structure of D-fructose
- Interconversions
- Structure of various disaccharides-maltose, lactose cellobiose and sucrose
- Deoxy sugars- 2-deoxy-D-ribose and D-ribose
- Polysaccharides-structure of cellulose and starch

1.1 INTRODUCTION

Carbohydrates are one of the essential food ingredients, which we all require. Everyone of us is quite familiar with the words glucose and cane sugar. They are simple carbohydrates. We consume carbohydrates in one form or the other every time we take our meals, whether it is breakfast, lunch or dinner. They are the primary source of energy in our body and constitute an important part of any well-balanced diet. We encounter carbohydrates at every turn of our lives. The paper we use for writing, the cotton clothes we wear and the wooden furniture around us are all made of cellulose. The bread we eat, rice, potatoes, peas, etc., all contain starch. The sweetening agents in fruits are nothing but simple carbohydrates.

1.2 OCCURRENCE AND DEFINITIONS

Carbohydrates are the most widespread organic compounds occurring in nature. They are present in almost all the plants and comprise about 80 per cent of the dry weight. The most important are cellulose—the chief structural material of plants, starches and the sugars—sucrose and glucose. In higher animals and human beings, glucose is an essential constituent of blood and takes part in various metabolic reactions. Carbohydrates are also present in biologically important compounds like adenosine triphosphate and nucleic acids.

Originally, the name carbohydrate was given to all such compounds having the general formula $C_x(H_2O)_y$ *i.e.*, they were considered to be hydrates of carbon. Now, we know that carbohydrates are not simply hydrates of carbon but have a variety of other structural features. They are usually defined as polyhydroxy aldehydes and ketones or substances that hydrolyse to yield polyhydroxy aldehydes and ketones. We shall see later that this definition is not entirely satisfactory as they exist primarily as hemiacetals and acetals or as hemiketals and ketals.

Polyhydroxy aldehyes are also called aldoses and polyhydroxy ketones are called ketoses. Here *ald-* is taken for aldehydes and *ket-* for ketones and *-ose* is the common suffix used for these.

Low molecular weight carbohydrates are also called as sugars or saccharides, *e.g.*, glucose and sucrose are simple sugars. More common sugars like glucose and fructose, and some less common sugars also occur in the combined state with various hydroxy compounds. Such derivatives are called glycosides and the non-sugar component is called 'aglycone'. When we have glucose as the sugar component, the compound is called glucoside, if fructose is present, it is a fructoside and so on.

Glycosides are widely distributed in plants and animals. Structurally, these compounds are related to simple methyl glucosides. When the sugar moiety forms an ether linkage with the aglycone, it is called *O*-glycoside and when there is a formation of a C–C bond, it is called *C*-glycoside.

1.3 FORMATION OF CARBOHYDRATES IN PLANTS

Carbohydrates are produced in green plants as a result of photosynthesis. It involves chemical combination or fixation of carbon dioxide and water by utilisation of light energy. In this process, water is oxidised to oxygen and carbon dioxide is reduced. The overall equation for photosynthesis can be written as follows:

$$x\,CO_2 + y\,H_2O + \text{solar energy} \rightarrow \underset{\text{Carbohydrate}}{C_x(H_2O)_y} + x\,O_2$$

Photosynthesis involves many individual enzyme-catalysed reactions, but all are not fully understood.

However, we know that photosynthesis begins with the absorption of light by the green plant pigment, chlorophyll. The chlorophyll has the ability to absorb sunlight in the visible region. This is due to the extended conjugated system present in chlorophyll. As photons of sunlight are trapped by the chlorophyll, energy becomes available to the plant in a chemical form that can be used to carry out series of reactions that reduce carbon dioxide to carbohydrates and oxidise water to oxygen. The first reduction product of carbon dioxide is not known definitely but seems to be closely related to D-glyceric acid (1).

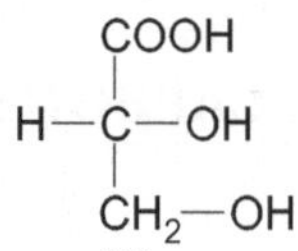

(1)

D-Glyceric acid

The plant carries out a series of enzyme-catalysed reactions which result in the synthesis of simple sugars like glucose and more complex ones like starch and cellulose. Glucose is the most imporant key product in the photosynthesis. It is usually stored in the plants in the form of starch to serve as food energy or to support the framework of the plant and is present as cellulose (Fig. 1.1).

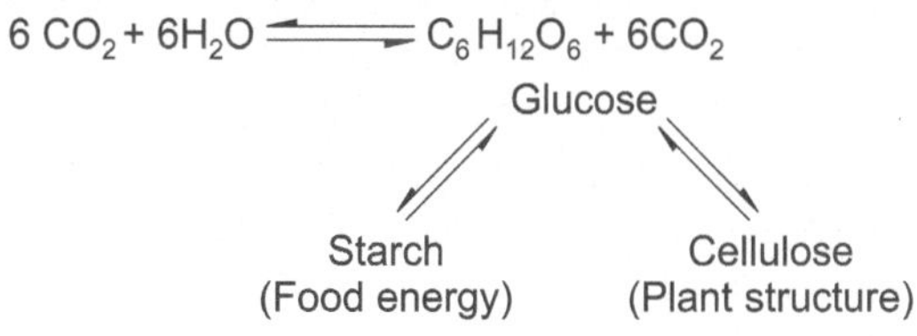

Fig. 1.1

Thus, in glucose nature stores the energy utilised in the photosynthetic process. This serves as the main source of energy in the living organisms. When we consume starch it is converted into glucose in our body and provides us energy. The remaining glucose in the body gets converted into animal starch, known as glycogen, which is stored in the muscle tissues. When we require energy at times when we are on fast, the glycogen is broken down once again into glucose molecules.

Besides providing energy, a major metabolic product of glucose is acetyl Coenzyme A, which acts as a starting point for the synthesis of fats, fatty acids, amino acids and a large number of other important compounds required by our body (Fig. 1.2).

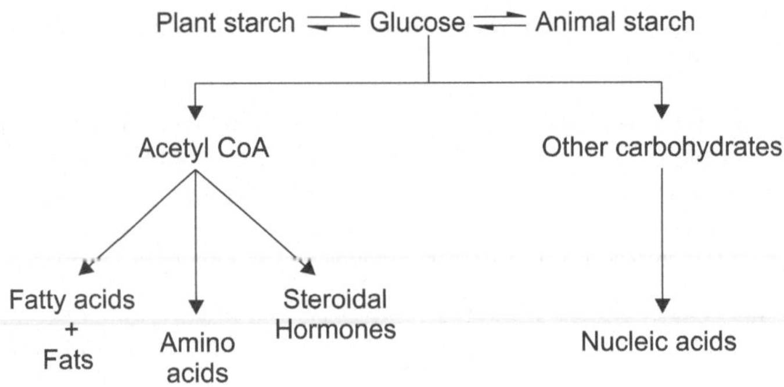

Fig. 1.2

1.4 CLASSIFICATION OF CARBOHYDRATES

Carbohydrates are classified on the basis of number of aldose and/or ketose units produced upon hydrolysis (summarised in Table 1.1).

1. Monosaccharides

The carbohydrates, which cannot be further hydrolysed into smaller simpler units, are called monosaccharides. Examples of aldose monosaccharides are arabinose, glucose and mannose. The most common ketose is fructose.

Monosaccharides are further characterised in terms of the number of carbon atoms present in an aldose or a ketose. Thus, pentoses and hexoses contain five and six carbons respectively. For example, glucose is a hexose but it is completely characterised as an aldohexose, which means it is a six carbon sugar containing an aldehyde group. Similarly, fructose is a ketohexose and arabinose an aldopentose. In aldoses, the aldehyde group is at the top of the chain and the aldehyde carbon is considered as carbon no. 1.

All naturally occurring ketoses have the keto group on the second carbon in the chain.

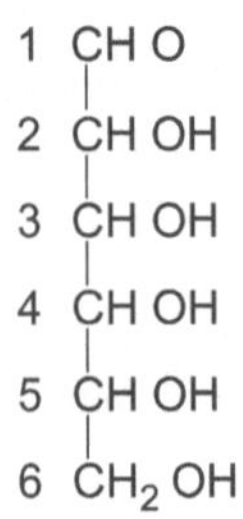

An aldohexose

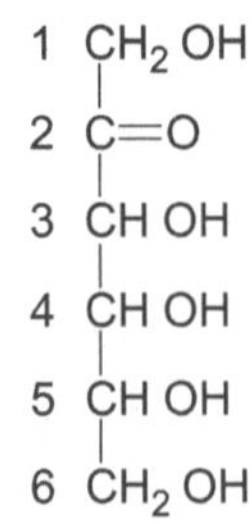

A ketohexose

The simplest sugars have three carbon atoms and are called trioses. The simplest aldose is glyceraldehyde, an aldotriose and the simplest ketose is dihydroxy acetone, a ketotriose.

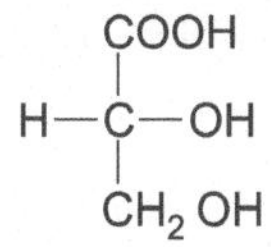

Glyceraldehyde

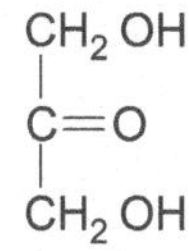

Dihydroxy acetone

Addition of an extra carbon atom to a triose gives successively a tetrose, a pentose, a hexose and so on. Most of the naturally occurring monosaccharides are either pentoses or hexoses. Trioses and tetroses do not occur in nature.

The most abundant pentoses are L-arabinose, D-ribose, 2-deoxy-D-ribose, and D-xylose, which are all aldopentoses (Table 1.2). The common hexoses are D-glucose, D-fructose, D-mannose and D-galactose. D-Fructose is a ketohexose whereas all others are aldohexoses (Table 1.3).

2. Oligosaccharides

These are low molecular weight condensation polymers containing 2-9 monosaccharide units. More often the monosaccharide units are hexoses.

Oligosaccharides can be further classified as disaccharides, trisaccharides, tetrasaccharides; etc., depending upon the number of monosaccharide units obtained on hydrolysis.

A. Disaccharides

A sugar that yields two monosaccharide units on hydrolysis is known as a disaccharide.

For example, sucrose, the common table sugar is a disaccharide because on hydrolysis it produces one mole of glucose and one mole of fructose. The disaccharides maltose and cellobiose, both produce two moles of glucose on hydrolysis. They differ in the linkage between the two units. Lactose has glucose and galactose as the two monosaccharide units.

$$\text{Sucrose} \xrightarrow{H_2O} \text{Glucose} + \text{Fructose}$$

$$\text{Maltose} \xrightarrow{H_2O} \text{Glucose} + \text{Glucose}$$

$$\text{Cellobiose} \xrightarrow{H_2O} \text{Glucose} + \text{Glucose}$$

$$\text{Lactose} \xrightarrow{H_2O} \text{Glucose} + \text{Galactose}$$

B. Trisaccharides

A trisaccharide yields three monosaccharide units on hydrolysis. Raffinose, is an important example of a trisaccharide, which on hydrolysis gives one mole each of glucose, fructose and galactose.

$$\text{Raffinose} \xrightarrow{H_2O} \text{Glucose} + \text{Fructose} + \text{Galactose}$$

3. Polysaccharides

Polysaccharides are generally high molecular weight polymers built up by repeated condensation of monosaccharide units. A polysaccharide on complete hydrolysis should furnish ten or more monosaccharide units. However, the natural polysaccharides consist of 100-3000 sub-units.

Polysaccharides are present both in plants and animals. Common examples of plant polysaccharides are cellulose, starches and the various gums. Glycogen is a polysaccharide present in animal tissues.

TABLE 1.1. Classification of Carbohydrates (Some Common Examples)

- Monosaccharides
  - Pentoses ($C_5H_{10}O_5$)
    - L-Arabinose
    - D-Ribose
    - 2-Deoxy-D-ribose
    - D-Xylose
  - Hexoses ($C_6H_{12}O_6$)
    - D- Glucose
    - D- Fructose
    - D- Mannose
    - D- Galactose
- Oligosaccharides
  - Disaccharides ($C_{12}H_{22}O_{11}$)
    - Sucrose (Glucose + Fructose)
    - Maltose (Glucose + Glucose)
    - Lactose (Glucose + Galactose)
  - Trisaccharides ($C_{18}H_{32}O_{16}$)
    - Raffinose (Glucose + Fructose + Galactose)
- Polysaccharides
  - Plants
    - Cellulose
    - Starch
    - Gums and Agar-Agar
  - Animals
    - Glycogen

TABLE 1.2 Some Common Aldopentoses, their Occurrence and Structure

Occurrence of Pentose	*Open Chain Structure*	*Cyclic Structure*
L-ARABINOSE It is of wide occurrence in plants. It is usually found combined with other monosaccharides in gums and as glycosides.	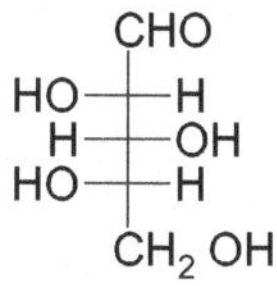	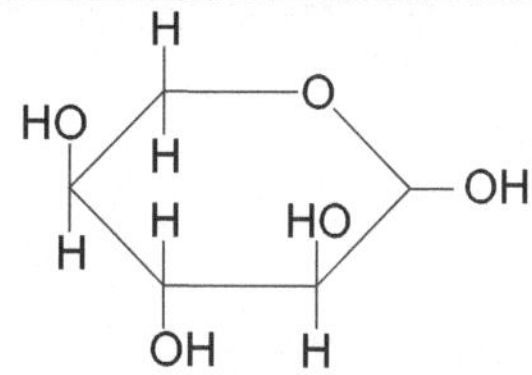
D-RIBOSE It is a component of the nucleic acid (RNA) and several vitamins and coenzymes. It is also a component of antibiotics Neomycin A and B.	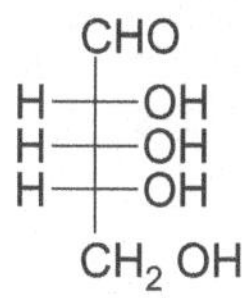	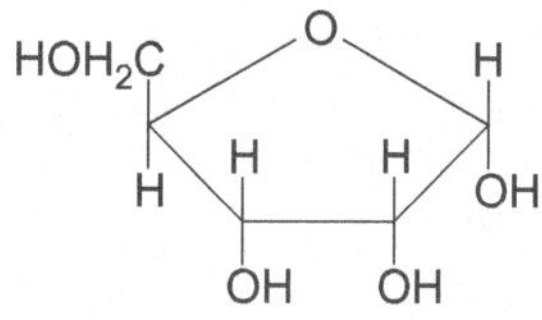 α-anomer
2-DEOXY-D-RIBOSE It occurs in cell nuclei as a component of deoxyribonucleic acid (DNA).	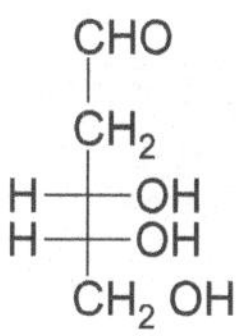	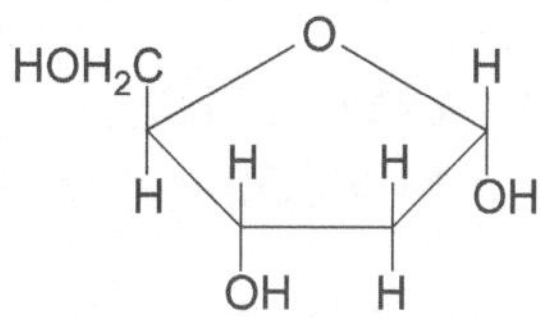 α-anomer
D-XYLOSE It is a component of the polysaccharides present in woody materials and of plant gums.	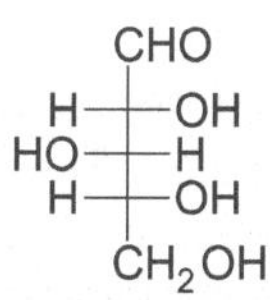	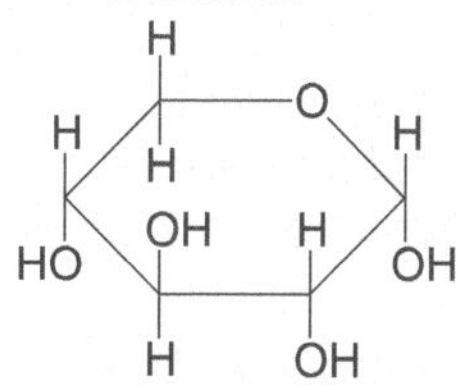 α-anomer

TABLE 1.3 Some Common Hexoses

Hexose	*Open Chain Structure*	*Cyclic Structure*
D-GLUCOSE It is the most abundant of all the hexoses. It occurs in the free state in many fruits, honey and plants, also in the combined state as glycosides. It is the basic unit of many polysaccharides like starch and cellulose. The most important disaccharide sucrose contains glucose as one unit.	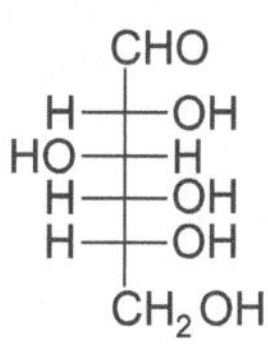	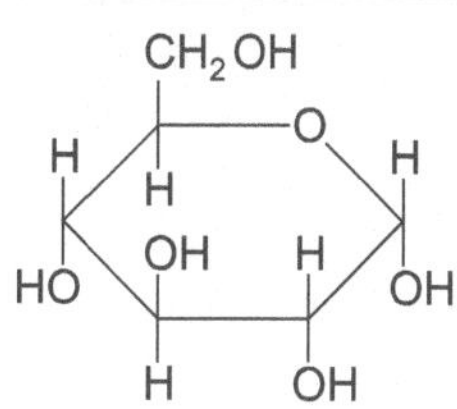 α-anomer
D-FRUCTOSE It occurs free in fruit juices and honey and in the combined state in sucrose and polysaccharides.	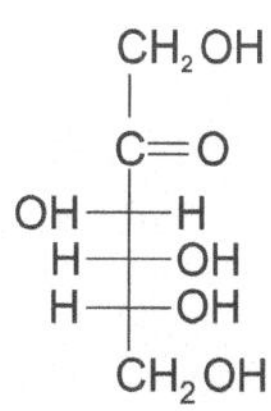	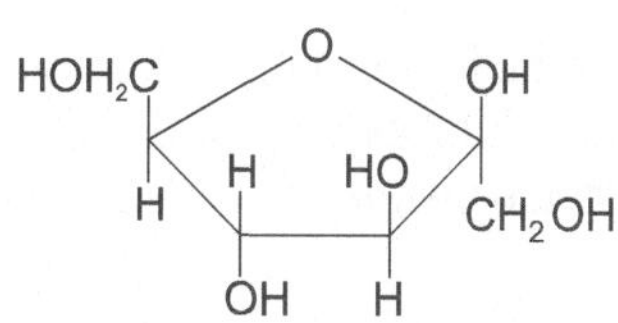 β-anomer

D-MANNOSE

It is reported to occur free in the peel of oranges and is a constituent of many polysaccharides.

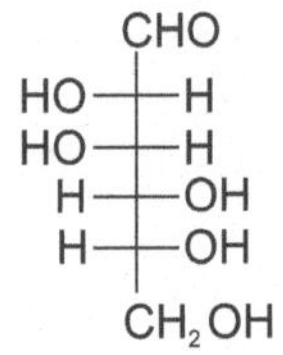

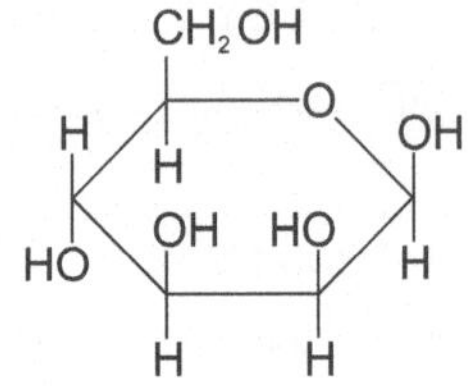

β-anomer

D-GALACTOSE

It is widely distributed in the animal and plant kingdom. Constituent of lactose and many polysaccharides (GUMS).

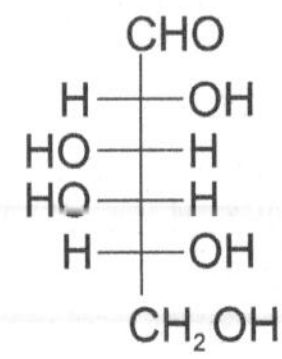

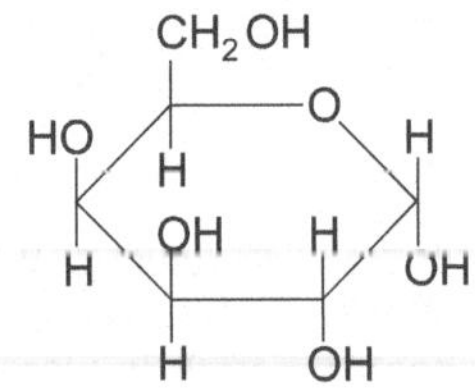

α-anomer

1.5 CONFIGURATIONAL RELATIONSHIP OF MONOSACCHARIDES

A. Correlation of the Configuration of D-glucose and D-glyceraldehyde

As mentioned above glyceraldehyde is the parent compound of the class of monosaccharides known as aldoses. Glyceraldehyde contains one chiral centre, therefore, there are two enantiomers. The configuration of these has been described by the R and S notation, which are represented by the Fischer projections as given below. However, in the carbohydrate chemistry, it is customary to use the older system of representation, *i.e.*, D and L. According to that the Fischer projection is orientated in such a manner so that the most oxidised carbon is on the top. Now if the group of highest priority is on the right it is called D and when this group is on the left, it is called L.

D(+)-Glyceraldehyde or R(+)-Glyceraldehyde

L(–)-Glyceraldehyde or S(–)-Glyceraldehyde

So under this convention all monosaccharides with the same configuration at the last chiral carbon atom as D-glyceraldehyde are called D-sugars. Those with opposite stereochemistry at the last chiral centre are called L-sugars.

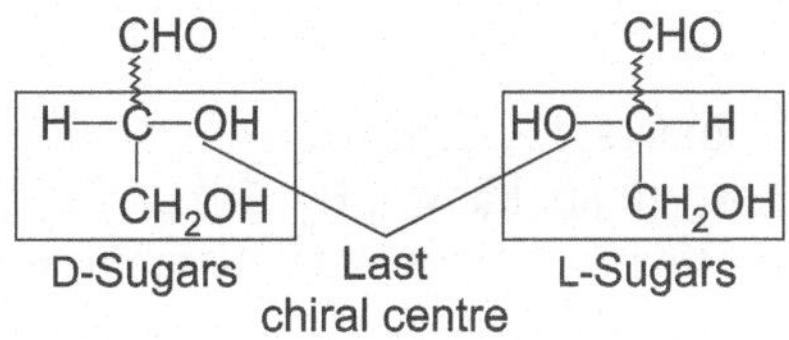

D-Glyceraldehyde was chosen as an arbitrary standard by Emil Fischer for correlating the configuration of higher sugars. But his guess was later shown to be correct by an X-ray diffraction analysis of the tartarate salts.

We can extend the aldose family starting from glyceraldehyde. As naturally occurring sugars generally belong to the D-family, so we will consider here only those belonging to the D-series, although there are an equal number of enantiomers for each class belonging to the L-series (Fig. 1.5).

B. Aldotetroses

D-Glyceraldehyde may be stepped up to give two isomeric tetroses (2 and 3) called D-erythrose and D-threose. They have the same configuration about the last chiral carbon as does D-glyceraldehyde. The question remains:- How to decide the configuration of the generated carbon atom or which is which?

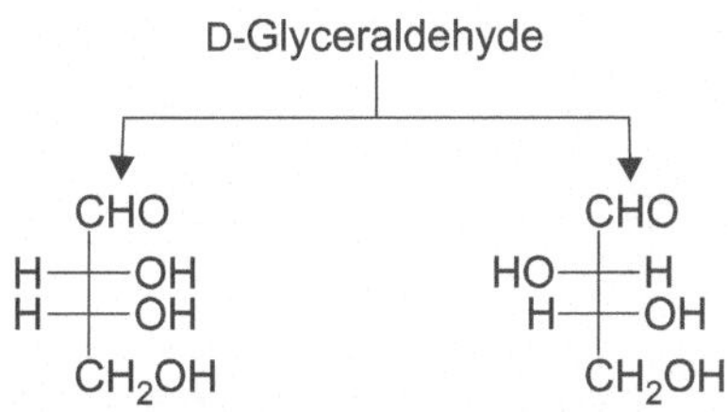

(2) (3)

This is decided by oxidising the two tetroses to the corresponding dicarboxylic acids and determining the optical rotation of these acids. Thus, on oxidation D-erythrose forms meso-tartaric acid; so its configuration must be (2). On the other hand, D-threose (3) gives optically active dicarboxylic acid.

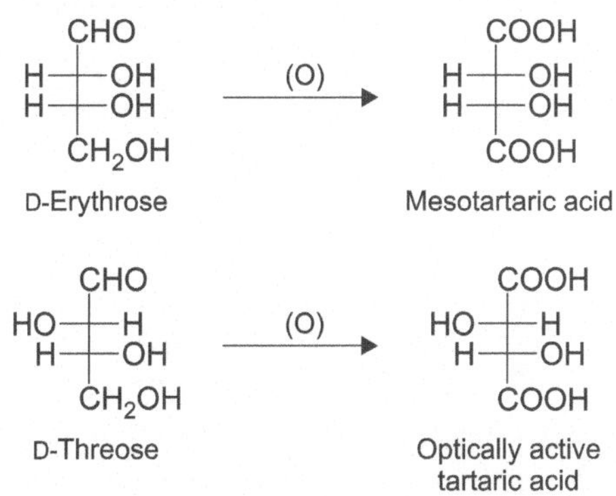

C. Aldopentoses

There are eight optical isomers, *i.e.*, four pairs of enantiomers. Four belong to D-series and four to L-series. All are known. By Kiliani-Fischer synthesis (page 62) D-erythrose can be converted to D-ribose and D-arabinose whereas D-xylose and D-lyxose are obtained from D-threose by the same reaction.

On oxidation with nitric acid D-arabinose gives an optically active dicarboxylic acid whereas D-ribose gives an optically inactive dicarboxylic acid. Similarly, D-xylose on oxidation gives optically inactive acid and the last D-lyxose affords an optically active dicarboxylic acid. Thus, the configuration of all aldopentoses belonging to D-family can be established (Fig. 1.3).

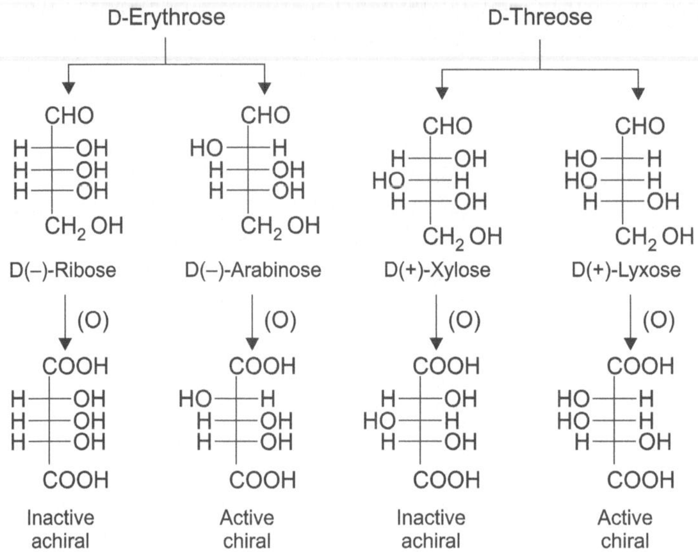

Fig. 1.3 Aldopentoses and their oxidation products.

D. Aldohexoses

Of the sixteen aldohexose stereoisomers, eight belong to the D series. The configuration of these sugars has been worked out by Fischer. They are represented in Fig. 1.4.

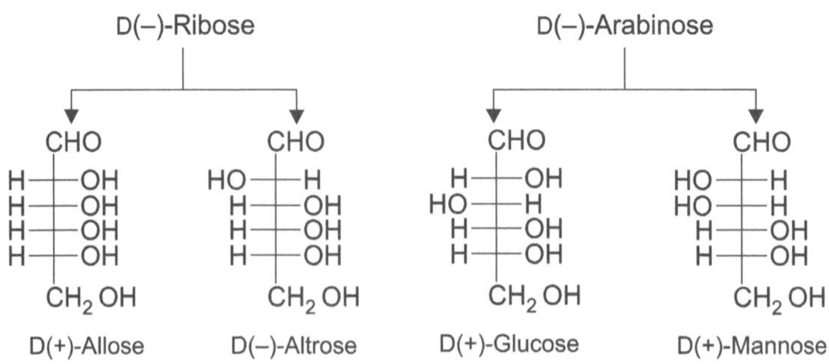

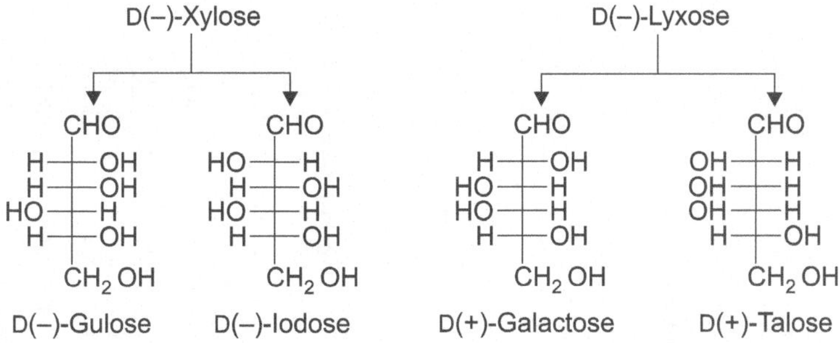

Fig. 1.4 Aldohexoses belonging to the D-series.

All these are diastereoisomeric. D-Arabinose gives D-glucose and D-mannose, which are epimers. Diastereoisomers which differ in configuration at one of the chiral centres are called epimers. Thus, D-glucose and D-galactose are also epimers.

1.6 STRUCTURE AND PROPERTIES OF D-GLUCOSE

Glucose is the most impotant monosaccharide. It is by far the most abundant monosaccharide which occurs in nature. There are more (+)–glucose units in nature than any other organic molecule.

(+)–Glucose is typical of all monosaccharides. The chemistry of all other carbohydrates revolves around glucose molecule. By learning about its structure and properties we can learn about the structure and properties of all other members of this family.

A. Open-Chain Structure of D-Glucose

The open-chain structure is derived at by the following facts:

1. Glucose has the molecular formula, $C_6H_{12}O_6$ as shown by elemental analysis and molecular weight determination.
2. *Presence of an aldehyde group:*

   That glucose molecule contains carbonyl group is shown by the formation of phenylhydrazone and oxime.

   Mild oxidation of glucose with Br_2 water gives a monocarboxylic acid, known as gluconic acid.
3. Glucose is a reducing sugar—is shown by the fact that it reduces Tollen's reagent and Fehling solution. Both these tests are given by aldehydes and not by ketones.
4. On reduction with sodium-alcohol glucose gives sorbitol, a hexa- hydric alcohol.

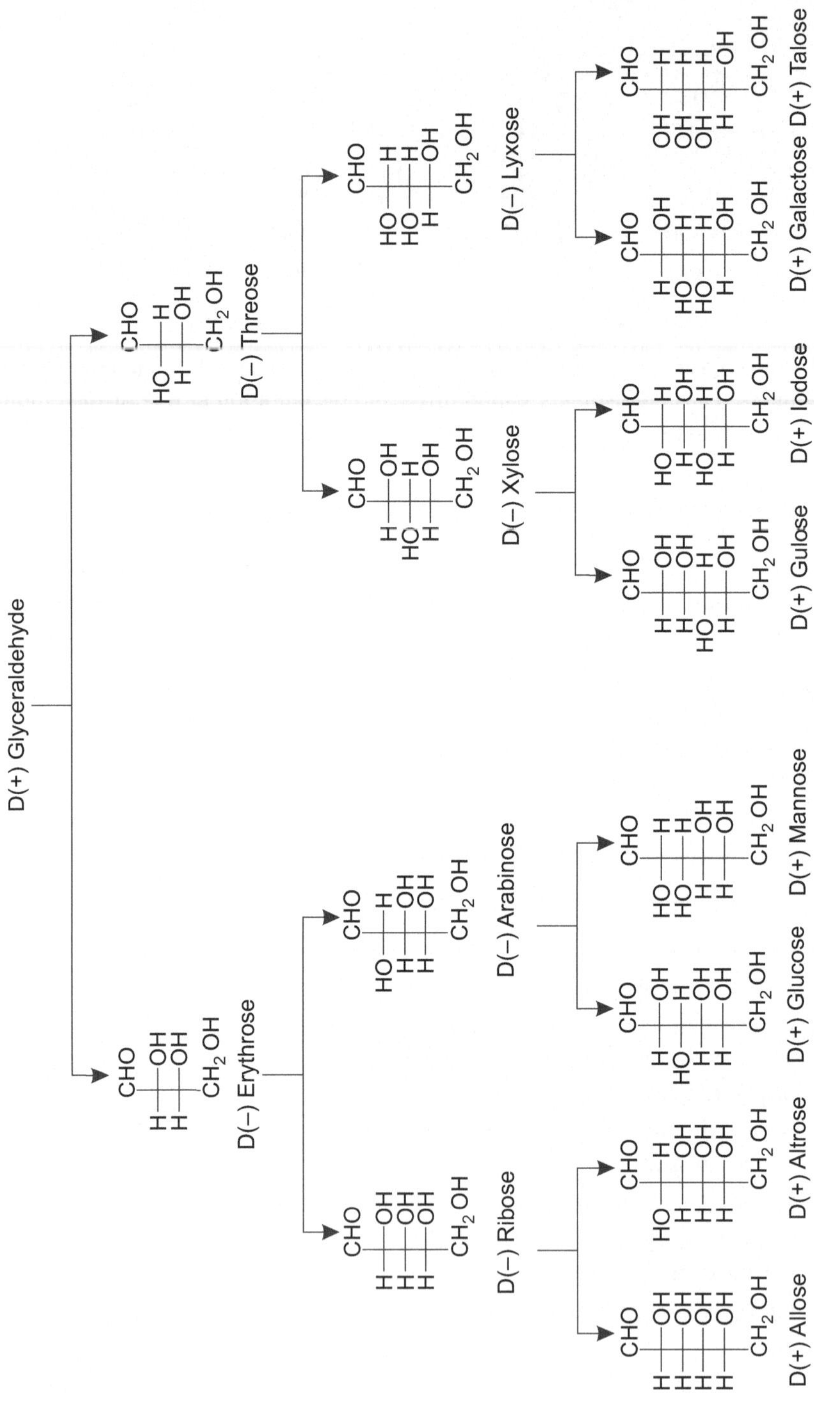

Fig. 1.5 Aldehexoses belonging to the D-series.

5. *Oxidation with nitric acid:*

   Dilute nitric acid oxidises both the—CHO group and the terminal—CH_2OH group of an aldose to —COOH groups. These dicarboxylic acids are known as aldaric acids.

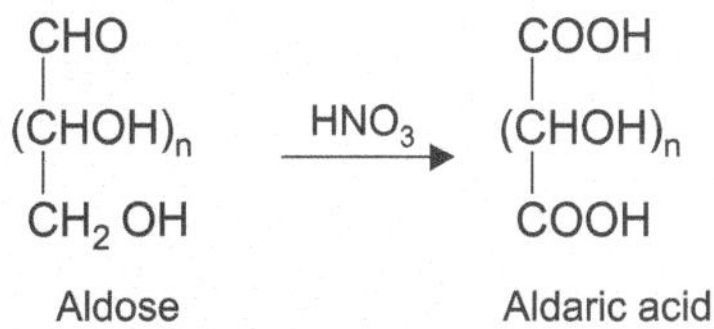

   Oxidation of glucose with dilute nitric acid gives a dicarboxylic acid known as D-glucaric acid or saccharic acid.

6. *Reaction with HI/P:*

   Reduction of glucose with hydriodic acid and red phosphorus gives *n*-hexane.

The cyanohydrin, obtained from glucose, on hydrolysis followed by treatment with HI/P gives *n*-heptanoic acid.

All the above facts suggest that glucose is a six-carbon, straight chain pentahydroxy aldehyde *i.e.*, glucose is an aldohexose. The structure of glucose is thus written as (4).

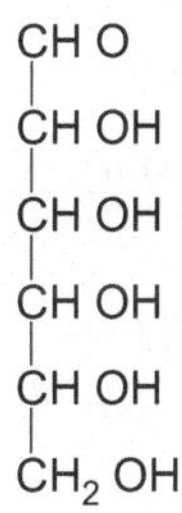

(4)

All the above reactions of glucose are summarised in Fig. 1.6.

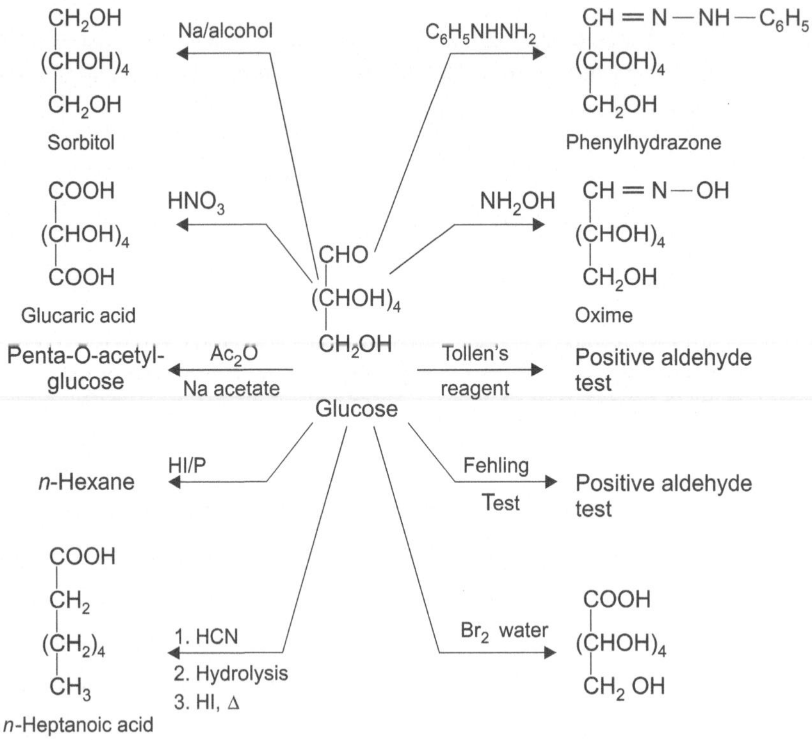

Fig. 1.6. Reactions of glucose.

B. Configuration of D (+)-Glucose

The above structure (4) of the (+)-glucose shows that there are four chiral centres in the molecule and hence there should be 2^4=16 stereoisomers, *i.e.*, eight pairs of enantiomers. All these 16 possible stereoisomers are now known. Some have been isolated from natural sources and others have been synthesised in the laboratory.

The question now is "which structure represents glucose out of these 16 stereoisomers"? The problem was solved by **Emil Fischer**. He began his work on the stereochemistry of (+)-glucose in 1888. He gave the configuration of (+)-glucose and was also able to synthesise and characterise all the other isomers. For this research work he was awarded the **Nobel Prize** in 1902.

Fischer had recognised that 16 stereoisomers consisted of two enantiomeric families. He arbitrarily guessed that (+)-glucose has got the same configuration at

C-5 as D(+) glyceraldehyde. The configuration of D(+)-glyceraldehyde was arbitrarily assigned as:

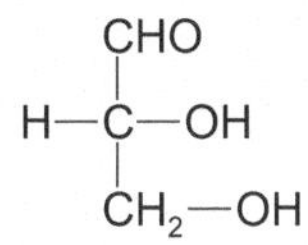

(5)

So all sugars having —OH at C-5 on the right were designated as D-sugars and those with —OH at C-5 on the left were designated as L-sugars. Fischer, therefore, rejected eight enantiomers retaining only those in which —OH at C-5 were on the right. It was not until 1957, when Bijuoet determined the absolute configuration of D(+)-tartaric acid (and, hence of D(+)-glyceraldehyde) that Fischer's arbitrary assignment of (+)-glucose to the D-family was known to be correct.

Fischer's assignment of structure (6) to (+)-glucose was based on the following arguments:

1. Out of the four possible aldopentoses belonging to the D-series, only D-arabinose could be converted to a mixture of D(+)-glucose and D(+)-mannose by Kiliani-Fischer Synthesis (*see* page 62). Here again Fischer had retained only those enantiomers in which the bottom chiral carbon atom carried OH on the right. This means that at C-3, C-4 and C-5 (+)-glucose has got the same configuration as that of arabinose.
2. Since (+)-glucose and (+)-mannose give the same osazone derivative, these sugars must have identical configurations at all chiral centres except C-2.

   Sugars which have different configurations at one of the chiral centres are called epimers. Thus, Fischer realised that (+)-glucose and (+)-mannose were epimeric at C-2 (Fig. 1.7).

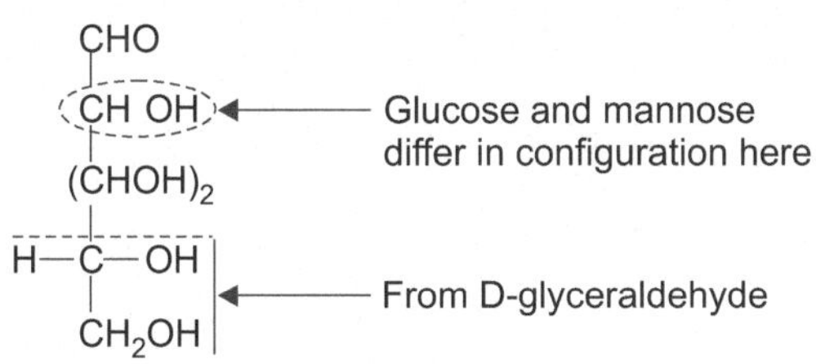

Fig. 1.7. Configurations of (+) glucose and (+) mannose.

3. Nitric acid oxidation of (–)-arabinose gives an optically active aldaric acid. This indicates that (–)-arabinose must have either of the two possible configurations (7 or 8). The other two aldopentoses (9 and 10) would have afforded an optically inactive compound, *i.e.,* meso aldaric acids.

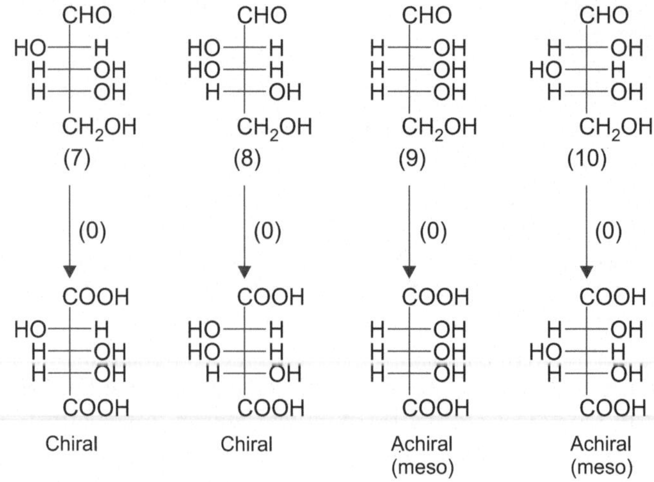

4. Nitric acid oxidation of (+)-glucose and (+)-mannose gives in each case an optically active aldaric acid. This means that —OH at C-4 in both these molecules is on the right. If it were on the left then at least one aldaric acid would have been achiral (meso). So, arabinose must have configuration (7) and not (8). This is illustrated by the following scheme (Fig. 1.8).

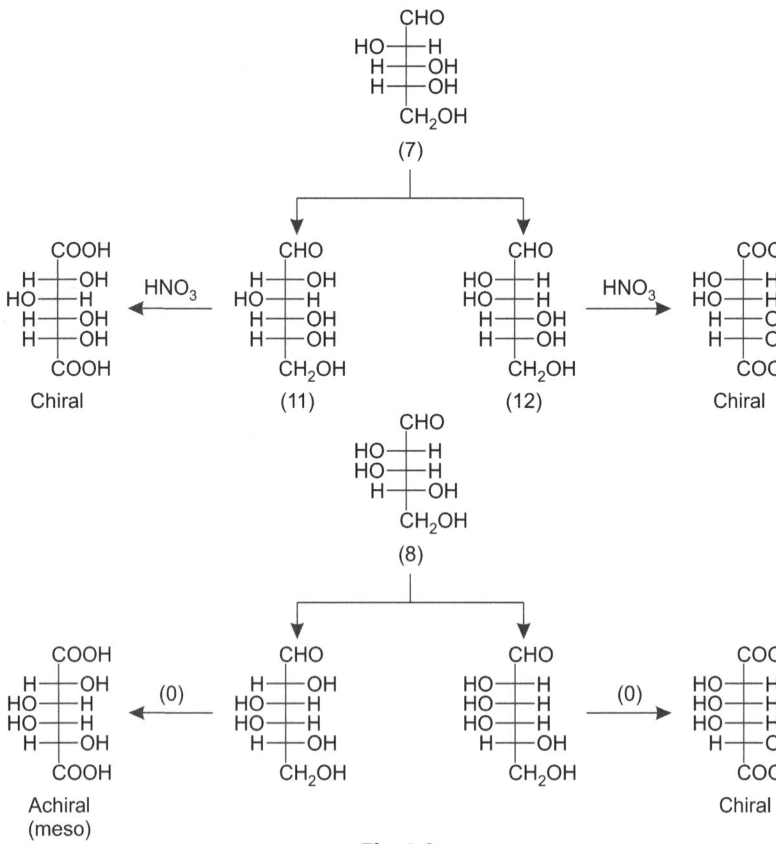

Fig. 1.8.

Once the configuration of (–)-arabinose is decided as (7), (+)-glucose and (+)-mannose should have structures (11) and (12). The question remains, however, which one is which.

5. Another aldohexose (+)-gulose on oxidation with nitric acid gives the same aldaric acid as (+)-glucose. Only acid (14) can be derived from two different hexoses. This means that (+)-glucose must have configuration (11) and (+)-gulose the inverted configuration (13). (+)-Mannose, therefore, has got configuration (12).

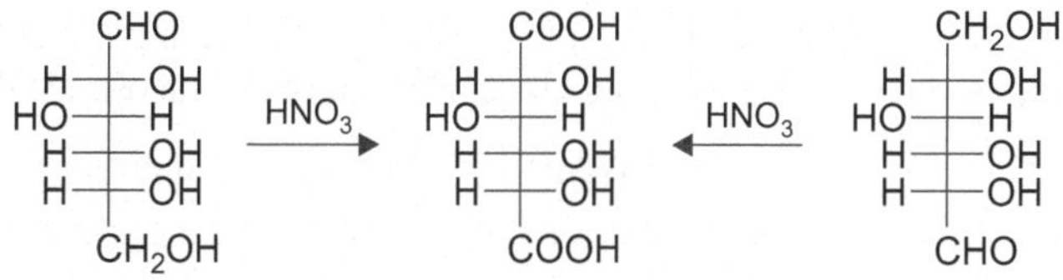

(11) (14) (13)

6. Fischer had already developed a method for effectively interchanging the two end groups —CHO and —CH_2OH of an aldose chain and with brilliant logic he arrived at structure of (+)-glucose as (11). So the problem can be viewed by considering the overall effect on glucose and mannose of a sequence of reactions which interchange the two end groups. If (+)-mannose has structure (12) an interchange of end groups will yield the same aldohexose.

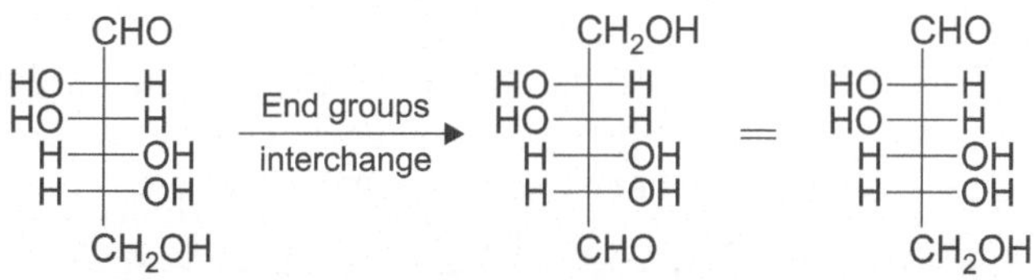

(12) (12)

On the other hand, if (+)-glucose has structure (11), an end group interchange will yield a different aldohexose (13), which is L-sugar and it is an enantiomer of D(–)-gulose.

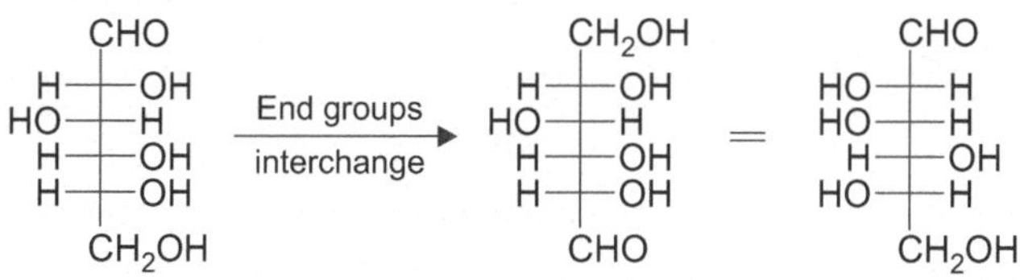

(11) (13), L(+) Gulose

C. Cyclic Structure of D(+)-Glucose

The above structure (11) for D(+)-glucose could not account completely for its chemical behaviour.

Facts that are still to be accounted for are the following:

1. D(+)-Glucose fails to undergo certain reactions typical of aldehydes, *e.g.*, it does not form a bisulphite addition product and gives a negative test with Schiff's reagent, although it is readily oxidised.
2. The penta-acetate and the pentamethyl ether derivatives of glucose are not oxidised by either Tollen's reagent or Fehling solution or also do not form phenylhydrazone or bisulphite addition product.

   One of the ether groupings in pentamethyl derivative is highly reactive. It undergoes hydrolysis in dilute acids to a tetramethyl derivative. The tetramethyl derivative reduces Tollen's reagent as effectively as glucose itself.
3. D(+)-Glucose exists in two isomeric forms, which undergo mutarotation. When crystals of ordinary glucose, m.p. 146°C, are dissolved in water, its specific rotation is +112°. On keeping this solution the value drops to +52.5°.

   When glucose is obtained from fresh hot concentrated solution, its specific rotation changes from an initial value of +19° to an equilibrium value of +52.5°. This form of D(+)-glucose is termed as β-D(+)-glucose (m.p. 150°C). The α-isomer is the previously used isomer. The change in rotation of these to the equilibrium value is called mutarotation.

$$\underset{\alpha\text{-isomer}}{\alpha\text{-D(+)-Glucose}} \longrightarrow +52.5^\circ \longleftarrow \underset{\beta\text{-isomer}}{\beta\text{-D(+)-Glucose}}$$

$$[\alpha]_D = +112^\circ \longrightarrow +52.5^\circ \longleftarrow [\alpha]_D = +19^\circ$$

4. D(+)-Glucose forms two isomeric methyl-D-glucosides; methyl α-D-glucoside and methyl-β-D-glucoside under conditions which normally convert an aldehyde to a hemiacetal or acetal derivative.

$$C_6H_{12}O_6 + CH_3\,OH \xrightarrow{HCl} \underset{\text{Methyl-}\alpha\text{- and }\beta\text{-D-glucoside}}{(C_6H_{12}O_5) - OCH_3}$$

We know aldehydes, react with alcohols in the presence of anhydrous hydrochloric acid to form hemiacetal or acetal.

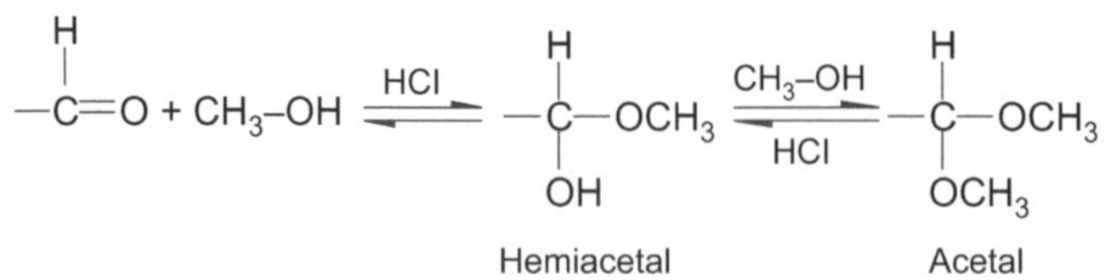

These glucosides do not reduce Tollen's reagent and Fehling solution and do not undergo mutarotation. That means they are comparatively quite stable under mild acidic or mild basic conditions. Furthermore, they have different m.p. and specific rotation. The isomer of higher positive rotation is called methyl-α-D-glucoside (m.p. 165°C, $[\alpha]_D$ = +158°) and the other methyl-β-D-glucoside has m.p. 107°C, $[\alpha]_D$ = –33°. (*See* Table 1.4.)

TABLE 1.4 Properties of D(+)-glucose and its methyl glucosides

Compound	m.p.	$[\alpha]_D$
α-D-(+) Glucose	146°C	+ 112°
β-D(+)-Glucose	150°C	+ 19°
Methyl-α-D-glucoside	165°C	+ 158°
Methyl-β-D-glucoside	107°C	- 33°

The above observations cannot be explained on the basis of the open-chain structure for the glucose molecule. It is obvious that the aldehyde group in the glucose molecule is not free but is tied up with one of the hydroxyl groups, which turns out to be at C-5 to form a hemiacetal structure as explained above. Thus, we get a cyclic structure for the glucose molecule.

A new chiral centre is created at C-1 by hemiacetal formation. There are, therefore, two stereoisomeric forms of D(+)-glucose. They are α-D(+)-glucose and β-D(+)-glucose, represented by structures (15) and (16). These are diastereoisomers differing in configuration at C-1. Such a pair of diastereoisomers are called **anomers**.

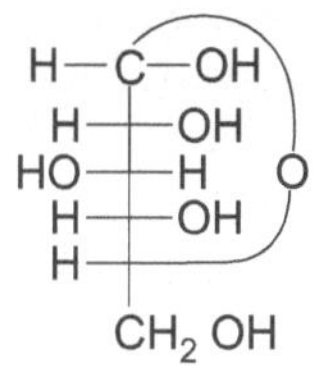

(15) α-D(+)-Glucose

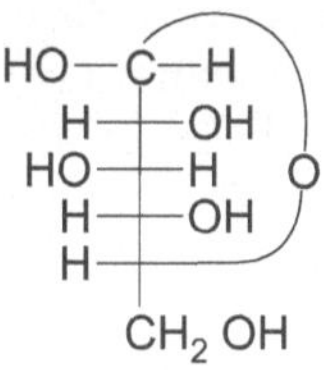

(16) β-D(+)-Glucose

X-ray analysis of D(+)-glucose indicates that in the α-anomer, the hydroxyl groups at C-1 and C-2 are *cis* to one another and *trans* in the β-anomer. The above representations are according to Fischer Projections. Hydroxyl groups on the same side in the Fischer Projection of the hemiacetal are *cis* to each other and those on the opposite side are *trans*. Thus, C-1 hydroxyl group is on the right in the Fischer projection of α-D(+)-glucose and on the left in β-D(+)-glucose.

1. Haworth Projections

Structures (15) and (16) drawn for the isomers of D(+)-glucose are not consistent with their molecular shapes. Haworth projections were given to overcome this difficulty. In this, hemiacetal structure is represented by a planar ring containing the oxygen atom.

To write the Haworth projection for a Fischer projection, we rotate the open-chain form in such a manner so that –OH at C-5 and the carbonyl group forming the hemiacetal are in close proximity. This is possible when C_4–C_5 bond is rotated. After rotating, hemiacetal formation results in giving Haworth structure. For example, we can write Haworth structure for the glucose molecule as given ahead.

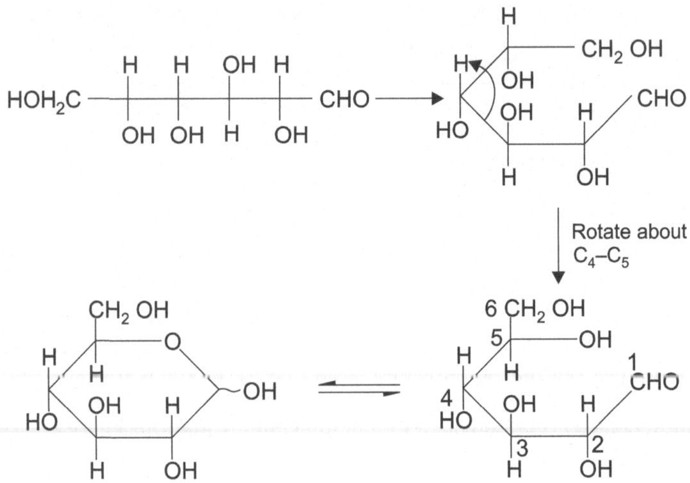

Haworth structure
of D-glucose

Thus, the Haworth structures for the α-D(+)-glucose and β-D(+)-glucose are represented by the structures (17) and (18) respectively. As they are derived from the parent heterocycle pyran, they are named, as α-D(+)-glucopyranose and β-D(+) glucopyranose. The corresponding five-membered cyclic hemiacetals are referred to as furanoses, named after furan.

In these projections, the $-CH_2OH$ group is written as up for D-sugars and down for L-sugars. For D-sugars the —OH at C-1 will be down in the α-form and up for the β-form.

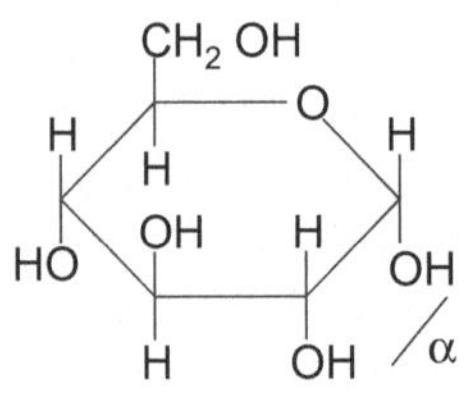

(17) α-D(+)-Glucopyranose

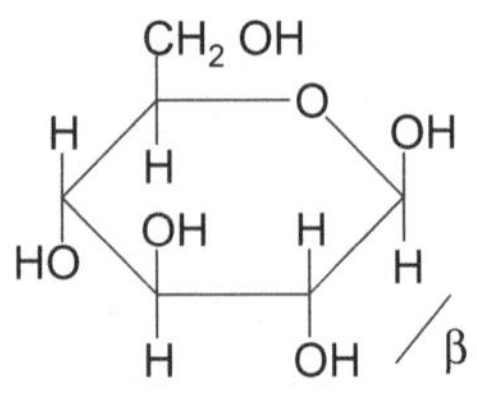

(18) β-D(+)-Glucopyranose

2. Correlation between Open-Chain and Ring Structures

α- and β-D(+)-Glucose are readily hydrolysed by water. In aqueous solution either anomer is converted *via* the open chain form into an equilibrium mixture containing both the cyclic isomers. Thus, mutarotation results from the ready opening and closing of the hemiacetal ring.

The typical aldehyde reactions of D(+)-glucose, for example, the oxidation of glucose to gluconic acid (reduction of Tollen's reagent and Fehling solution), the formation of oxime or osazone and the conversion of glucose to a methyl glucoside are presumably due to open-chain structure of the glucose. This is present in very low concentratins at equilibrium (~ 0.01%). As the aldehyde is consumed by the chemical reaction, it is rapidly replaced by opening of the α- or β-anomers. The concentration of this open-chain structure, is, however, too low for certain easily reversible aldehyde reactions like bisulphite addition and the Schiff's test.

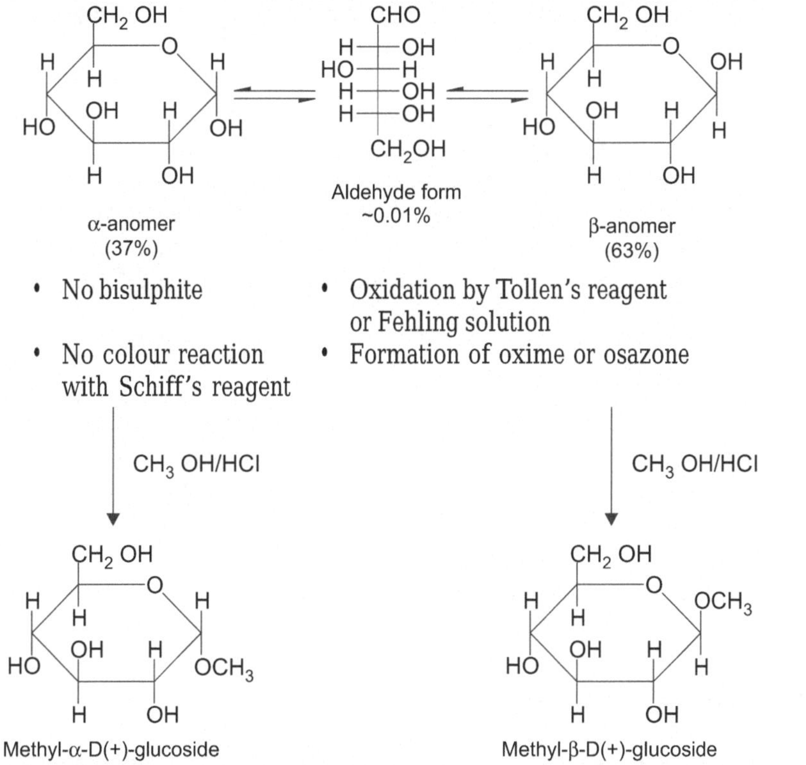

D. Conformations of D(+) Glucose

Although Haworth structures describe sugars as five and six membered rings, they do not represent the most important structural feature, *i.e.*, conformations. We know that cyclohexane is not planar but has got chair conformations. There is least strain and minimum repulsion in these conformations.

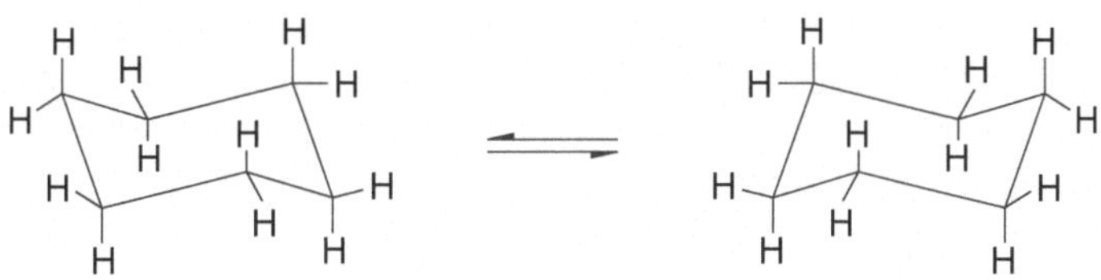

By analogy with cyclohexane we expect the six-membered cyclic oxide structure of glucose to exist preferentially in a chair conformation. To avoid interactions, bulky substituents prefer to be in the equatorial rather than the axial positions. X-ray studies have in fact established that crystalline α-D-glucose has the chair conformation (19) and crystalline β-D(+)-glucose has structure (20) in which all the substituents can be seen to occupy equatorial positions.

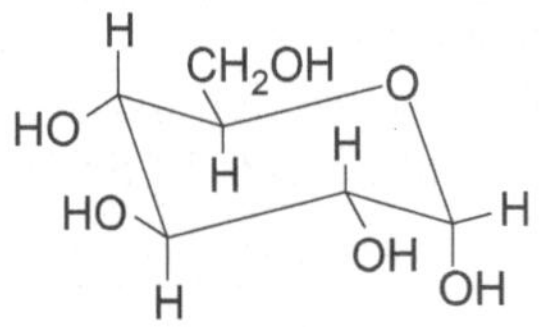

(19) α-D(+)-Glucose

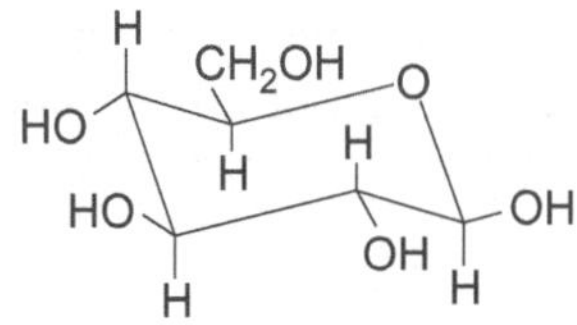

(20) β-D(+)-Glucose

Due to small difference in energy between the boat and chair forms, it is uncertain whether these conformations are preserved when D(+)-glucose is in solution. However, when the hydroxyl groups are acetylated to give α- and β-D-glucose-2, 3, 4, 6-tetra-acetate, it has been found that a chair conformation is preferred with largest number of bulky substituents occupying the equatorial positions. The —OH at C-1 is axial in the α-form and equatorial in the β-form.

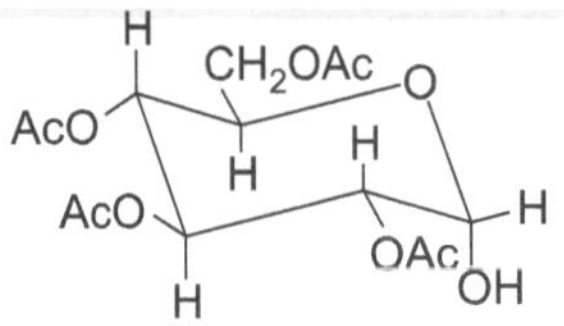

α-isomer

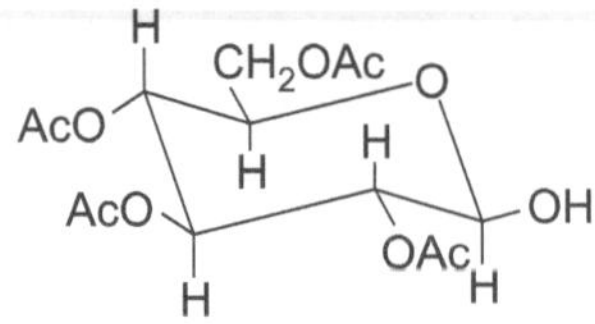

Ac = – $COCH_3$

β-isomer

As it is clear that in β-D(+)-glucose the —OH at C-1 occupies equatorial position; the β-isomer is more stable than the α-isomer. It is, therefore, understandable that β-D(+)-glucose is the most predominant structure occurring in nature.

1.7 MUTAROTATION

We have already seen that glucose can be isolated in two different crystalline forms. These forms are α-D(+)-glucose and β-D(+)-glucose. They have different melting points and different specific rotations. The α-form is soluble in water to the extent of 82 g/100 ml at 25°C and β-isomer is soluble in water to a greater extent, 154 g/100 ml at 15°C. Both these crystalline forms of α- and β-D(+)-glucose are quite stable. In solution (water) each form slowly changes into an equilibrium mixture of both. The process can be readily observed as a decrease in the specific rotation of the α-anomer (+112°) or an increase for the β-anomer (+19°) to an equilibrium value of +52.5°. This phenomenon is known as mutarotation and is commonly observed for reducing sugars. Aldopentoses, aldohexoses and ketohexoses all undergo mutarotation.

We have seen that mutarotation takes place in water solution. However, mutarotation can take place in any solvent, which is amphiprotic in nature, *i.e.*, a solvent which can function both as an acid as well as a base. It has been observed that mutarotation does not take place when pyridine or cresol are used as solvents separately but it takes place in a mixture of the two. That is, mutarotation is particularly effective if both an acid and a base are present in the same solution.

We know that water acts both as an acid as well as a base. Therefore, mutarotation takes place readily in water.

Mechanism of Mutarotation

The phenomenon of mutarotation indicates that the two forms of D(+)-glucose are interconvertible and they have a common intermediate. This change presumably takes place *via* the open-chain structure of glucose. There is sufficient evidence for the existence of this open-chain form. The generally accepted mechanism is virtually the same as described for acid and base-catalysed hemiacetal and hemiketal formation of aldehydes and ketones.

Acid Catalysed Hemiacetal Formation:

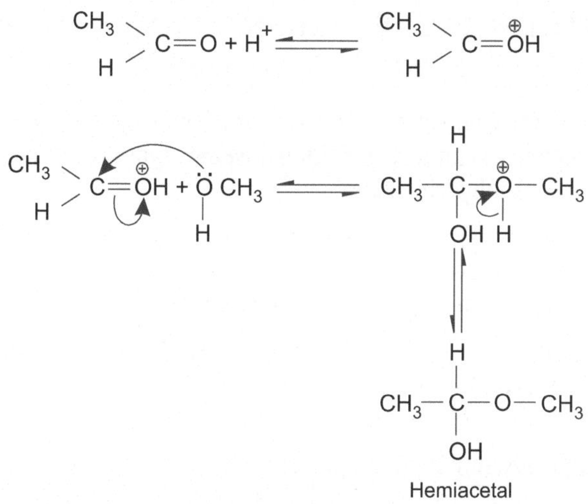

Base Catalysed Hemiacetal Formation

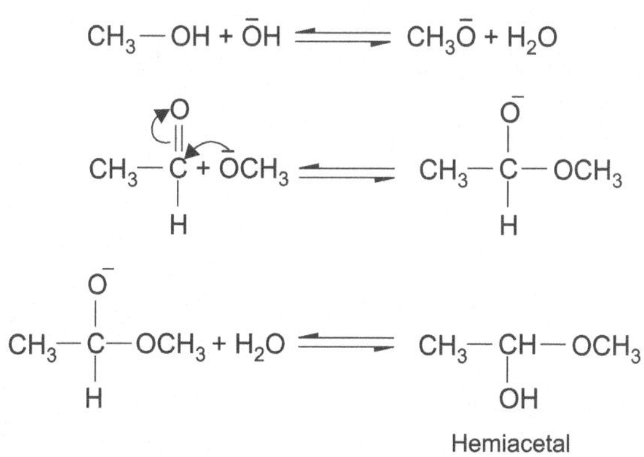

When mutarotation takes place, the ring of one form opens up and then recloses in the inverted position to give the 2nd form.

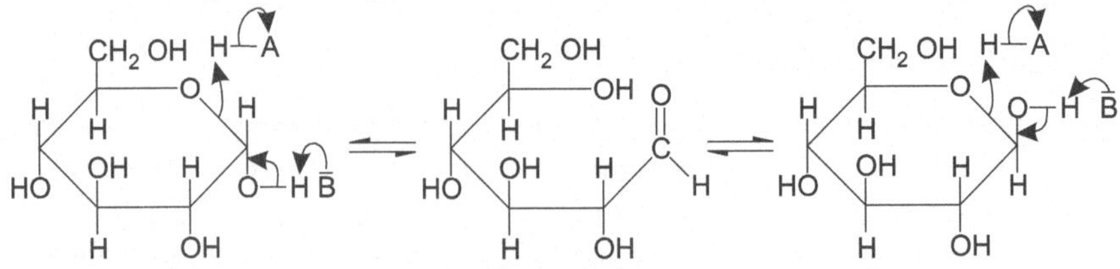

As already mentioned at equilibrium about 63% of the β-anomer and 37% of the α-anomer are present. The amount of free aldehyde form is very small (~ 0.01%).

1.8 LENGTHENING THE CARBON CHAIN OF ALDOSES- THE KILIANI-FISCHER SYNTHESIS

Aldose chains may be lengthened by one carbon through the cyanohydrin formation. The reaction sequence is known as **Kiliani-Fischer** Synthesis. A new chiral centre is created at the cyanohydrin stage.

Thus, we get a pair of epimeric stereoisomers at this site. A pair of aldoses can be recognised as epimers not only by their conversion into the same osazone, but also by their formation in the same Kiliani-Fischer synthesis.

Heinrich Kiliani at the **Technische Hochschule** in **Munich**, showed that an aldose can be converted into the glyconic acids of the higher sugars by addition of HCN and hydrolysis of the resulting cyanohydrins.

At the same time **Fischer** reported that reduction of a glyconic acid, in the form of its lactone, can be controlled to yield the corresponding aldose.

Thus, a lower sugar can be converted to two epimeric higher sugars and the total reaction is called Kiliani-Fischer synthesis.

If we subject D(+)-glyceraldehyde to the Kiliani-Fischer synthesis, two stereoisomeric tetroses, erythrose and threose, are formed. The reaction sequence is shown in Fig. 1.9.

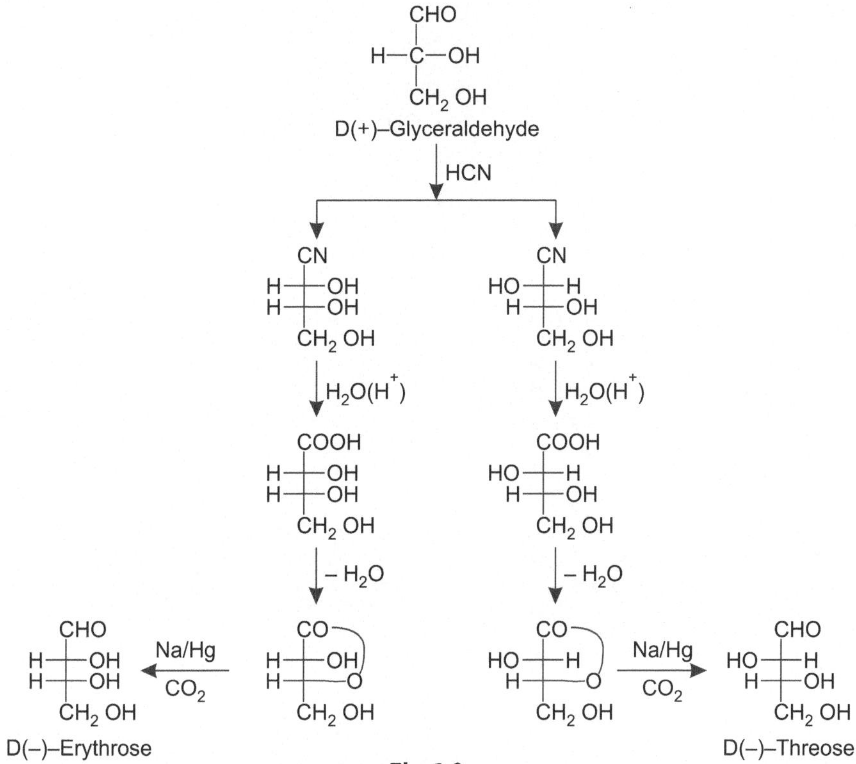

Fig. 1.9.

Similarly, D(–)-arabinose an aldopentose, can be converted into two epimeric aldohexoses, D(+)-glucose and D(+) mannose (Fig. 1.10). The separation can be achieved at the diastereoisomeric glyconic acids. We would expect these acids to form lactones under acidic conditions. These lactones are finally reduced to the corresponding aldoses. Reduction can also be carried out selectively by using $NaBH_4$.

1.9 SHORTENING THE CARBON CHAIN OF ALDOSES

Aldose chains may be selectively shortened from aldehyde end by one carbon atom by a number of methods.

A. The Wohl Degradation

This is the reverse of Kiliani-Fischer synthesis. The aldose is converted to its oxime, which undergoes simultaneous acetylation and dehydration to the poly-*O*-acetyl nitrile when treated with acetic anhydride in the presence of anhydrous sodium acetate. Loss of HCN together with deacetylation occurs when the nitrile is treated with ammoniacal silver oxide to give the lower sugar.

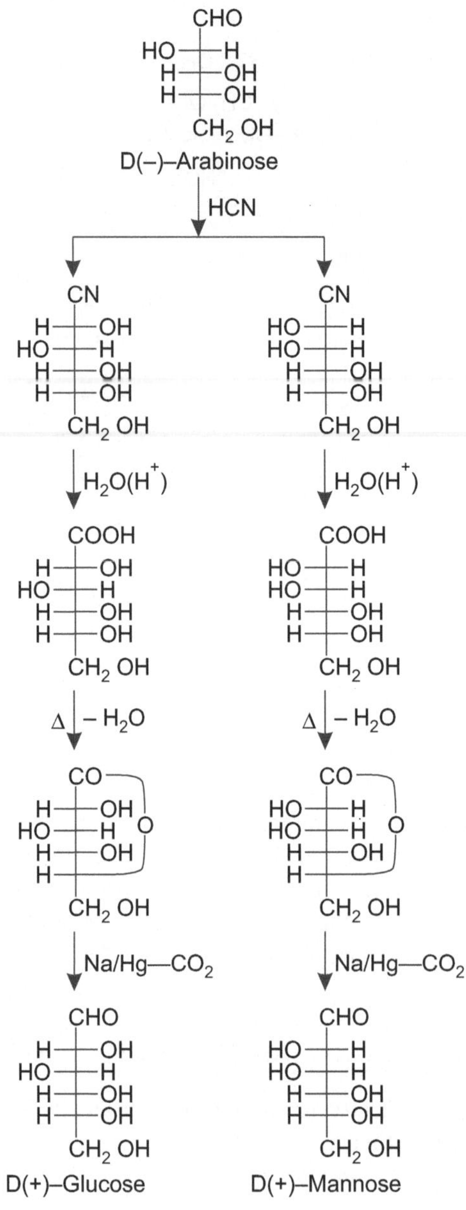

Fig. 1.10.

The reaction sequence can be illustrated by taking D-glucose as an example. D-Arabinose is the product (Fig. 1.11).

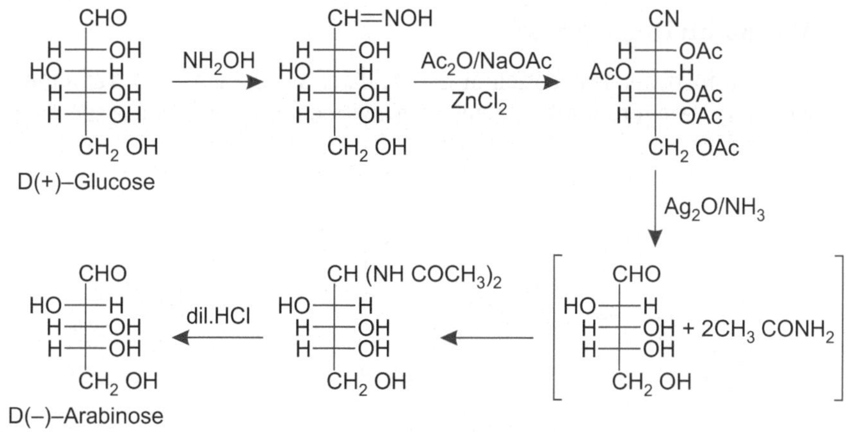

Fig. 1.11. The Wohl Degradation.

B. The Ruff Degradation

This is the most widely used method. In this, a soluble salt of an aldonic acid undergoes oxidative decarboxylation with H_2O_2 in the presence of Fe^{3+} ions to give the lower aldose sugar. It is assumed that the intermediate product is an α-keto acid which undergoes decarboxylation.

D(+)-Glucose is converted into D(+)-arabinose in the following manner (Fig. 1.12):

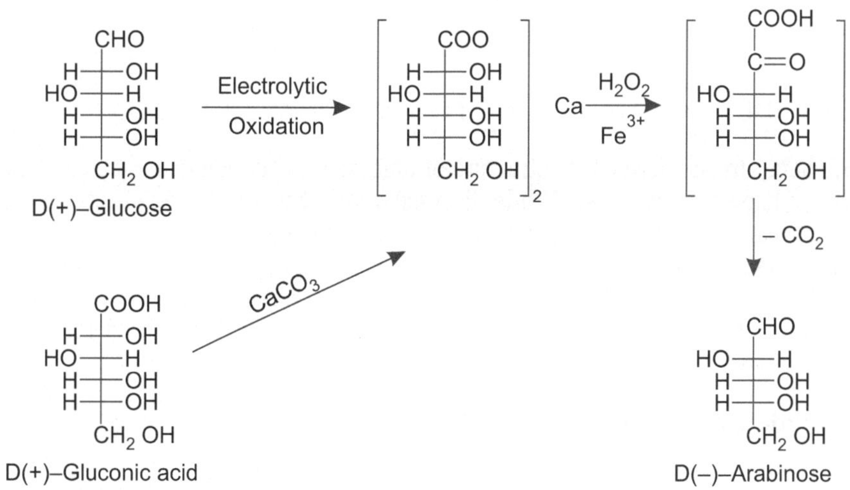

Fig. 1.12. Ruff Degradation

Roy L. Whistler improved this method by oxidising glucose with 3 moles of hypochlorite at pH 11 to gluconic acid and then oxidising this with 1.4 moles of the reagent at pH 4.5–5.0.

C. Weermann Degradation

This method involves degradation of the amide of a sugar acid by a variation of the Hofmann reaction. On treatment with hydrochlorite or bromite the amide undergoes Hofmann degradation to the intermediate isocyanate, which then disproportionates to the lower aldose. This method is not used extensively (Fig. 1.13).

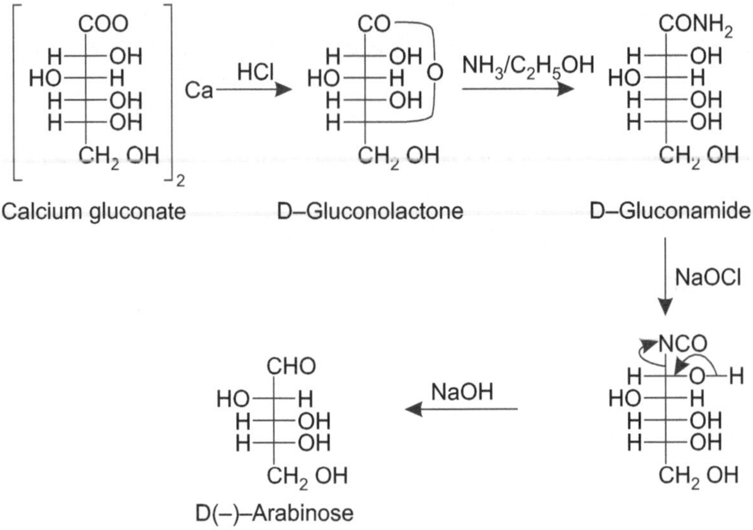

Fig. 1.13. Weermann Degradation.

1.10 DETERMINATION OF THE RING SIZE OF GLUCOSE

We have already derived the structure of glucose as a six-membered cyclic oxide that is known as pyranose. While discussing this structure it was presumed that hemiacetal formation takes place between the aldehyde group and —OH at C-5. This pyranose structure for the glucose molecule has been proved by a number of methods including the X-ray analysis and periodate oxidations. Here we give the standard chemical method *i.e.*, by methylation studies which is used for determining the ring size of any sugar.

1. Evidence for Pyranose Structure of D(+)-Glucose

The method was developed by Haworth and Hirst in 1927.

D(+)-glucose is refluxed with methanol in the presence of a small amount of hydrochloric acid. The methyl glucoside so obtained is methylated with excess of dimethyl sulphate in the presence of sodium hydroxide to give methyl-tetra-*O*-methyl-D-glucoside. This on hydrolysis with dilute hydrochloric acid gives tetra-*O*-methyl-D-glucose. A lactone is isolated when the tetra-*O*-methyl-D-glucose is oxdised with bromine at 90°C. Finally, the lactone on further oxidation with HNO_3 affords trimethoxy glutaric acid.

One —COOH group of this acid comes from the lactone group that is combined with the hydroxyl in the formation of the lactone ring in gluconolactone. The second carboxyl is derived from the non-methylated carbon *i.e.*, from the —CHOH group that is involved in the ring formation of the sugar. So, there are three methoxyl groups in the lactone ring and thus C-5 hydroxyl must be involved in the ring formation. Therefore, the lactone is 2,3,4,6-tetra-*O*-methyl gluconolactone. Working backwards in this way D(+)-glucose must have structure as D(+)-glucopyranose. This is true with both α- and β-isomers.

The reactions involved are given below:

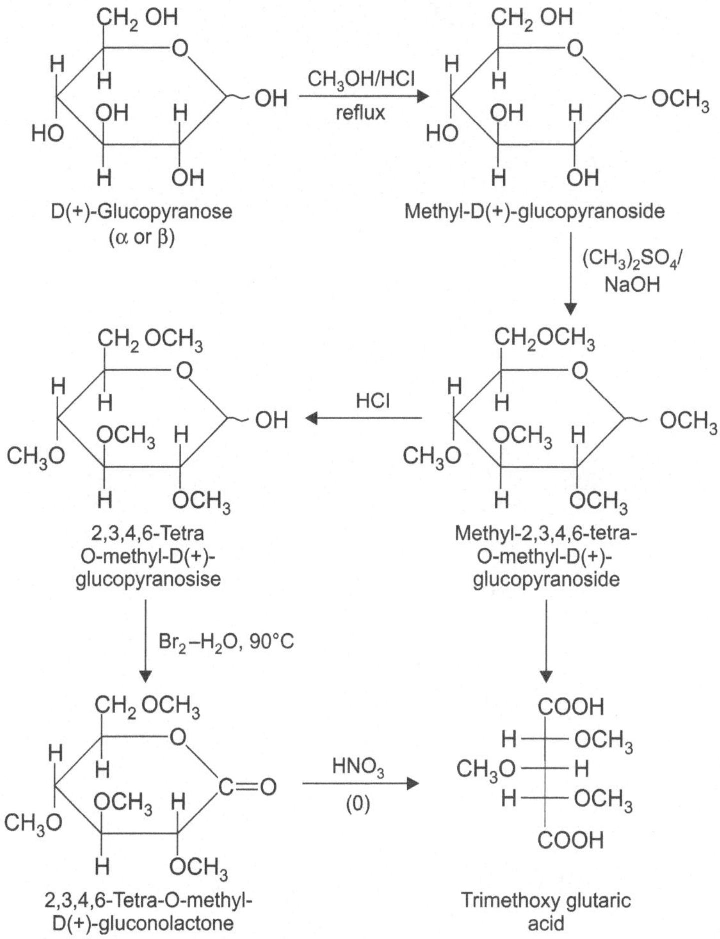

2. Evidence for the Furanose Structure of D(+)-Glucose

Fischer in 1914 prepared the methyl glucoside in a different way. He dissolved glucose in methanol and added to that 1%-hydrochloric acid. The mixture was allowed to stand at 0°C. He obtained a syrup instead of a crystalline product as obtained in the earlier case. This syrup was subsequently shown to be a mixture of α- and β-glucofuranosides. The procedure adopted here was just the same as given above. The end product in this case comes out to be dimethyl tartaric acid.

The formation of dimethyl-D-tartaric acid indicates that there are two methoxy groups within the ring *i.e.*, lactone is a γ-lactone. Therefore, its structure is given as 2,3,5,6-tetra-*O*-methyl gluconolactone and hence the glucose has got furanose structure named as D-glucofuranose. It must be noted here that D-glucofuranose has not been isolated. The structure is deduced on the basis of isolation of methyl-D-glucofuranoside (*See* Fig. 1.14).

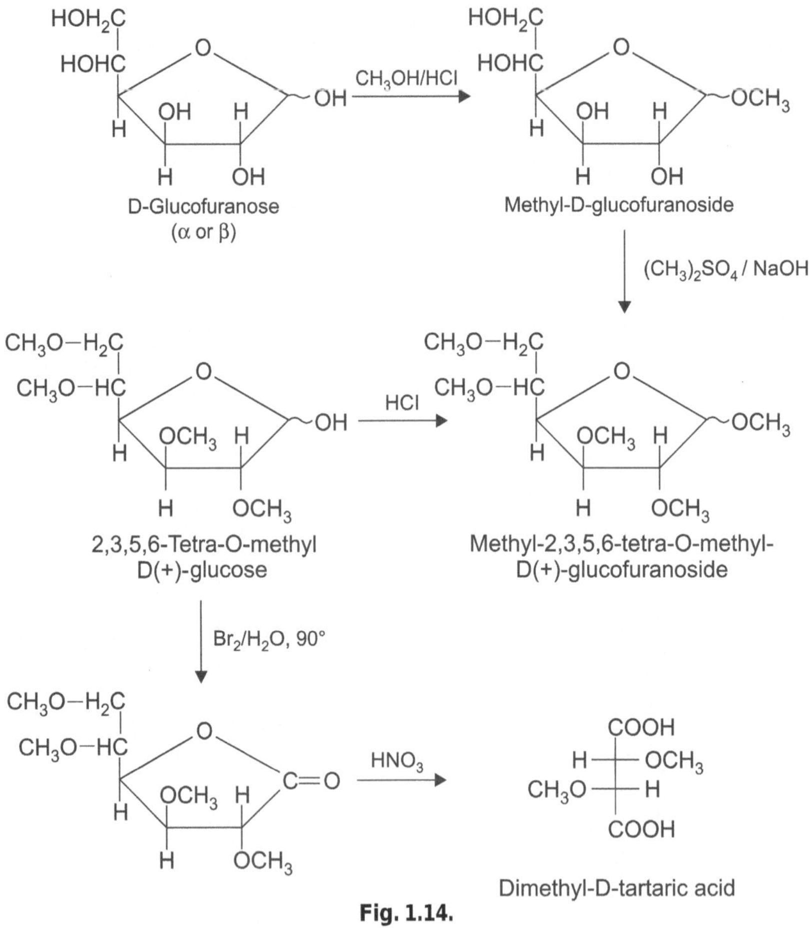

Fig. 1.14.

Both the above schemes favour different ring sizes for the glucose molecule. So, it is possible to decide the actual ring size on the basis of methylation studies.

Methylation at different temperatures give different methyl glucosides. At refluxing temperatures a pyranoside is formed whereas at 0°C the formation of a furanoside is favoured. The important question now is: What is the actual size of the ring in the original sugar? Oxidation of an aldose with hypobromite produces δ-lactone. This is the first product of oxidation.

Further investigations have proved that δ-lactone obtained by the oxidation of glucose is quite unstable. That slowly gets converted into more stable γ-lactone. It, therefore, follows that the size of the ring in normal sugars is pyranose.

McDonald *et al.* examined α-D-glucose by the X-ray analysis and confirmed the presence of six-membered ring.

3. Determination of Ring Size by Oxidations with Periodic Acid

This is an elegant method for the determination of the ring structures of glycosides and hence of sugars. We have already seen the methods adopted in the methylation studies are quite laborious and structures are deduced in an indirect manner. In contrast to that here the procedures are facile and direct. The consumption of the oxidant and liberation of products such as CO_2, formaldehyde, and formic acid can be determined quantitatively on a semi-micro scale.

It was discovered by **Malaprade** in 1928 that vicinal diols undergo cleavage of their C–C bond when treated with aqueous solutions of periodic acid or its salts. This cleavage is quantitative and can be used for the structure elucidations.

The mechanism of the periodic oxidation has been studied in detail. It involves the formation of a cyclic diester of the periodic acid. Decomposition of the diester yields the two carbonyl fragments and iodic acid as shown below:

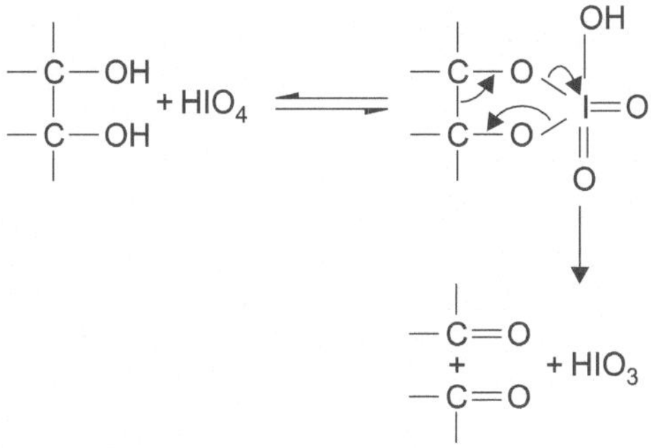

Sugars and their methyl glycosides are also cleaved like vicinol diols by periodic acid. The above reaction can be applied to glycosides for determining the ring size *i.e.*, whether they have pyranose or furanose structure. Under mild conditions (0.01 M to 0.05 M of HIO_4 at 15°C for 23 hours) the reaction can be controlled and the amount of periodic acid used is determined quantitatively.

Thus, methyl-α- and β-D-glucopyranoside each reacts with only two moles of periodate and give one mole of formic acid and no formaldehyde. This proves that the pyranoside structure assigned from methylation studies were correct. Furanoside, however, reacts with two equivalents of the reagent but yields only one equivalent of formaldehyde and no formic acid.

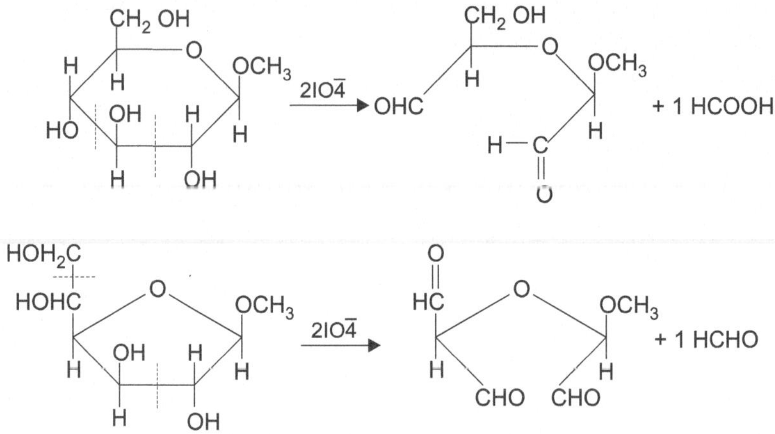

Glucose itself under strong conditions (1% aqueous $NaIO_4$ in 1 N H_2SO_4) consumes five equivalents of HIO_4 and yields five equivalents of formic acid and one equivalent of formaldehyde. It appears that with free sugars, it is the open-chain form that is oxidised and complete oxidative cleavage takes place.

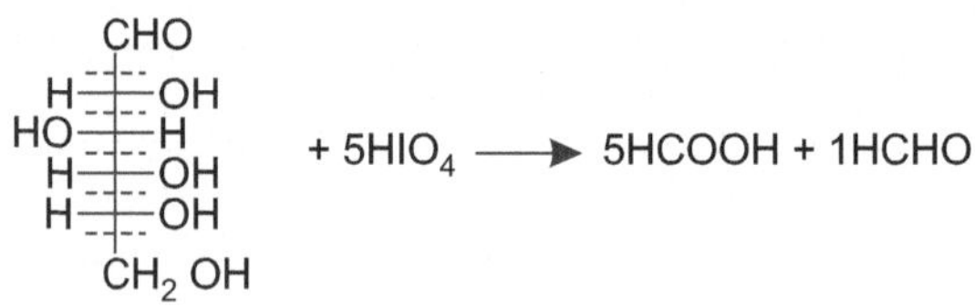

1.11 OSAZONE FORMATION

The reaction of glucose with phenylhydrazine is quite interesting. With excess of phenylhydrazine glucose forms a crystalline derivative known as glucose-phenyl osazone or glucosazone. The characteristic feature is that all aldoses form these derivatives, called osazones. Osazone formation requires the oxidation of the carbon group adjacent to the carbonyl carbon. Under normal conditions of the reaction phenylhydrazone is formed first which then reacts with excess of phenylhydrazine to form the osazone.

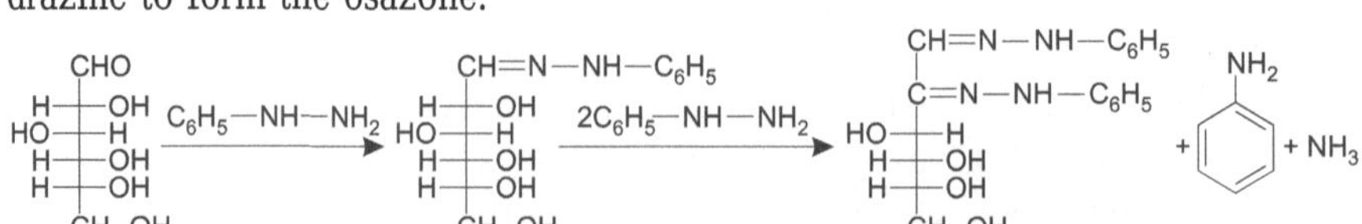

D-Glucose — D-Glucose phenyl hydrozone m.p. 160°C — D-Glucose phenyl osazone m.p. 205°C

Osazones are crystalline derivatives which can be handled very easily. Hence, they are used for characterising the sugars through their characteristic melting points.

Osazones are also useful for establishing the stereochemical relationships in isomeric sugars. Osazone formation destroys the chirality at C-2 position of an aldose. Therefore, epimeric sugars give the same osazone, *e.g.*, D-glucose and D-mannose form the same osazone. D-Fructose also gives an osazone identical with the one obtained from D-glucose. This indicates that two sugars, D-glucose and D-fructose have the same configuration at C-3, C-4 and C-5.

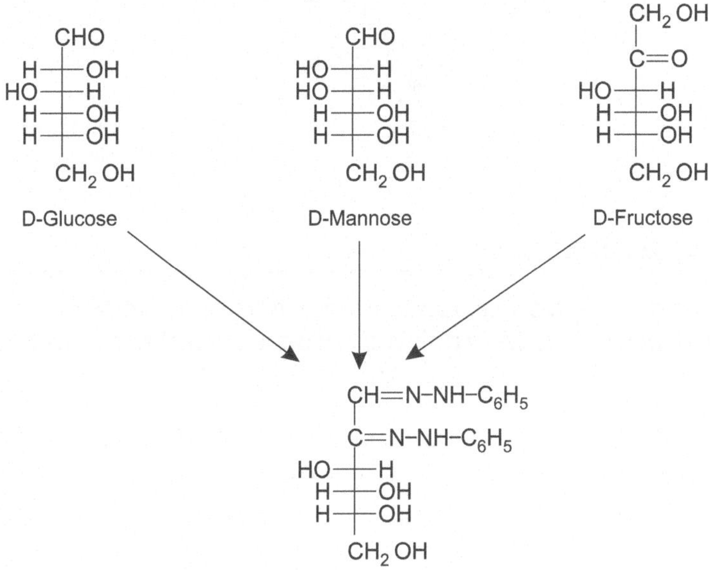

D-Glucose phenylosazone

Mechanism of Osazone Formation

In the osazone formation, we have seen that three molecules of phenyl- hydrazine are used for 1 mole of glucose. Two molecules are incorporated into one mole of glucose and the third is converted to aniline and ammonia. To explain this phenomenon a number of mechanisms were proposed. But the most expected was given by Shemyakin *et al.* in 1965.

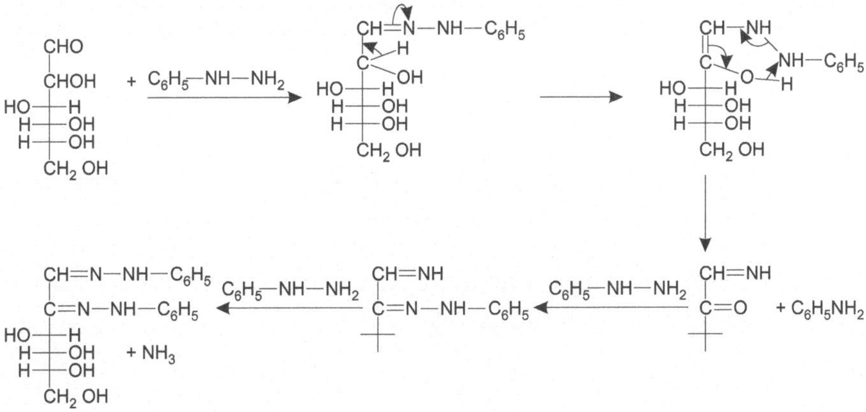

When there is ketohexose; we have the following mechanism:

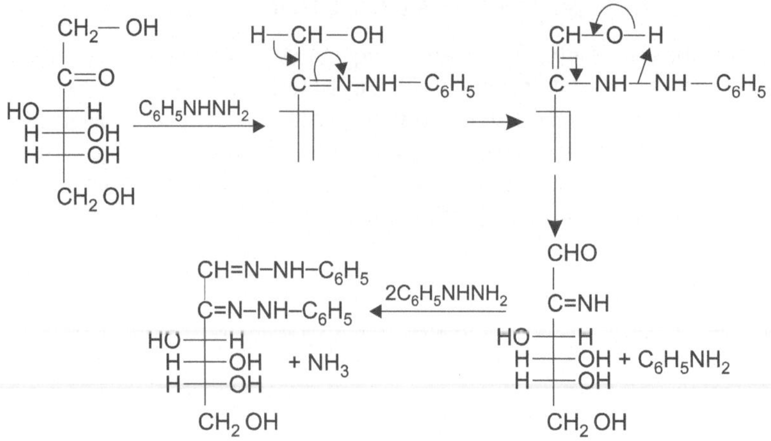

1.12 KETOHEXOSES

All the ketohexoses that occur in nature have the keto group adjacent to a terminal —CH_2OH group. Thus, there are three chiral carbon atoms present in the structural formula of a ketohexose. There are eight stereoisomeric forms possible, *i.e.*, four pairs of enantiomers. Six are known. These are D-(–) and L(+)-fructose, D(+) and L(–)-sorbose, D(+)-tagatose and L(–)-psicose. Out of these only D(–)-fructose, L(–)-sorbose and D(+)-tagatose occur in nature.

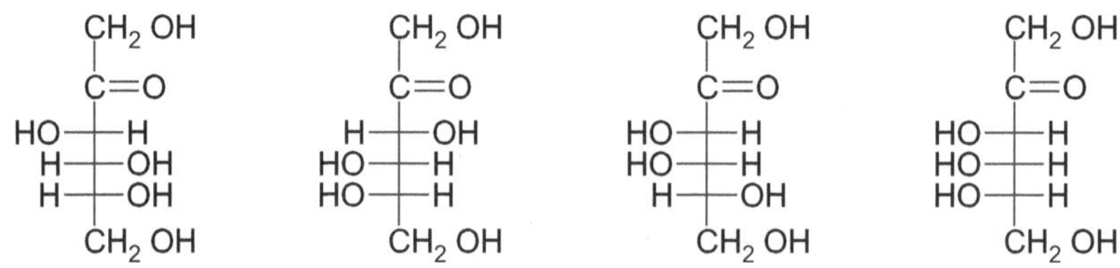

D(–)-Fructose L(–)-Sorbose D(+)-Tagatose L(–)-Psicose

1.13 STRUCTURE OF D(–)-FRUCTOSE

D(–)-Fructose is present in the free state in some fruits and honey and in the combined form with glucose in the most important disaccharide, sucrose. Fructose is also quite sweet. Naturally occurring fructose is laevorotatory and is, therefore, called as laevulose.

Open-chain structure of D(–)-fructose is derived on the basis of the following observations:

1. D(–)-Fructose has got the molecular formula $C_6H_{12}O_6$ as deduced from elemental analysis and molecular weight determination.
2. Fructose forms the oxime when treated with hydroxylamine and a cyanohydrin is formed with hydrogen cyanide. These reactions indicate the presence of a carbonyl group.

3. Fructose is a reducing sugar. It reduces Tollen's reagent and Fehling solution.
4. Fructose is reduced with sodium-amalgam and water to a mixture of hexahydric alcohols, sorbitol and mannitol. The alcohols on treatment with hydriodic acid and red phosphorus at 100°C give a mixture of *n*-hexane and 2-iodohexane.
5. Fructose forms the penta-acetate when treated with acetic anhydride and anhydrous sodium acetate indicating the presence of five hydroxyl groups.
6. Fructose is not oxidised by bromine water. However, oxidation with nitric acid affords a mixture of trihydroxy glutaric acid, tartaric acid and glycollic acid. Since all these acids contain fewer number of carbon atoms than the fructose molecule, the carbonyl group in the fructose must be present as keto group.

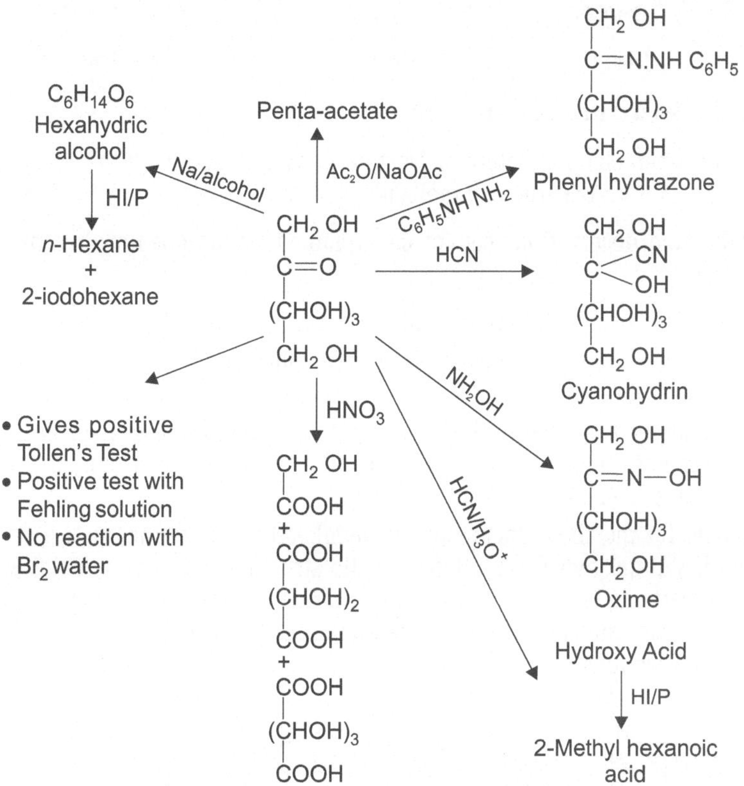

Fig. 1.15. Reactions of Fructose.

The above reactions (given in Fig. 1.15) show that fructose has got a six carbon straight chain structure as given below:

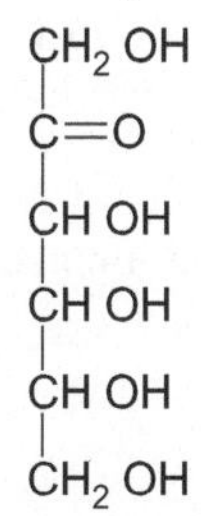

Fructose

A. Configuration of D-Fructose

We have already seen that D-fructose gives the same osazone as obtained from D-glucose. This means that the configuration at C-3, C-4 and C-5 is same in the two sugars, D-glucose and D-fructose. Thus, the configuration of D-fructose is established.

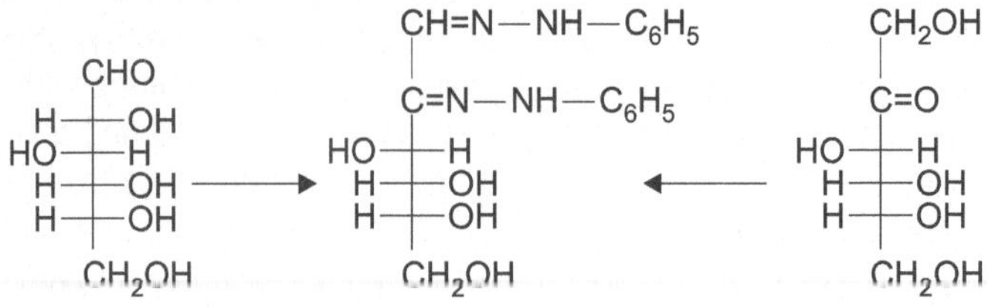

D(+)-Glucose Osazone D(–)-Fructose

B. Cyclic Structure of D-Fructose

The above open chain structure of fructose does not explain all the observations/ reactions given by the fructose molecule.

The following observations cannot be explained by the open-chain structure of D-fructose:

1. Fructose does not form a bisulphite addition product.
2. Fructose undergoes mutarotation in aqueous solutions, *e.g.*, $[\alpha]_D = -133°$ changes to –92°.
3. As has been shown in the case of glucose, fructose also gives two methyl glycosides, known as methyl fructosides when treated with methyl alcohol in the presence of hydrochloric acid.

To account for the above facts a cyclic hemiketal structure for fructose has been proposed. We get two forms of this cyclic structure with the generation of an anomeric carbon. Thus, we can have two structures for the fructose designated as α-D-fructopyranose and β-D-fructopyranose represented by the following Haworth projections:

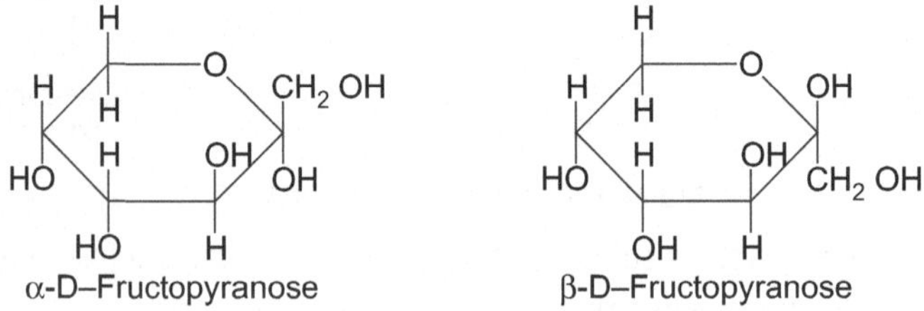

α-D–Fructopyranose β-D–Fructopyranose

In the crystalline form fructose probably exists in the pyranose form with β-isomer predominating (α-form has not been isolated).

However, fructose always occurs in the furanose form when combined in natural products. There is good evidence that a considerable amount of furanose form is present in solution of fructose.

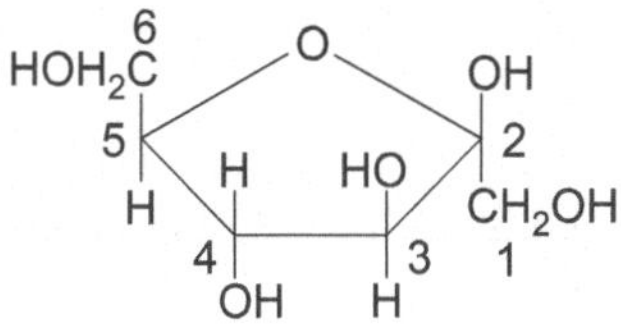

β–D-Fructofuranose

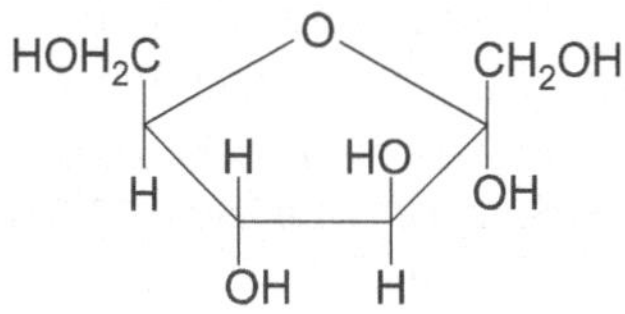

α–D-Fructofuranose

TABLE 1.5 Derivatives of Fructose

Companies	m.p.	$[\alpha]_D$
β-D-Fructopyranose penta-acetate	109°C	- 121°
α-D-Fructo pyranose penta-acetate	70°C	+ 35°
Methyl β-D- Fructopyranoside	120°C	- 172°
Methyl-α-D-Fructopyranoside	102°C	+ 90°.

1.14 INTERCONVERSIONS

1. Conversion of an Aldose to the corresponding Ketose

An aldose is converted to the corresponding ketose *via* the osazone. We know that an aldose and its corresponding ketose give the same osazone. So, an aldose is first converted to an osazone using excess of phenylhydrazine. The osazone is hydrolysed with dilute hydrochloric acid to give osone, which on reduction with zinc and glacial acetic acid gives the corresponding ketose. This is due to the fact that zinc reduces aldehyde group in preference to the ketonic group. For example, D-glucose may be converted into D-fructose in the following manner:

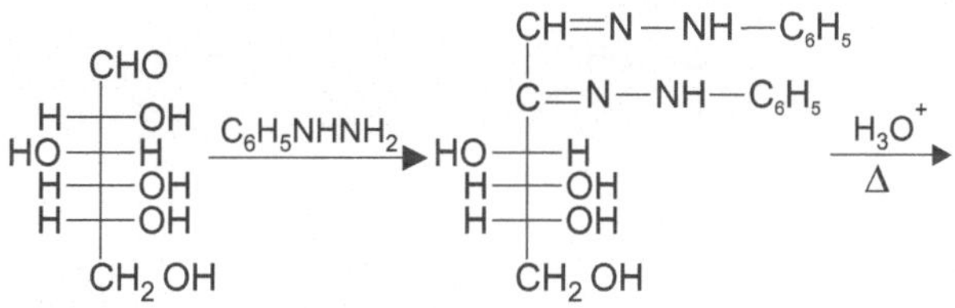

D-Glucose Osazone

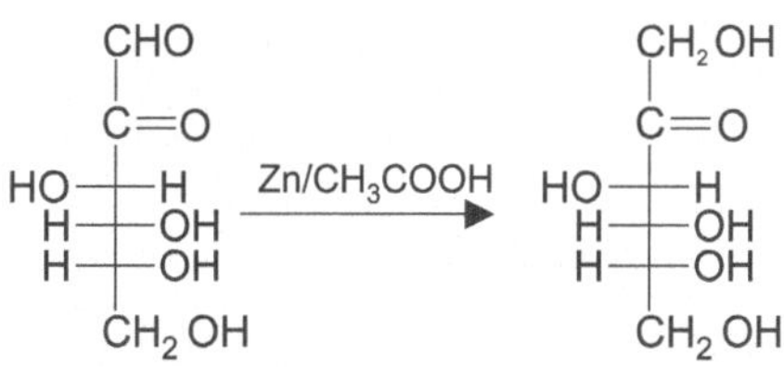

Osone D-Fructose

2. Conversion of a Ketose into the corresponding Aldose

The ketose is reduced catalytically (H_2/Ni) to the corresponding polyhydric alcohol. As a new chiral centre is created, this process results in the formation of two different polyhydric alcohols. These are oxidised together with nitric acid to give the corresponding aldonic acids. These acids on heating give a mixture of γ-lactones, which are separated by fractional crystallisation. The lactones are then reduced with sodium amalgam in a weakly acidic solution to give aldoses, which are isomeric with the original ketose. Thus, D-fructose may be converted into D-glucose and D-mannose as shown below.

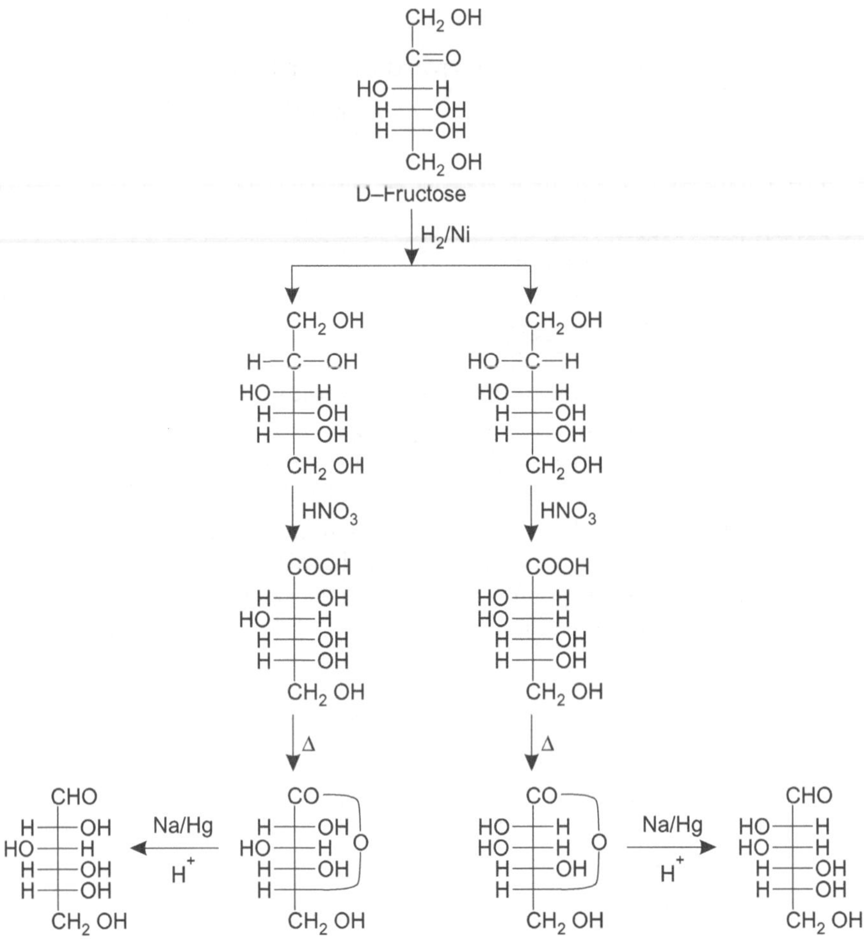

3. Conversion of an Aldose to its Epimer

The aldose is oxidised with bromine water to give the corresponding aldonic acid, which on treatment with pyridine gives a mixture of the original acid and its epimeric aldonic acid. These acids are converted into the corresponding lactones, separated and reduced with sodium amalgam to give the original aldose and its C-2 epimer. Thus, D-glucose may be converted into D-mannose as shown below:

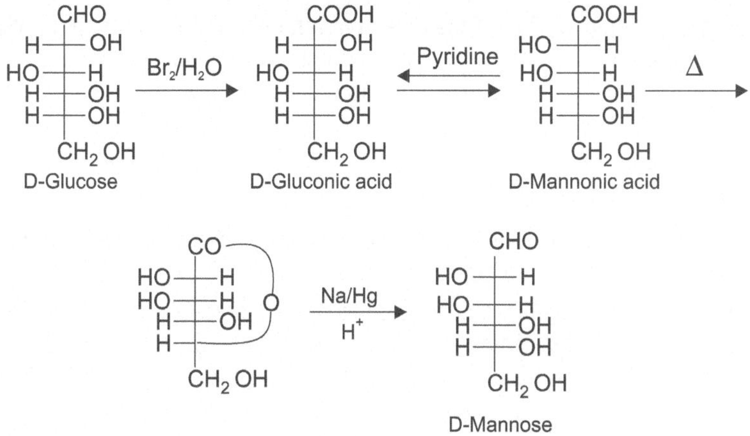

1.15 MALTOSE

Maltose is a simple disaccharide, which is produced by the enzymatic hydrolysis of starch. Diastase (α- and β-amylases) acts on starch to give maltose in about 80% yield. Maltose is also found free in small quantities in some plants, *e.g.*, barley grains.

Maltose is readily fermented by yeast. Under ordinary conditions, it is first converted to glucose by the enzyme, maltase, but at pH 4.5 (conditions which inhibit the action of maltase) direct fermentation occurs.

Structure of Maltose

The structure of maltose is established on the basis of the following observations:

1. Maltose is hydrolysed by dilute acids to yield two moles of D(+)-glucose.
2. Maltose is a reducing sugar, *i.e.*, it gives positive tests with Fehling, Benedict and Tollen's reagents. That means in maltose one of the glucose units has its anomeric hydroxyl free. It may, therefore, undergo normal carbonyl reactions just as the monosaccharides do.

   Maltose also reacts with phenylhydrazine to form a phenylosazone. The reaction goes in a similar way as we have seen in case of glucose. It takes up three moles of phenylhydrazine.
3. Maltose is oxidised by bromine water to give maltobionic acid.
4. Maltose exists in two anomeric forms:

   $$\alpha\text{-Maltose}, \ [\alpha]_D^{25} = +168°$$

   $$\beta\text{-Maltose}, \ [\alpha]_D^{25} = +112°$$

   The anomers undergo mutatoration in water to give an equilibrium mixture of α- and β-forms, $[\alpha]_D^{25} = +136°$.

The above considerations establish clearly that in maltose one of the glucose units is present in a hemiacetal form; the other unit must be present as a glucoside.

5. Complete methylation of maltose gives an octa-methyl derivative, which is non-reducing. This hepta methyl-methyl maltoside on hydrolysis yields one mole of 2,3,4,6-tetra-O-methyl-D-glucose and one mole of 2,3,6-trio-O-methyl-D-glucose.

   The free —OH at C-5 in the second product indicates that it must have been involved in the oxide ring.

   The free —OH at C-5 in the second product indicates that it must have been involved in the oxide ring.

6. Methylation of maltobionic acid gives methyl-octa-O-methyl maltobionate, which is hydrolysed under acidic conditions to give 2,3,4,6-tetra-O-methyl-D-glucose and 2,3,5,6-tetra-O-methyl-D-gluconic acid (Fig. 1.16).

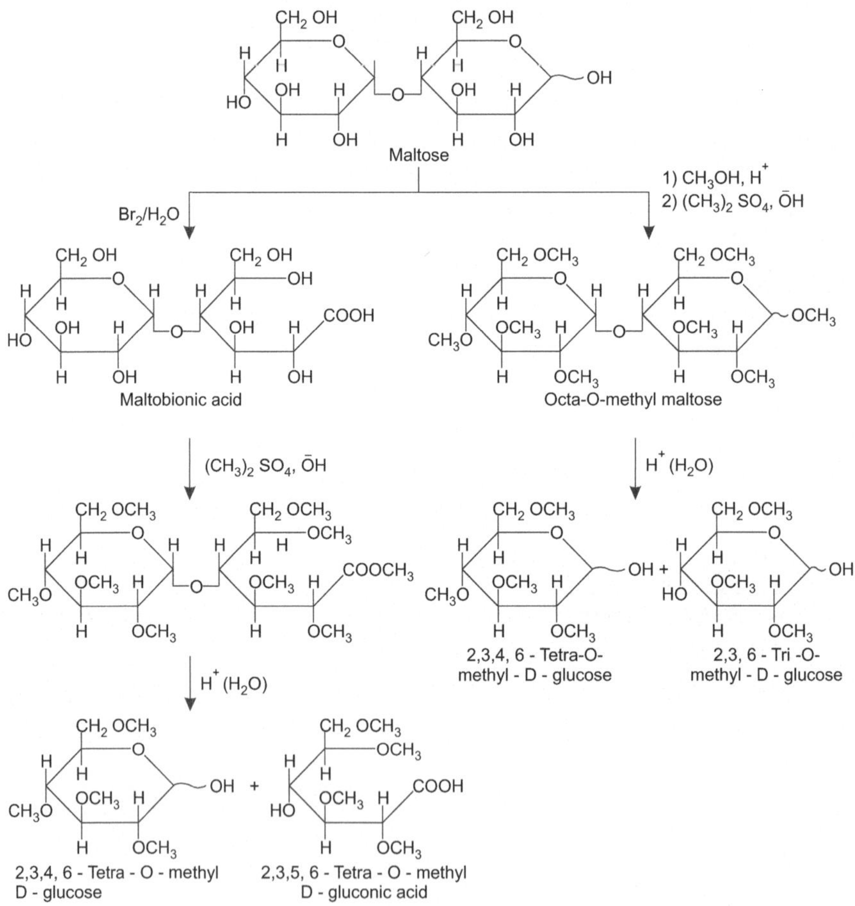

Fig. 1.16

The second product has free hydroxyl at C-4, which must have been involved in the glycosidic linkage.

These facts show that both the glucose units in maltose are in the pyranose form and C-4 hydroxyl of the reducing sugar is involved in a glycosidic linkage with C-1 hydroxyl of the non-reducing glucose.

7. The only remaining question is the stereochemistry of the glycosidic linkage in maltose. This is established by the fact that maltose is hydrolysed by maltase, an enzyme specific for α-glucosides whereas β-glucosidase does not affect the disaccharide. Thus, the glycosidic linkage in maltose is α. The high specific rotation of maltose and its derivatives is consistent with the above facts.

Maltose is, therefore, designated as 4-0-(α-D-glucopyranosyl)-D-glucopyranose. Thus, we have two anomers of maltose as shown below (Fig. 1.17).

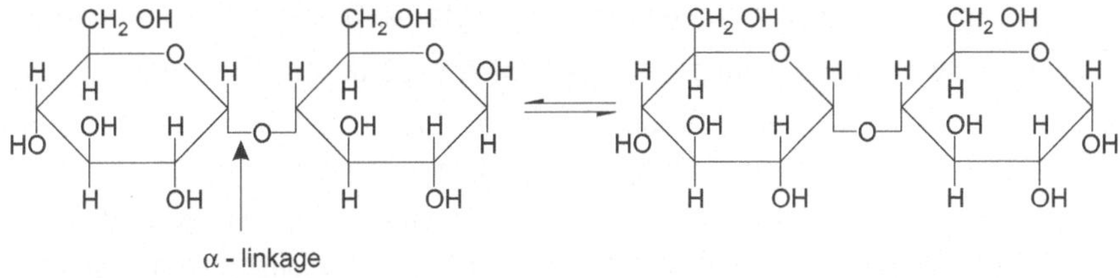

β - anomer
4-0-(α-D-glucopyranosyl)-
β-D-glucopyranose

α - anomer
4-0-(α-D-glucopyranosyl)-
α-D-glucopyranose

Fig. 1.17 Structure of maltose

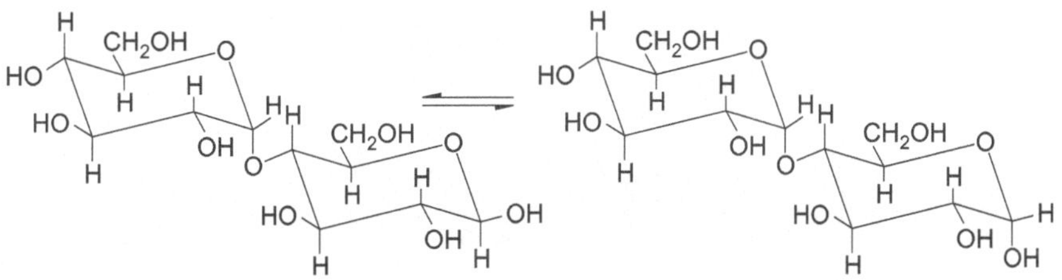

β - anomer α - anomer

Fig. 1.18 Conformations of maltose

1.16 LACTOSE

Lactose is known as milk sugar. It is an example of a disaccharide in which two monosaccharide units are different. Lactose occurs in the milk of mammals, both free and in the combined state in oligosaccharides. The concentration varies from 2% to 9%, the average being about 5% by weight of mammalian milk. It also occurs in a number of fruits. Crystals of lactose are gritty to the tongue and only faintly sweet. The disaccharide is readily converted to lactic acid by the lactic bacillus. This process is responsible for the souring of milk.

Lactose is produced commercially from whey, which is obtained as a byproduct in the manufacture of cheese. Evaporation of whey at temperatures below 95°C causes the less soluble α-anomer to precipitate out. Large amounts of lactose are being used for the preparation of homoeopathic medicines.

Structure of Lactose

The following facts establish the structure of lactose as 4-0-(β-D-galactopyranosyl)-D-glucopyranose:

1. Hydrolysis of lactose affords one mole of galactose and one mole of glucose.
2. Lactose is a reducing disaccharide. It responds to various tests of reducing sugars as given by reducing monosaccharides. It also forms an osazone.
3. Lactose exists in α- and β-forms, which undergo mutarotation.

   The α-isomer is more stable, has m.p. 252°C, $[\alpha]_D$ = +90° changes to +55.4° (water), β-D(+)-lactose has $[\alpha]_D$ = +35°.
4. That the galactose moiety is in the non-reducing part of the molecule is clearly shown by the hydrolysis of lactobionic acid to galactose and gluconic acid (Fig. 1.19).

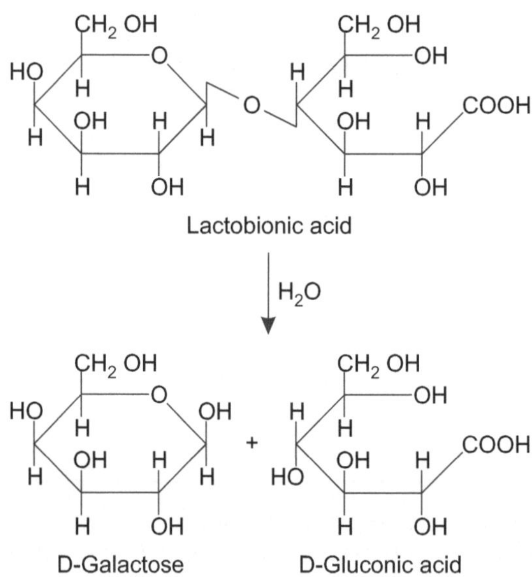

Fig 1.19

5. The glycosidic linkage between galactose and glucose in the lactose molecule is shown by application of the methylation technique.

Hydrolysis of octa-O-methyl lactose affords 2,3,4,6-tetra-O-methyl-D-galactopyranose and 2,3,6-tri-O-methyl-D-glucopyranose. This means that C-4 hydroxyl of the glucose moiety is bound in the glucosidic linkage (Fig. 1.20).

6. Further complete methylation of lactobionic acid gives methyl-octa-O-methyl lactobionate, which on acid hydrolysis yields 2,3,4,6-tetra-O-methyl-D-galactopyranose and 2,3,5,6,-tetra-O-methyl-D-gluconic acid. (Fig. 1.19).

   The above facts show that anomeric hydroxyl (C-1) of the galactose is bound in the form of glycosidic linkage to C-4 of the glucose moiety.

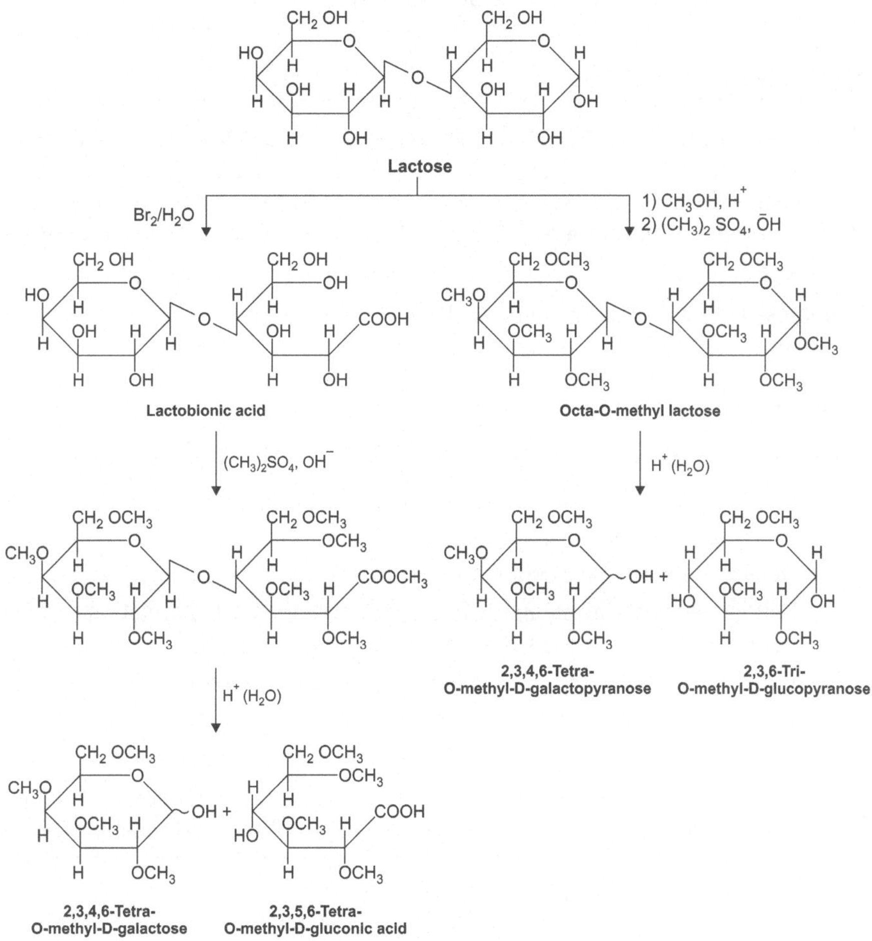

Fig. 1.20

7. Lastly, the hydrolysis of lactose is catalysed by an enzyme called β-galactosidase which is specific for β-galactoside linkages.

Therefore, structure of lactose is given as in Fig. 1.21:

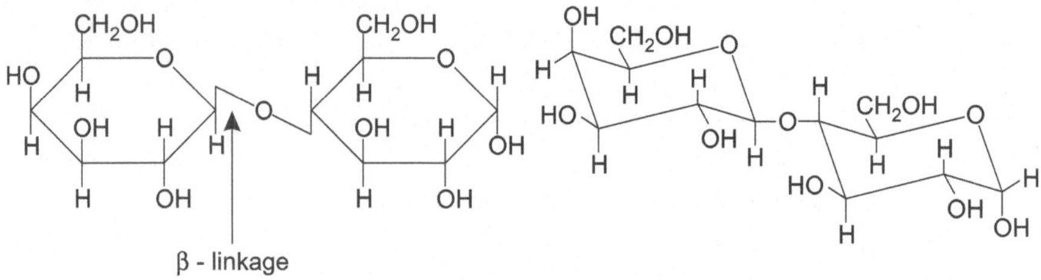

Fig. 1.21 Structure of lactose

1.17 CELLOBIOSE

We have already mentioned that cellobiose is obtained as a result of partial hydrolysis of cellulose. It is generally isolated as the octa-acetate when cellulose is treated with acetic anhydride and sulphuric acid. Acetylation and hydrolysis takes place simultaneously in this process. Alkaline hydrolysis of the octa-acetate gives cellobiose.

Cellobiose with molecular formula $C_{12}H_{22}O_{11}$ resembles maltose in every respect except one and that is the configuration of its glycosidic linkage.

Structure of Cellobiose

The structure of cellobiose is deduced on the basis of the following observations:

1. Like maltose cellobiose also yields two moles of D-glucose on acid hydrolysis.
2. It is a reducing sugar and forms an osazone. On oxidation with bromine water cellobiose gives cellobionic acid.
3. Cellobiose exists in α- and β-forms, which undergo mutarotation. As we have seen in other sugars the β-form of cellobiose is more stable.

   It has m.p. 225°C, $[\alpha]_D$ = +14° changes to +35° in water.

   α-D(+)-Cellobiose, $[\alpha]_D$ = +72°

   β-D(+)-Cellobiose, $[\alpha]_D$ = +14°
4. Methylation of cellobionic acid gives octa-O-methyl derivative, which on hydrolysis affords the same products, *i.e.*, 2,3,4,6-tetra-O-methyl-D-glucose and 2,3,5,6-tetra-O-methyl-D-gluconic acid as hydrolysis of octa-O-methylmaltobionic acid.

   That means cellobiose is isomeric with maltose.
5. The glycosidic linkage in cellobiose is, however, decided on the basis of enzymatic hydrolysis. The hydrolysis of cellobiose is catalysed by β-glucosidase and not by α-glucosidase.

Therefore, the structure of β-D-cellobiose is given as 4-0-(β-D-glucopyranosyl)-β-D-glucopyranose (Fig. 1.22).

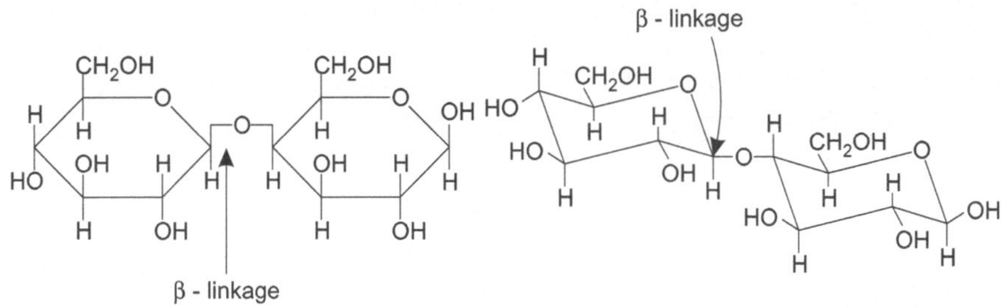

Fig. 1.22 Structure of cellobiose

1.18 SUCROSE

We are quite familiar with the ordinary table sugar, which is a disaccharide called sucrose. Large amounts of it are being consumed daily the world over.

Sucrose is one of the most widespread sugars in nature and is one of the few pure organic compounds produced industrially in very large amounts. It occurs almost universally in the plant kingdom and is found in all parts of the plant. The richest sources are sugar cane (*Saccharum officinarum*) and the sugar beet (*Beta vulgaris*). In sugar cane, it occurs to the extent of about 20% of the juice. Sucrose is commercially obtained from these sources.

Sucrose exists in two crystalline forms, sucrose A, m.p. 184°–185°C, which is the stable form obtained on recrystallisation from most solvents and sucrose B, m.p. 169°–170°C, obtained by recrystallisation from methanol. Molten sugar on cooling forms a glass which slowly becomes crystalline and loses its transparency. That is how we get various preparations of sucrose. At 195°–200°C sucrose changes into a brown uncrystallisable product.

Sucrose is dextrorotatory having $[\alpha]_D = +66°$.

D-Glucose is also dextrorotatory. The equilibrium mixture of α- and β-anomers of glucose has $[\alpha]_D = +52.5°$. But D-fructose is laevorotatory, the equilibrium mixture having $[\alpha]_D = -92.4°$.

Sucrose is hydrolysed by invertase or on warming with dilute acids to give equimolar mixture of D-glucose and D-fructose. In the process of hydrolysis the dextrorotatory sucrose solution becomes laevorotatory because an equilibrium mixture of D-glucose and D-fructose has $[\alpha]_D = -20°$. The commonly encountered mixture is called 'invert sugar'. The name is given from the process of inversion in the sign of rotation that occurs during its formation.

$$\text{Sucrose} \xrightarrow[\text{or Invertase}]{\overset{+}{H}(H_2O)} \alpha\text{- and }\beta\text{-D-glucose} + \alpha\text{- and }\beta\text{-D-fructose}$$

$[\alpha] = +66°$ $\qquad$ $[\alpha]_D = +52.5°$ $\qquad$ $[\alpha]_D = -92.4°$

$$\downarrow$$

$[\alpha]_D = -20°$

A number of organisms including the honey bee have enzymes that catalyse the hydrolysis of sucrose. These enzymes are usually called invertases and are specific for the β-fructofuranoside linkages. Honey is largely a mixture of D-glucose, D-fructose, and sucrose.

Acetylation of sucrose gives an octa-acetate, $[\alpha]_D = +59.6°$, which exists in two crystalline forms having m.p. 69–70°C and 75°C respectively.

Structure of Sucrose

The structure of sucrose is based on the following observations:

1. Acid hydrolysis of one mole of sucrose yields one mole of D-glucose and one mole of D-fructose.
2. Sucrose is a non-reducing sugar. It does not reduce Tollen's reagent and gives negative test with Benedict's solution.
3. Sucrose does not form an osazone. However, on treatment with dilute hydrochloric acid inversion takes place. The mixture then deposits osazone on treatment with phenylhydrazine. It results in inversion of the sign of rotation. The resulting mixture is known as invert sugar.
4. Sucrose does not undergo mutarotation.

   These facts suggest that there is no hemiacetal or hemiketal group present in the glucose or fructose portion of sucrose. The two units are linked together by a glycosidic linkage between the two anomeric carbons. Thus, C-1 of D-glucose is linked to C-2 of D-fructose.
5. The ring size of D-glucose and D-fructose in sucrose is established on the basis of usual methylation studies. Thus, methylation of sucrose gives an octamethyl derivative which on acid hydrolysis gives 2,3,4,6-tetra-*O*-methyl glucose and 1,3,4,6-tetra-*O*-methyl fructose. The formation of these products indicates that in sucrose, glucose is in the pyranose form and fructose is in the furanose form (Fig. 1.23).
6. The configuration at the glycosidic linkage follows from studies with enzymes.

Sucrose is hydrolysed by an α-glucosidase obtained from yeast and not by β-glucosidase. This shows α-configuration at the glucoside portion. It is unaffected by emulsion (β-glucosidase).

Sucrose is also hydrolysed by sucrase, an enzyme specific for β-fructofuranosides. This indicates that fructose has got β-configuration.

$$\text{Sucrose} \xrightarrow[\text{NaOH}]{(CH_3)_2SO_4} \text{Octamethyl sucrose} \xrightarrow{H_3O^+}$$

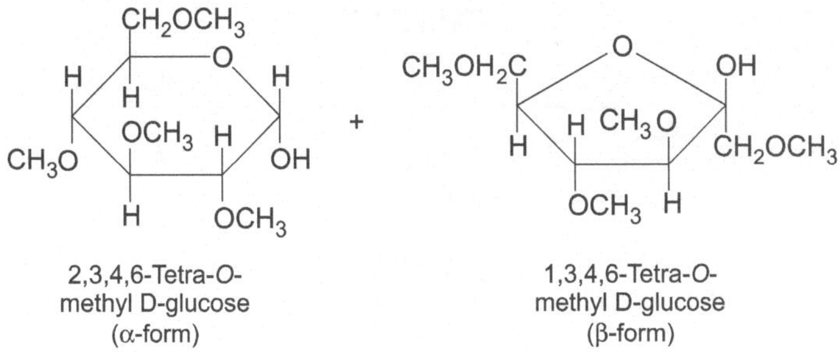

Fig. 1.23.

Thus, structure of sucrose is (21) α-D-glucopyranosyl-β-D-fructofuranoside.

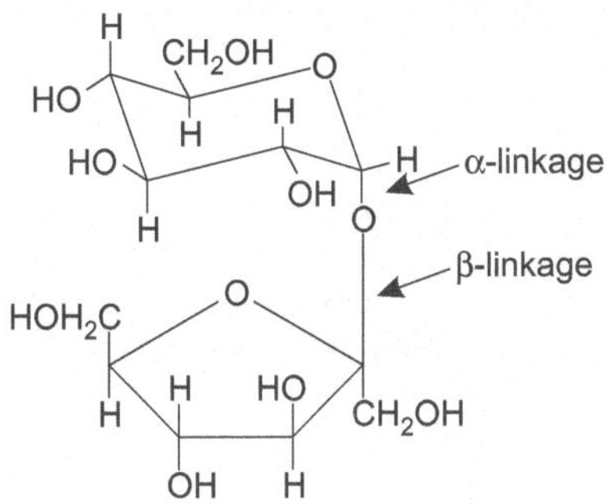

Fig. 1.24 (21) Structure of Sucrose

1.19 DEOXY SUGARS

Deoxy sugars are those compounds in which one or more hydroxyl groups have been replaced by hydrogens. Mono, di- and trideoxy sugars occur naturally. The 6-deoxyaldohexoses occur widely in nature *e.g.*, 6-deoxy-L-mannose, known as L-rhamnose and 6-deoxy-L-galactose (L-fucose) are quite common.

1. 2-Deoxy-D-Ribose

The most important 2-deoxyaldose is 2-deoxy-D-erythropentose or 2-deoxy-D-ribose as it is more commonly, but incorrectly called. It is of great importance because of its occurrence in the cell nuclei as a component of deoxyribonucleic acid (DNA) in which it exists in the furanosyl form.

The melting points and specific rotations of the two anomers and their acetates are given in Table. 1.5.

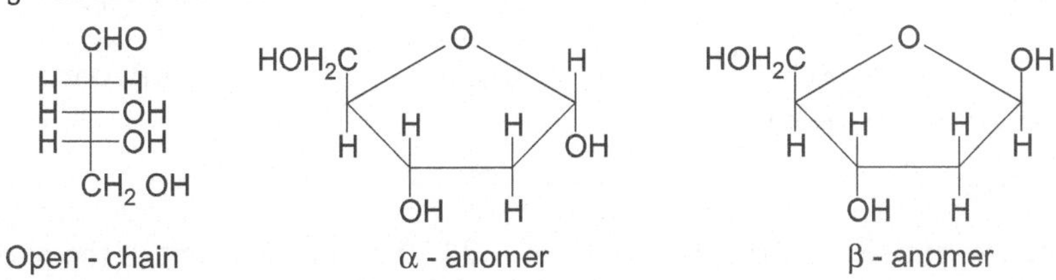

TABLE 1.5

2-Deoxy-D-ribose	*m.p.*	*[α]$_D$*
α-nomer (furanose)	78-82°C	- 55°
β-anomer (furanose)	96-98°C	-91° → - 58° (mutarotates)
Tri-O-acetate (β-furanose)	98-99°C	- 139°
Tri-O-acetate (β-pyranose)	90-91°C	- 169°.

Crystalline 2-deoxy-β-D-erythropentose has been shown by X-ray studies to exist in the seemingly unfavourable pyranose conformation.

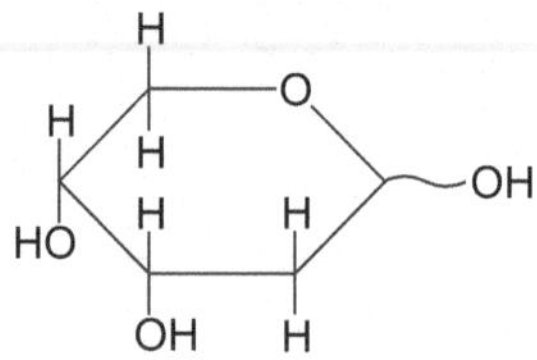

Haworth structure

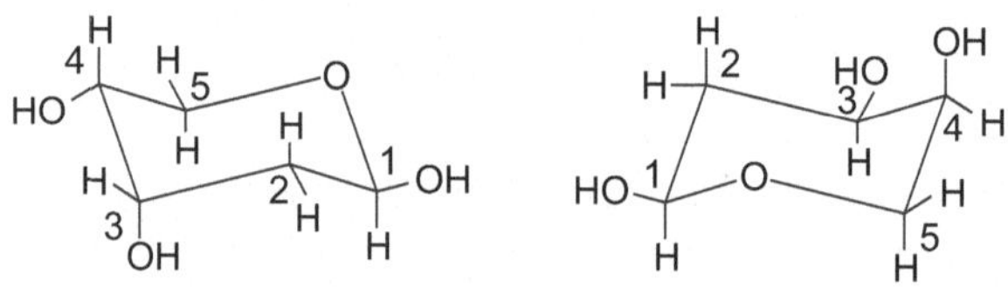

Synthesis of 2-Deoxy-D-ribose

2-Deoxy-D-ribose has been synthesised by the nitro methane condensation method as given below:

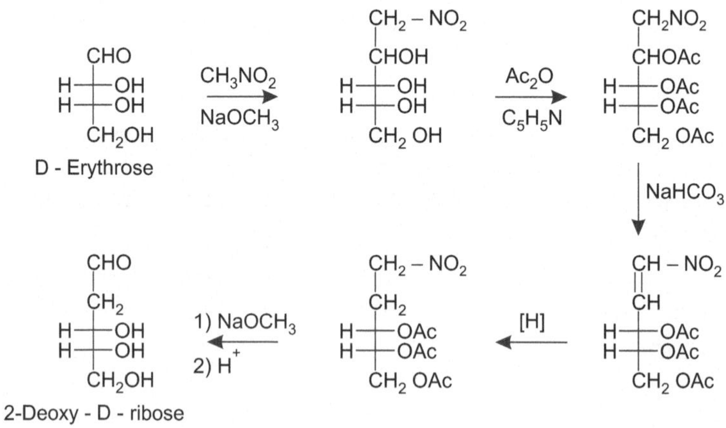

2. D-Ribose

D-Ribose, m.p. 87°C, $[\alpha]_D$ = – 23.1° → – 23.7°, is a component of the nucleic acid, RNA and several vitamins and coenzymes. Therefore, it is present in the cells of all animal and plant material. It is also a component of the antibiotics

Neomycins A and B and Paromomycin, D-Ribose arises in nature from D-glucose as the 6-phosphate as oxidative decarboxylation. The pentose may be prepared by the partial hydrolysis of yeast nucleic acids either with magnesium oxide or enzymes to nucleosides which are then cleaved by acid hydrolysis to give D-ribose after the removal of the aglycone.

The sugar occurs in natural products in the furanosyl form. In aqueous solution, the free sugar exhibits a complex mutarotation and considerable quantities of the furanose and straight chain modifications are thought to be present. D-ribose is not fermentable by ordinary yeasts. Commercial D-ribose is best purified *via* its anilide.

Mutarotation of D-Ribose

Evidence for the pyranose, furanose and aldehydo forms of D-ribose in aqueous solution has been obtained by NMR studies in D_2O, but none of the other aldopentoses showed significant amounts of furanoses. It was estimated that at equilibrium at 35°, ribose yields 6% of α-furanose, 18% β-furanose, 20% α-pyranose and 50% β-pyranose forms. Ribose forms about 8.5% of the aldehydo form whereas glucose forms only 0.01% as revealed by polarography.

The conformation of the furanose ring has an important bearing upon the equilibrium composition. Both pyranose and furanose forms of D-ribose contain unfavourable non-bonded interactions which are probably responsible for the abnormal concentration of the aldehydo form.

The concentration of aldehydo and furanose forms in solution tend to increase with conformational instability of the pyranose form. As axial-hydroxyl on the stable chair conformations as in D-galactose and D-arabinose results in greater amounts of aldehydo forms in solution than are produced by all equatorial D-glucose and D-xylose which exist as expected predominantly in the favourable pyranose chair conformations.

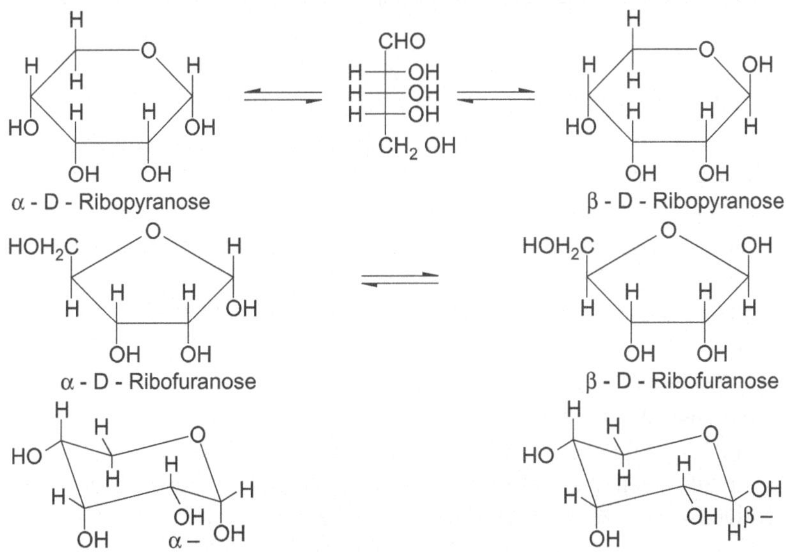

It has been shown by NMR that there is a conformational equilibrium between the chair forms of β-D-ribopyranose tetra-acetate.

1.20 POLYSACCHARIDES

A polysaccharide is a naturally occurring sugar polymer. As the name implies the polymer contains many sugar units, the actual number runs into thousands. Molecular weights have been reported as high as 1,000,000. In spite of the large number of —OH groups, only the smaller polymers exhibit water solubility. The bulk of the polysaccharides are insoluble in water. They may be broken into soluble derivatives only by enzymes or by acid or base catalysed hydrolysis.

Starch and cellulose are the most important of the polysaccharides. We have already mentioned that starch is the reserve food supply for plants while cellulose is the structural unit which provides both rigidity and form. We use starch and cellulose in the same way as nature does, starch as food and cellulose as basic structural material.

1. Cellulose

As mentioned earlier, cellulose is the main constituent of cell walls of plants to support the framework. Thus, cellulose is the chief component of wood and plant fibres and is very widely distributed in nature. Cotton, for example, is almost 100% pure cellulose. The cell walls in wood, are only 50% cellulose. The chief impurities which are associated with raw cotton are waxes and fats. These impurities are removed by washing the raw cotton with alcohol. It is then treated with hot caustic soda solution, washed with water and dried to give pure amorphous cellulose. It is insoluble in water and is tasteless. We use cellulose for its structural properties. We use wood for houses and furniture, as cotton for clothing, as paper for communications and packaging.

Structure of Cellulose:

1. Cellulose is a polysaccharide with molecular formula $(C_6H_{10}O_5)_n$.
2. It is composed entirely of D-glucose units as complete hydrolysis by acid yields D(+)-glucose as the only monosaccharide. That means, D(+)-glucose is the monomer of cellulose.
3. Hydrolysis of completely methylated cellulose gives a high yield of 2,3,6-tri-*O*-methyl-D-glucose. Therefore, cellulose is made up of chains of D-glucose units, each unit joined by a glycoside linkage to C-4 of the next.
4. On treatment with acetic anhydride and sulphuric acid, cellulose yields octa-*O*-acetyl-cellobiose. Therefore, cellulose may be considered as a polymer of cellobiose.
5. Further, cellulose on hydrolysis with enzymes yields cellobiose. We have already seen that cellobiose has β-glycosidic linkage. That means all glycosidic linkages in cellulose are β-linkages.

Therefore, structure of cellulose is represented as in Fig. 1.25.

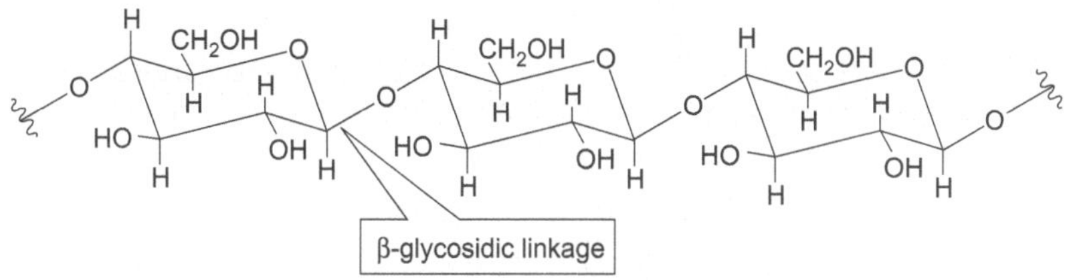

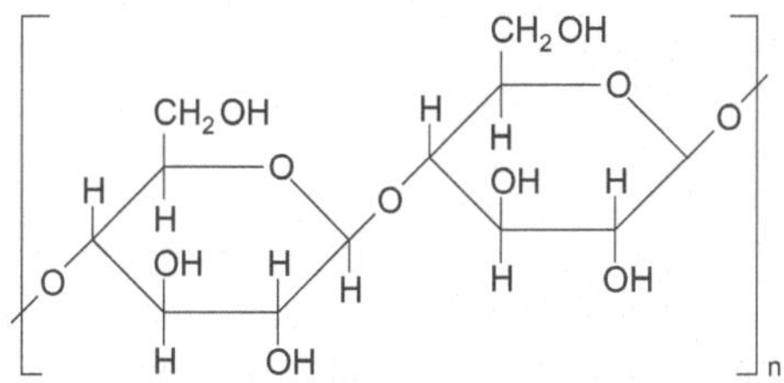

Fig. 1.25 Structure of cellulose

The above structure of cellulose indicates that it is made up of a long chain of cellobiose units in which glucose is linked alternately through the carbon atoms in positions 1 and 4. The linkage is β and conformation is completely equatorial. Each glucose unit contains three free hydroxyl groups except at the ends of chain. One end contains a free or potential aldehyde group and the other end has an extra hydroxyl group at position 4.

Cellulose molecule has been shown to have molecular weights ranging from 2,50,000 to 1,000,000. That means there are at least 1500 glucose units present in cellulose. These long chains lie side by side in bundles, which are held together by hydrogen bonds between the neighbouring —OH groups. These bundles are twisted together to form rope like structures which are grouped to form the fibres. In wood, these cellulose ropes are embedded in lignin to give the compact structure.

This compact structure of cellulose gives another interesting and important fact about cellulose. We, human beings, cannot digest cellulose. The digestive enzyme of humans cannot attack its β, 1:4 linkages. Hence, cellulose cannot serve as a food source for us as can starch. However, cows and other animals can use cellulose of grass as a food source because their digestive juices contain β-glucosidases.

2. Starch

We have already mentioned above that starch makes up the reserve food supply of plants and occurs chiefly in seeds. It is more water soluble than cellulose, more easily hydrolysed and hence more readily digested. We use starch as a food. The chief sources are potatoes, corn, wheat, rice, barley etc. Starch occurs as granules in the roots, tubers and seeds of plants which vary in size and shape depending upon their plant source.

Starch is not a pure compound. It is a mixture of two polysaccharides – amylose and amylopectin. Amylose is water-soluble fraction and gives a deep blue colour with iodine. In general, starch contains 20% of amylose and 80% of amylopectin, which is water-insoluble and gives no colour with iodine.

Both these forms of starch correspond to different carbohydrates of high molecular weight having the general formula $(C_6H_{10}O_5)_n$. When starch is treated with dilute acid or appropriate enzymes, it is slowly hydrolysed to maltose and then to D-glucose. That means both amylose and amylopectin are made up of D(+)-glucose units, but differ in molecular size and shape.

A. Structure of Amylose

1. Hydrolysis of amylose gives maltose as the only disaccharide and D(+)-glucose as the only monosaccharide. Therefore, amylose is made up of chains of many D(+)-glucose units. As maltose is also the product of hydrolysis of amylose, that means each glucose unit is joined by a-glycosidic linkage to C-4 of the next one.
2. When amylose is methylated and hydrolysed, we get mainly 2,3,6-trio-O-methyl-D-glucose. That is expected when each D(+)-glucose unit in amylose is joined to two others, it contains only three free —OH groups which are methylated to give the tri-methyl product (Fig. 1.26).

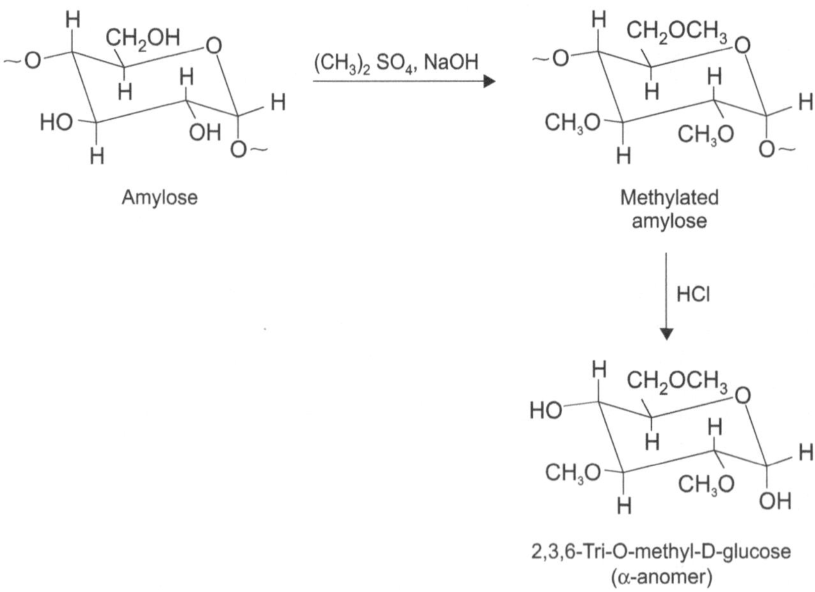

Fig. 1.26.

However, we also get a small amount of 2,3,4,6-tetra-O-methyl-D-glucose (Fig. 1.27), which is about 0.2 – 0.4% of the total product.

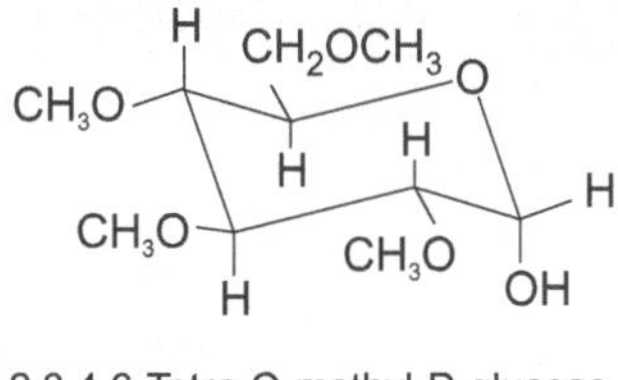

2,3,4,6-Tetra-O-methyl-D-glucose
(α-anomer)

Fig. 1.27.

Therefore, we conclude that each D-glucose unit in amylose is attached to two other D-glucose units, one through C-1 and the other through C4. The formation of the above products indicates that amylose is a straight chain polymer with one end having free aldehyde group. The other end has a D-glucose unit that has a free —OH at C4.

Thus, each molecule of completely methylated amylose on hydrolysis should give one molecule of 2,3,4,6-tetra-O-methyl-D-glucose along with a number of tri-O-methyl-D-glucose molecules. The formation of 0.2% of tetra-O-methyl derivative indicates the chain length of about 500 units. But physical methods suggest that the chains are longer, 1000 to 4000 glucose units per molecule of amylose. That means some degradation results during methylation and hydrolysis of the amylose molecule.

Therefore, amylose is a linear polymer of D-glucose units joined together by α-linkages as in maltose. There is no branching of the chain (Fig. 1.28). Amylose differs from cellulose only in the stereochemistry of the glycosidic linkage.

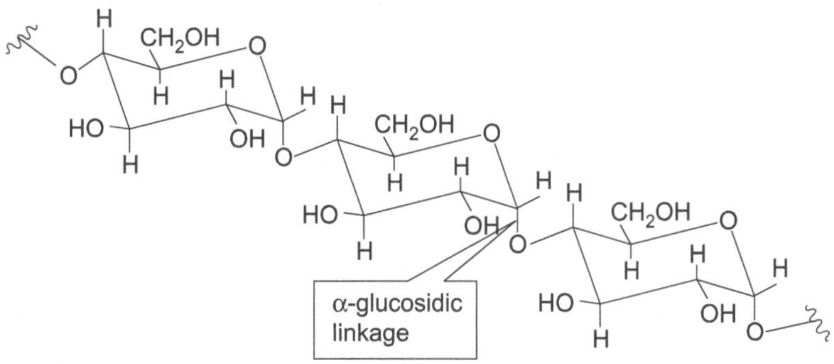

Fig. 1.28 Amylose

Amylose is the fraction of starch that gives the intense blue colour with iodine. X-ray analysis shows that chains of amylose are coiled to give a helix like structure.

The inside of the helix has just the right space to accommodate an iodine molecule. This is the basis of the starch iodide test for oxidising agents. The material to be tested is added to an aqueous solution of starch and potassium iodide. If the material is an oxidiser, some of the iodide is oxidised to iodine, which forms a blue complex with amylose of the starch.

B. Structure of Amylopectin

1. It is the insoluble fraction of starch.

   Amylopectin is also hydrolysed to the single disaccharide, maltose.

2. On methylation and hydrolysis of amylopectin, we get mainly 2,3,6-tri-*O*-methyl-D-glucose. Therefore, amylopectin is made up of chains of D-glucose with each unit joined by a-glycoside linkage of C-4 of the next unit.

3. Hydrolysis of methylated amylopectin also gives about 5% of 2,3,4,6-tetra-*O*-methyl-D-glucose along with an equal amount of 2,3-di-O-methyl-D-glucose (Fig. 1.29).

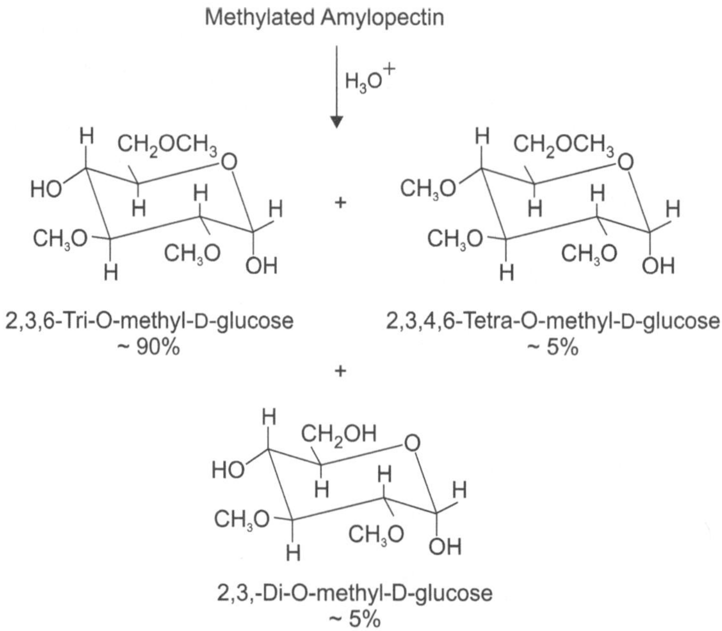

Fig. 1.29.

Formation of 5% of tetra-*O*-methyl-glucose indicates that there are 20 end D-glucose units per chain and the formation of dimethyl-glucose suggests that amylopectin has a highly branched structure consisting of several hundred short chains of about 20-25 D-glucose units. One end of each of these chains is joined through C-1 to C-6 on the next chain.

Therefore, amylopectin is a branched α-1, 4′ polymer of glucose with α-1,6′ linkage that provides the attachment point for another chain (Fig. 1.30).

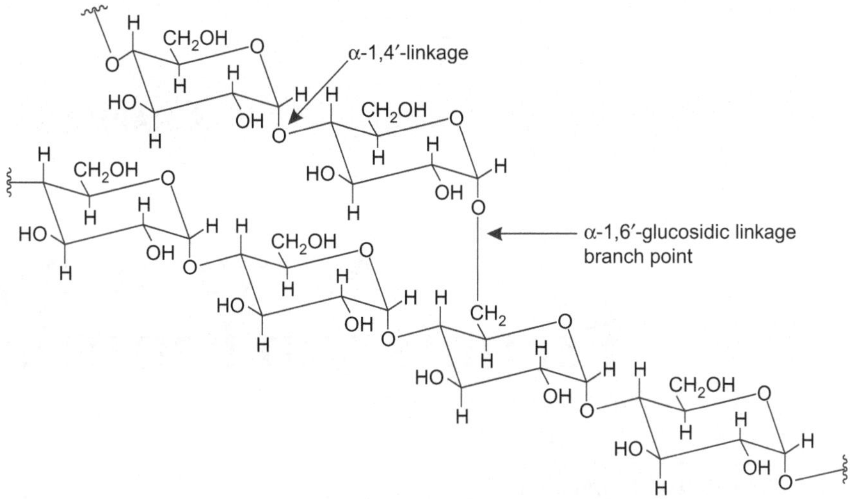

Fig. 1.30 Structure of Amylopectin

❑❑❑

CHAPTER 2

AMINO ACIDS, PEPTIDES AND PROTEINS

Learning Objectives

In this chapter, we will study:

- Amino acids-classification and stereochemistry
- The essential amino acids
- Synthesis and reactions of amino acids
- Amino acids as dipolar ions and as acids and bases.
- Isoelectric point
- Separation of amino acids
- Polypeptides-Nomenclature and synthesis
- Solid phase peptide synthesis
- End group analysis-identification of N-terminal and C-terminal amino acids
- Proteins-importance, classification
- Structure of proteins:primary, secondary, tertiary, quaternary structures
- Denaturation of proteins

2.1 INTRODUCTION

The three important groups of biopolymers are polysaccharides, proteins and nucleic acids. We already know the importance of polysaccharides. They act as food reserves in animals and human beings and in plants as structural materials. Proteins are the most abundant organic molecules in animals and human beings. We consume proteins primarily for growth and maintenance. Any balanced diet must contain adequate amount of proteins. They perform a variety of functions in living organisms. They are the principal material of muscles, skin, tendons, nerves and blood. As enzymes and hormones, proteins catalyse and regulate the reactions that occur in the body; as haemoglobins they transfer oxygen to its most remote corners. Even in plants, where carbohydrates are more abundant as structural materials, proteins are present in those parts that are responsible for growth and reproduction.

The fundamental structure of proteins is simple. Proteins are biopolymers of α-amino acids. They consist of long chains of α-amino acids bonded to each other by amide bonds also known as peptide linkages between the carboyxlic group of one amino acid with amino group of another. Thus, proteins are high molecular weight polypeptides. Their molecular weight may range from 5000–1,00,000 units. A single protein molecule contains hundreds or even thousands of amino acid units. About twenty different amino acids are the building blocks of proteins. The number of different combinations, that is the number of different protein molecules that are possible, is almost infinite.

2.2 AMINO ACIDS – THE BUILDING BLOCKS OF PROTEINS

We have seen above that α–amino acids are the building blocks of proteins. Hydrolysis of proteins with acid or base yield a mixture of different amino acids. Naturally occurring proteins on hydrolysis may yield twenty-two different amino acids. These amino acids have an important structural feature in common. All are α–amino acids *i.e.*, amino group is bonded to the α–carbon atom next to the carboxyl group. The simplest amino acid is glycine. All others can be derived from glycine.

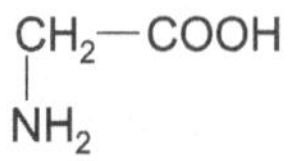

Glycine

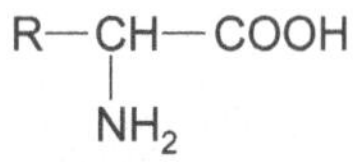

A substituted amino acid

where R group is different for different amino acids.

With the exception of glycine, all naturally occurring amino acids have the L-configuration at the α–carbon. That means they have the same relative configuration as L-glyceraldehyde.

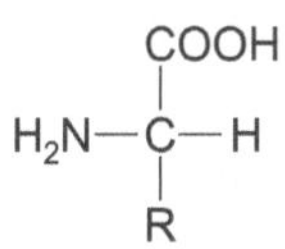

An L-amino acid

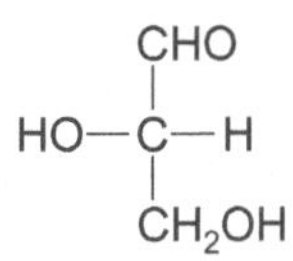

L-glyceraldehyde

As the two functional groups present in amino acids are acidic and basic, all amino acids are amphoteric and actually exist as zwitterions. Thus, the simplest amino acid glycine exists in the form shown below rather than aminoacetic acid.

$$\underset{\text{Aminoacetic acid}}{H_2N{-}CH_2{-}COOH} \rightleftharpoons \underset{\text{Zwitterion}}{H_3\overset{+}{N}{-}CH_2{-}CO\bar{O}}$$

2.3 CLASSIFICATION OF AMINO ACIDS

The 22 α–amino acids that are obtained from proteins can be classified into four different groups on the basis of the structures of their side chains:

1. Neutral Amino Acids
2. Acidic Amino Acids
3. Neutral Amino Acids with polar side chains
4. Basic Amino Acids

These are listed with their symbols, abbreviations and structures in Table 2.1.

Only 20 of the 22 amino acids listed in Table 2.1 are actually used by cells when they synthesise proteins. Hydroxyproline is synthesised from proline and cystine is synthesised from cysteine.

Asparagine and glutamine are derived from aspartic acid and glutamic acid.

TABLE 2.1 Amino Acids commonly found in proteins

1. Neutral Amino Acids			
Name	**Symbol**	**Abbreviation**	**Structure**
Glycine	G	Gly	$CH_2(NH_2){-}COOH$
Alanine	A	Ala	$CH_3{-}CH(NH_2){-}COOH$
Valine	V	Val	$(CH_3)_2CH{-}CH(NH_2){-}COOH$ e
Leucine	L	Leu	$(CH_3)_2CH{-}CH_2{-}CH(NH_2){-}COOH$ e
Isoleucine	I	Ile	$(CH_3)_2CH{-}CH(CH_3){-}CH(NH_2){-}COOH$ e
Proline	P	Pro	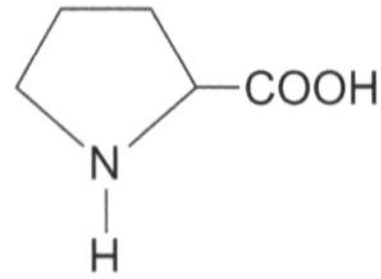

Contd...

Name	Symbol	Abbreviation	Structure	
Phenylalanine	F	Phe	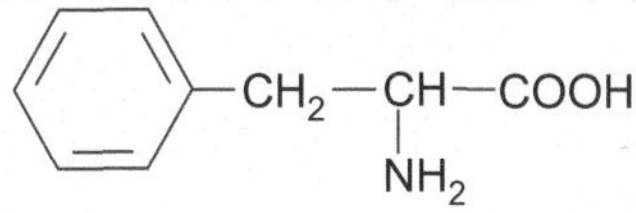	e
Tryptophan	W	Trp	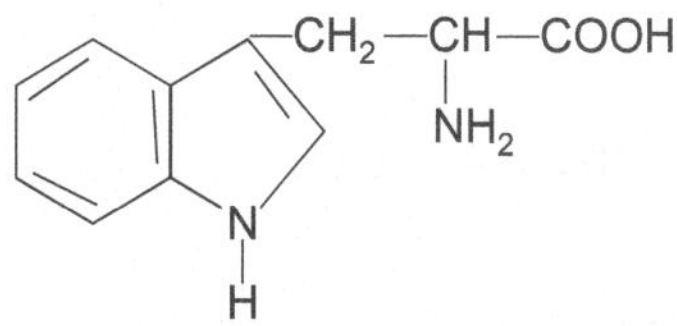	e
Methionine	M	Met	$CH_3-S-CH_2-CH_2-CH(NH_2)-COOH$	e
Cystine		$(Cys)_2$	$(-S-CH_2-CH(NH_2)-COOH)_2$	
2. Acidic Amino Acids				
Aspartic acid	D	Asp	$HOOC-CH_2-CH(NH_2)-COOH$	
Glutamic acid	E	Glu	$HOOC-CH_2-CH_2-CH(NH_2)-COOH$	
3. Neutral Amino Acids with polar side chains				
Serine	S	Ser	$HO-CH_2-CH(NH_2)-COOH$	
Threonine	T	Thr	$CH_3-CH(OH)-CH(NH_2)-COOH$	e
Cysteine	C	Cys	$HS-CH_2-CH(NH_2)-COOH$	
Tyrosine	Y	Tyr	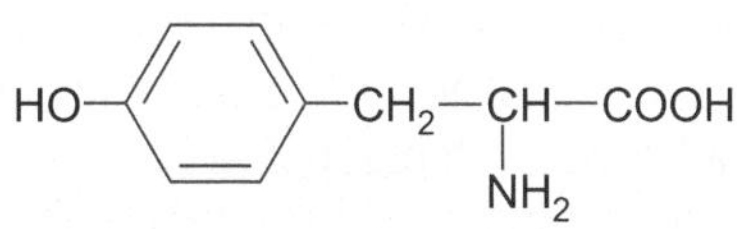	
Asparagine	N	Asn	$H_2N-C(=O)-CH_2-CH(NH_2)-COOH$	
Glutamine	Q	Gln	$H_2N-C(=O)-CH_2-CH_2-CH(NH_2)-COOH$	

Contd...

Name	Symbol	Abbreviation	Structure
Hydroxyproline		Hyp	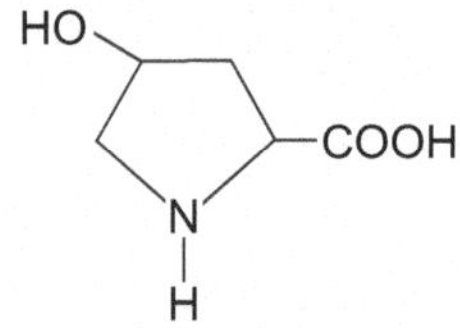
4. Basic Amino Acids			
Lysine	K	Lys	$CH_2(NH_2)-CH_2-CH_2-CH_2-CH(NH_2)-COOH$ **e**
Histidine	H	His	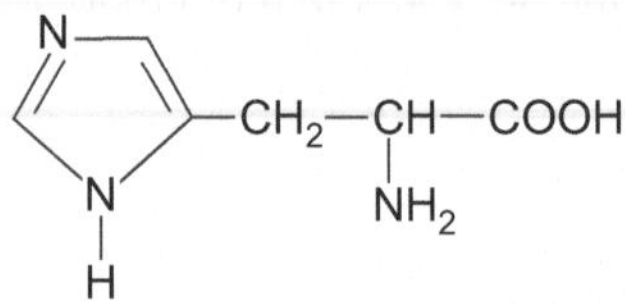 **e**
Arginine	R	Arg	$H_2N-C(=NH)-NH-CH_2-CH_2-CH_2-CH(NH_2)-COOH$ **e**

2.4 THE ESSENTIAL AMINO ACIDS

Amino acids are synthesised by all living organisms, animals and plants for their protein requirements. The biosynthesis of proteins requires the presence of all the constitutent amino acids. If one of the 20 amino acids is missing or in short supply, protein biosynthesis is inhibited. Some organisms, such as *E.Coli* can synthesise all the amino acids that they need. However, human body is unable to synthesise some of these amino acids required for making proteins. These are called **essential amino** acids. For adult humans there are ten amino acids, which have been designated with the superscript **e** in the Table 2.1. They must be included in our diet. We eat proteins, break them down in our body to their constituent amino acids and then use some of these amino acids to build up other proteins which we require to maintain good health. So, for the requirement of essential amino acids we depend upon external sources, *i.e.*, the proteins we take. The requirement of each essential amino acid for an adult human body is in the range of 1 to 2 grams per day. An adequate protein supplies all the essential amino acids in the quantities needed for growth and repair of body tissues. Not all proteins contain all the essential amino acids. Amino acids are not stored and dietary sources of essential amino acids are needed at regular intervals.

Most proteins from plant sources are deficient in one or more essential amino acids. Zein, the corn protein, is deficient in two essential amino acids lysine and tryptophan. People whose diet consists chiefly of corn may suffer from malnutrition. A protein deficiency disease, called **Kwashiorkar**, is quite common in parts of Africa where corn is the major food. The problem in this disease, which is particularly severe in growing children, is not simply starvation but the breakdown of body's own proteins. Protein from rice is short of lysine and threonine. Gliadin, the wheat protein lacks in lysine and soya protein does not have the essential amino acid methionine. A complete protein is one that provides all essential amino acids in appropriate amounts for human survival. These amino acids cannot be synthesised by humans but they are needed for the biosynthesis of proteins. Lysine and methionine are two essential amino acids that are frequently in short supply in plant proteins. Most proteins from animal sources contain all the essential amino acids in adequate amounts. Meat, milk, fish, eggs and cheese are complete proteins. For example, casein, the protein from milk, contains nineteen of the common amino acids and all of the essential ones. Therefore, it is a complete protein. A balanced diet should consist of about 20% protein by weight. It must include all the essential amino acids in sufficient quantity. If we consume proteins from vegetable sources only, we may have deficiency of one or the other essential amino acid. Because grains such as rice and corn are usually poor in lysine and because beans are usually poor in methionine, vegetarians are at risk for malnutrition unless they eat grains and beans together. This leads to the concept of complementary proteins, *i.e.*, mixtures that provide all the essential amino acids. Extreme vegetarianism is, therefore, dangerous. It is best to eat proteins from a variety of sources to ensure an adequate supply of all the essential amino acids in the diet.

2.5 STEREOCHEMISTRY OF α-AMINO ACIDS

With the exception of glycine, all the common amino acids listed in Table 2.1 are chiral molecules. Therefore, there are two (optically active) stereoisomers for each of the amino acids. As with carbohydrates, it is customary to use the D- and L-nomenclature with the amino acids. All the naturally occurring amino acids have the L-configuration, *i.e.*, they have the same relative configuration as L-glyceraldehyde. Applying Fischer representation to these amino acids, D-amino acids are those in which $-\overset{+}{N}H_3$ group is on the right side of the chiral centre, $-COO^-$ being at the top and the carbon chain at the bottom. Similarly, L-amino acids have $-\overset{+}{N}H_3$ group on the left of the chiral centre as shown below:

D-Amino acid L-Amino acid

The stereo structures of L-alanine and L-serine in relationship with L-glyceraldehyde are shown in Fig. 2.1.

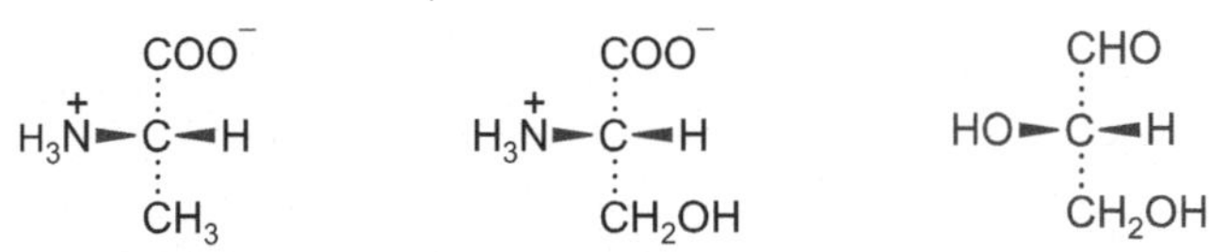

L-Alanine L-Serine L-Glyceraldehyde

Fig. 2.1 The relationship of L-alanine and L-serine to L-glyceraldehyde.

Some D-amino acids are found as components of the cell walls of bacteria and certain antibiotics. However, only L-amino acids are incorporated in proteins. These D- and L-designations have nothing to do with the actual optical rotation of amino acids, which is determined experimentally.

2.6 SYNTHESIS OF α-AMINO ACIDS

1. Amination of α-Halogeno Acids

This is the first method that has been developed for the synthesis of α-amino acids. With its various modifications, this method is quite useful, although it can not be applied for the synthesis of all the amino acids.

In this method, α-chloro or α-bromo acids are subjected to direct ammonolysis by treating with a large excess of concentrated aqueous ammonia. The necessary α-halo acids are prepared by the Hell-Volhard Zelinksy reaction of the unsubstituted acids or through the appropriately substituted malonic ester. Generally the bromo compounds show greater reactivity.

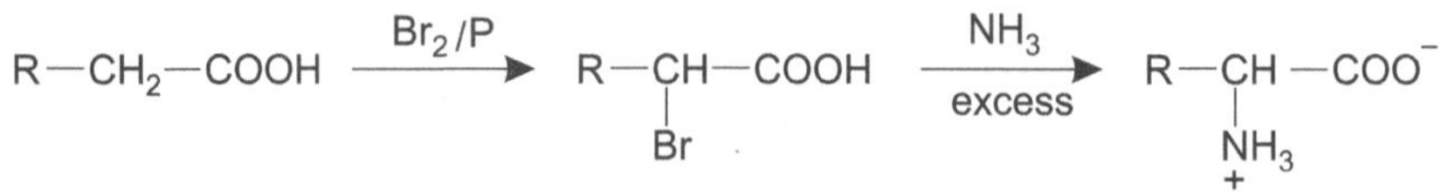

For example, alanine can be prepared from propionic acid in the following manner:

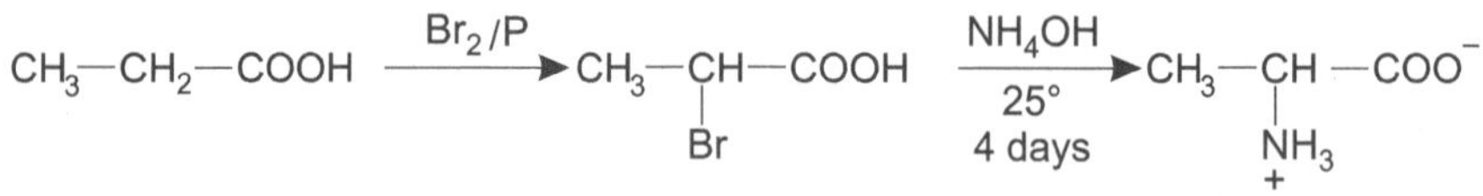

The general scheme for the synthesis of α–halo acids from malonic ester is outlined below:

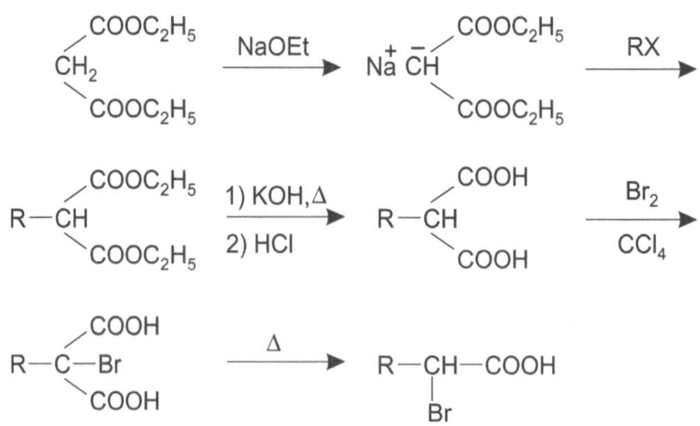

Phenylalanine may be synthesised starting from malonic ester by the following route:

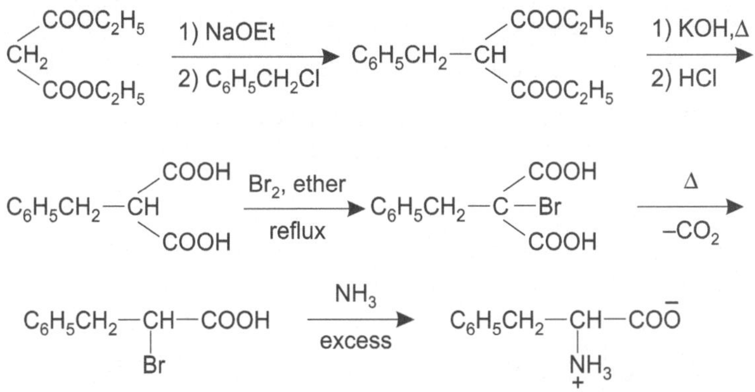

Phenylalanine

This route has also been used for the synthesis of histidine. Other amino acids which can be easily synthesised by the above method are glycine, serine, threonine, valine, leucine and isoleucine.

2. Gabriel Phthalimide Synthesis

Better yields of amino acids are obtained by the Gabriel Phthalimide Synthesis. α–Halo-esters are used instead of α–halo acids. The halogen in the α–halo esters is replaced by a phthalimido group either by fusion or by gentle heating with potassium phthalimide in dimethyl formamide. The intermediate ester yields the desired amino acid by drastic hydrolysis or better in two steps by treatment first with 4N alkali followed by acid-hydrolysis. The basic scheme is outlined below:

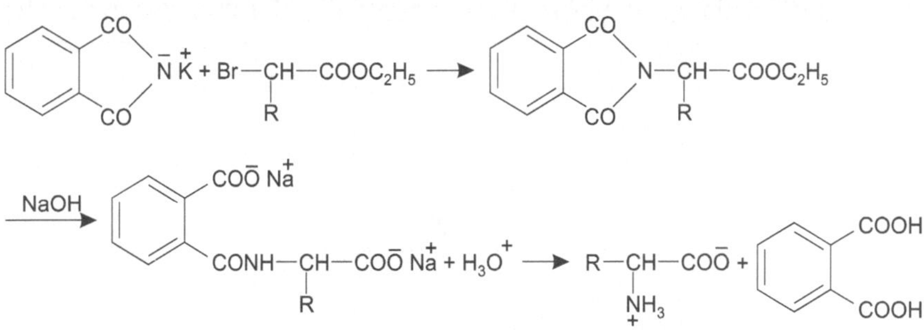

The above method is generally used for the preparation of glycine and leucine.

3. Strecker Synthesis

In this method, α–aminonitriles obtained by the treatment of aldehydes with ammonia and HCN are hydrolysed to α–amino acids. The method was first developed by Strecker to synthesise alanine and is known after his name.

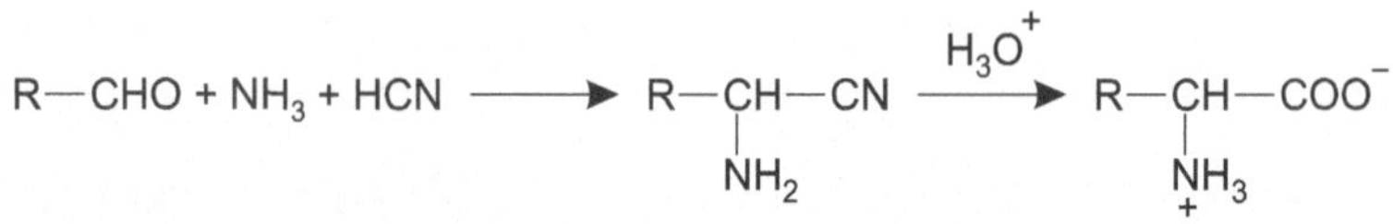

Use of acetaldehyde in the above scheme results in the formation of alanine.

The mechanism of formation of the α–aminonitrile probably involves the addition of HCN to the imine, which is formed by condensation of the aldehyde with ammonia.

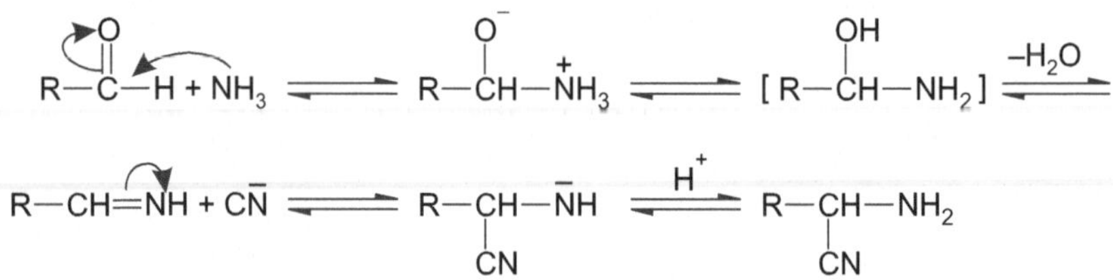

Imine α-Aminonitrile

In a modification, an equimolecular mixture of ammonium chloride and potassium cyanide is used instead of the more dangerous HCN. This method is useful for the synthesis of glycine, alanine, serine, valine, methionine, glutamic acid, leucine, isoleucine and phenylalanine. For example, glycine has been synthesised by the following route:

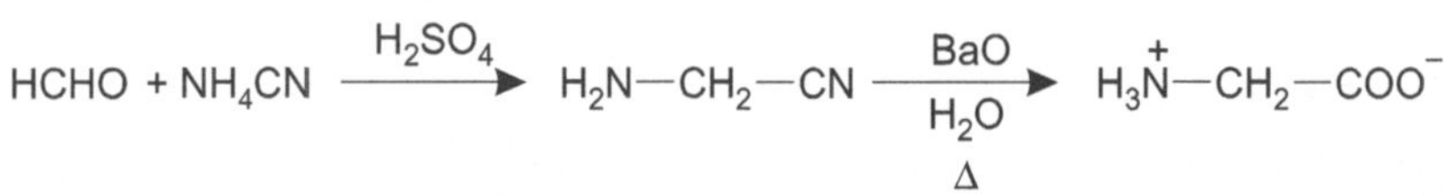

A modification of the chemical procedure was given by Buchrer for improving the yields. On heating the cyanohydrins with urea or better ammonium carbonate in alcoholic solution, high yields of the corresponding 5-substituted hydantoins are formed, which provide the desired amino acids after hydrolysis with alkali or concentrated acid. The cyanohydrin may also be obtained *via* the bisulphite addition compound.

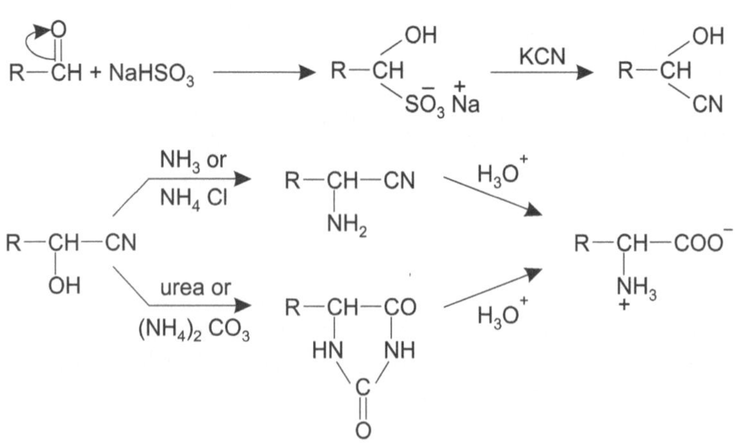

4. Malonic Ester Synthesis

By variations of the malonic ester synthesis, useful general methods have been developed for the synthesis of a large number of α-amino acids.

A. From Phthalimidomalonic Esters

In the earlier synthesis, malonic ester has been used for the preparation of α-halo acids. If we couple malonic ester synthesis with Gabriel phthalimide scheme, we get a useful method for the synthesis of a number of amino acids. This gives us Phthalimidomalonic Ester method.

N-Phthalimidomalonic ester is prepared by condensing potassium phthalimide with monobromo malonic ester. The ester may be alkylated by a variety of alkyl halides or condensed with α,β-unsaturated carbonyl compounds. Vigorous acid hydrolysis causes hydrolysis of both ester groups and the phthalimido group. Decarboxylation of the resulting malonic acid gives the desired amino acid. The product is a racemic modification and may be resolved.

The basic scheme is outlined below:

Phthalic acid is formed as a byproduct.

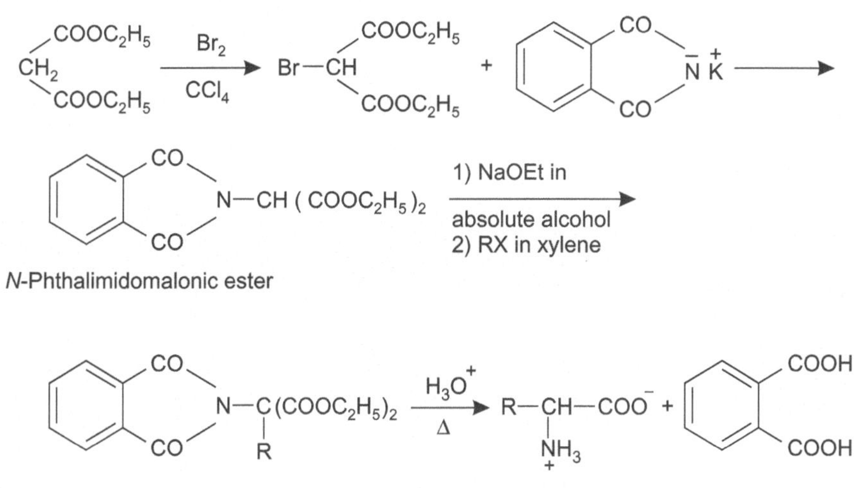

The above method has been successfully used for the synthesis of aspartic acid, phenylalanine, glutamic acid, cysteine, proline, methionine, serine, tyrosine and lysine.

Some specific examples are given below:

Aspartic acid

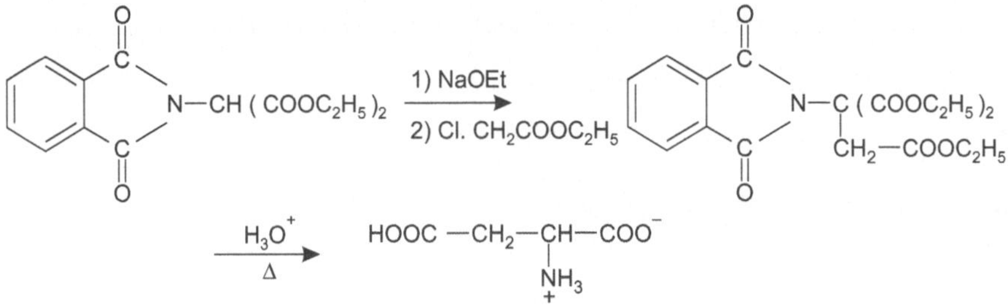

Aspartic acid

Phenylalanine

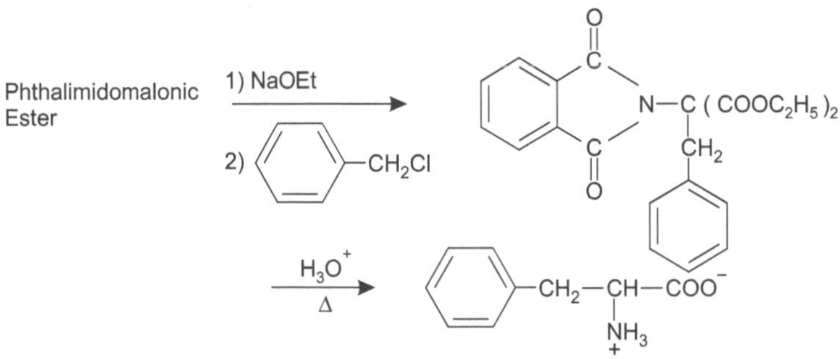

Phenylalanine

Glutamic acid

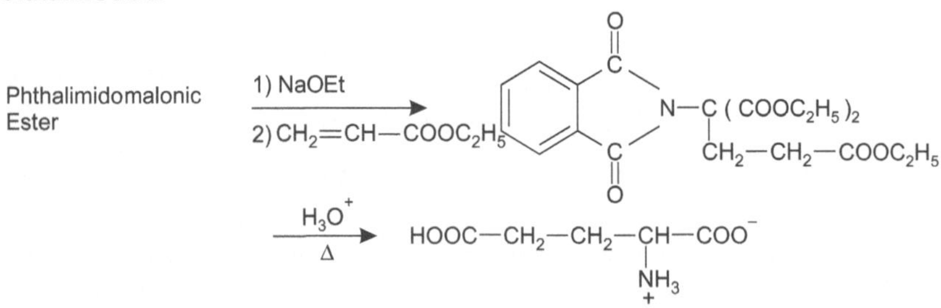

Glutamic acid

Methionine

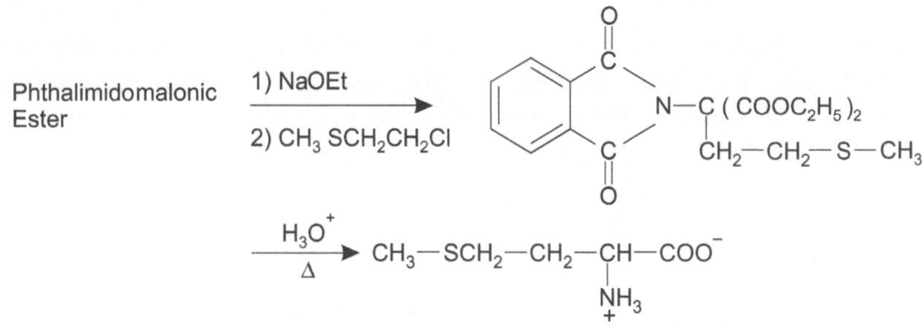

Methionine

Proline

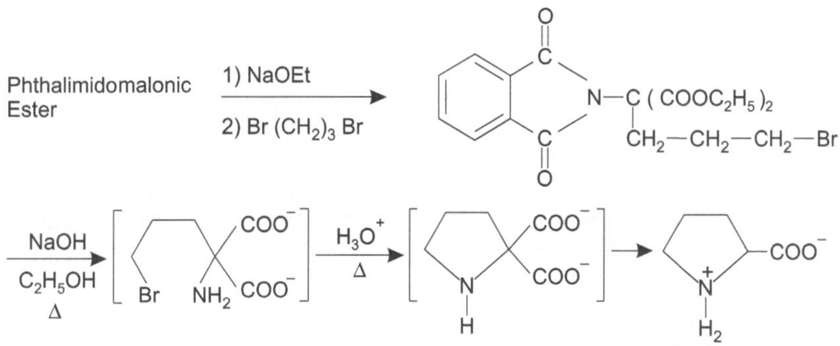

Proline

Cysteine-Cystine

Benzylthiomethyl chloride is prepared first and condensed with *N*-phthalimidomalonic ester. The product is hydrolysed and reduced to obtain cysteine. Air oxidation of cysteine gives cystine.

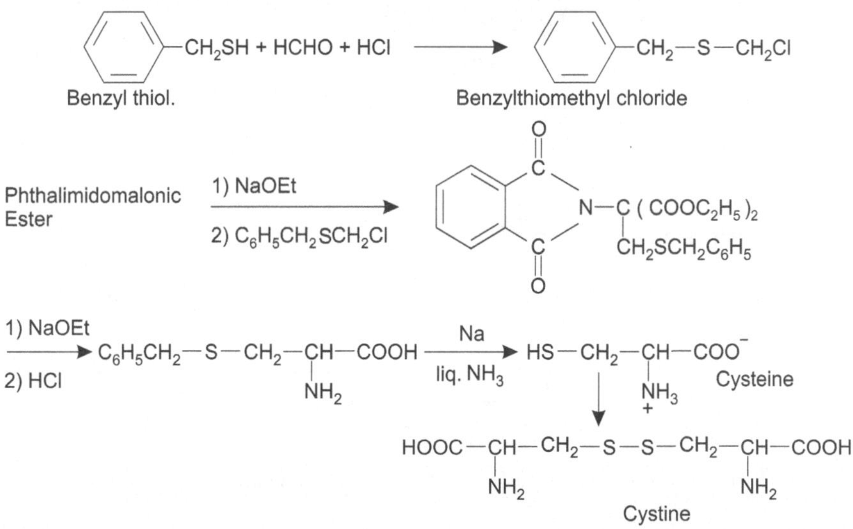

B. From Acetamidomalonic Esters

Acetamido derivatives of malonic ester may also be used to synthesise amino acids. *N*-Acetamidomalonic ester may be prepared by treating the malonic ester with nitrous acid to give a nitroso derivative which tautomerises to the oxime and reduction of the oxime using zinc/acetic acid in the presence of acetic anhydride. Catalytic hydrogenation of the oxime in acetic anhydride solution also affords *N*-acetamidomalonic ester.

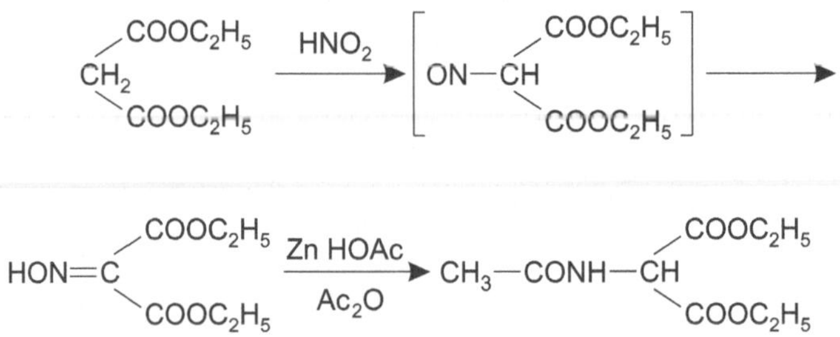

N-Acetamidomalonic ester is alkylated in the usual manner and the resulting product is hydrolysed and decarboxylated to give the desired amino acid. Some specific examples are given below:

Serine

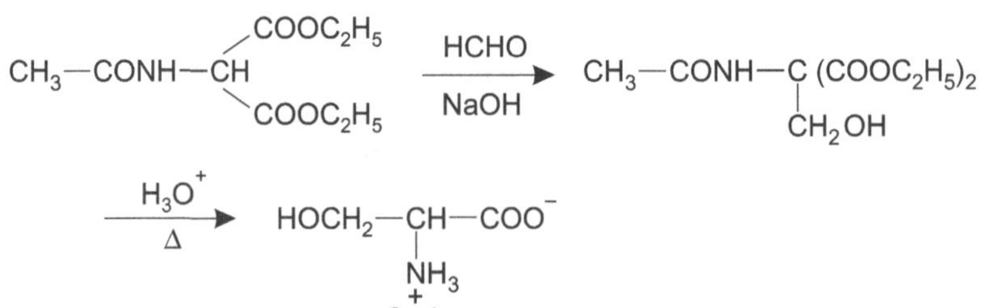

Serine

Leucine

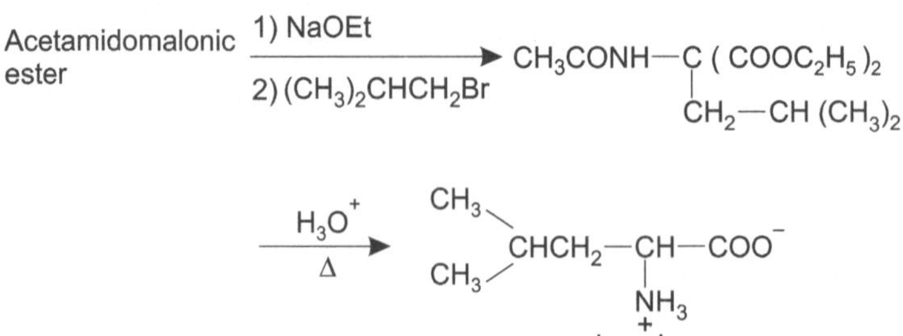

Leucine

Histidine

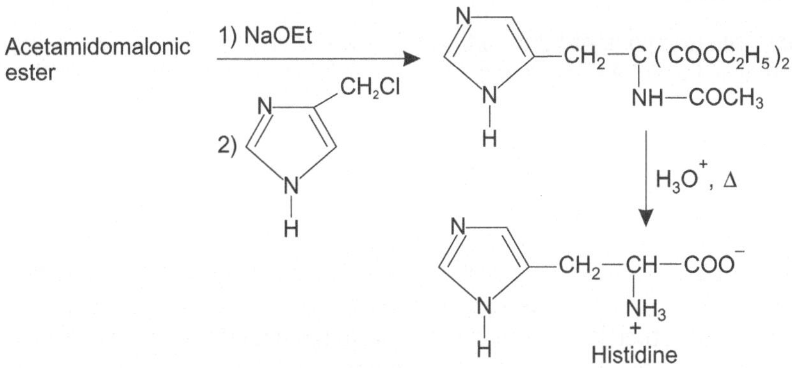

C. From Benzamidomalonic Esters

Diethylaminomalonate can be prepared from diethyl malonate by the following reactions:

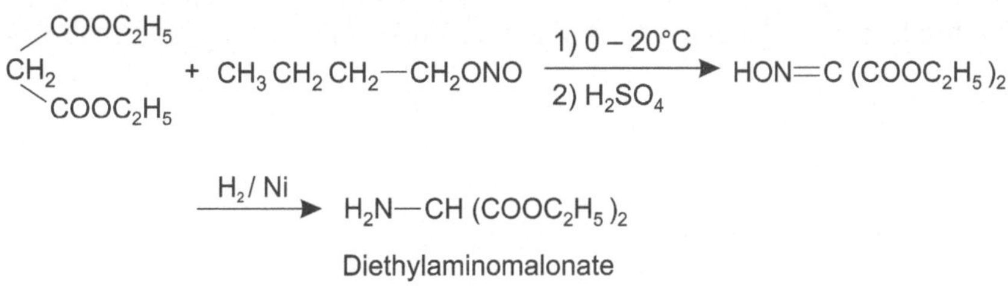

Diethylaminomalonic ester is now benzoylated to give benzamidomalonic ester in good yields which is a useful reagent for the synthesis of amino acids.

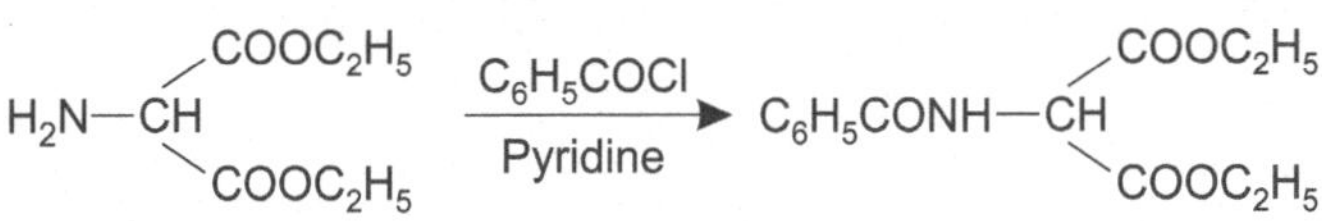

Benzamidomalonic ester so obtained is alkylated in the usual manner and the product so obtained on acid hydrolysis and decarboxylation affords the desired amino acid. For example, leucine may be synthesised in the following manner:

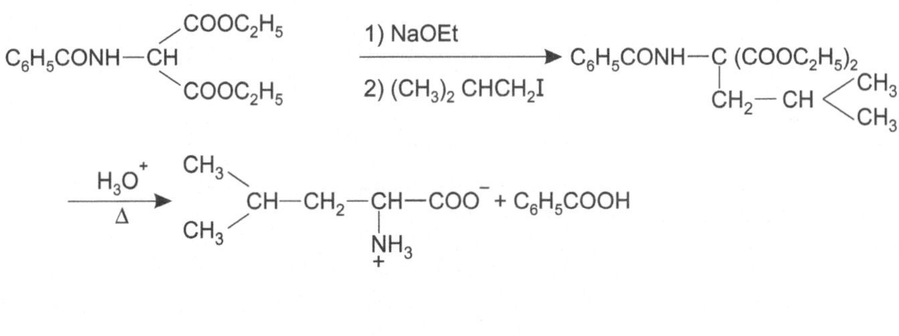

5. Aldehyde Condensations with Glycine Derivatives (Erlenmeyer Azlactone Synthesis)

From structural point of view all the amino acids can be considered as derivatives of glycine in which one of two α–hydrogen atoms is substituted by the side chain.

$$\boxed{H}-\underset{\displaystyle NH_2}{\underset{|}{CH}}-COOH \longrightarrow \boxed{R}-\underset{\displaystyle NH_2}{\underset{|}{CH}}-COOH$$

It is possible to introduce any group by reacting aldehydes with the α–hydrogens of glycine. For this it is necessary to protect the amino group, which would otherwise react first. In the Erlenmeyer azlactone synthesis, hippuric acid (benzoyl glycine) is used for condensing with aromatic aldehydes. The amino group is thus protected by benzoyl group.

An azlactone is formed by boiling the mixture of an aromatic aldehyde and hippuric acid in the presence of acetic anhydride and sodium acetate. Partial hydrolysis of the azlactone by warming with 1% sodium hydroxide solution opens up the ring. The product on reduction with sodium amalgam followed by acid hydrolysis gives α–amino acid. Also heating the azlactone with a mixture of red phosphorus and hydriodic acid converts it in one step into the amino acid.

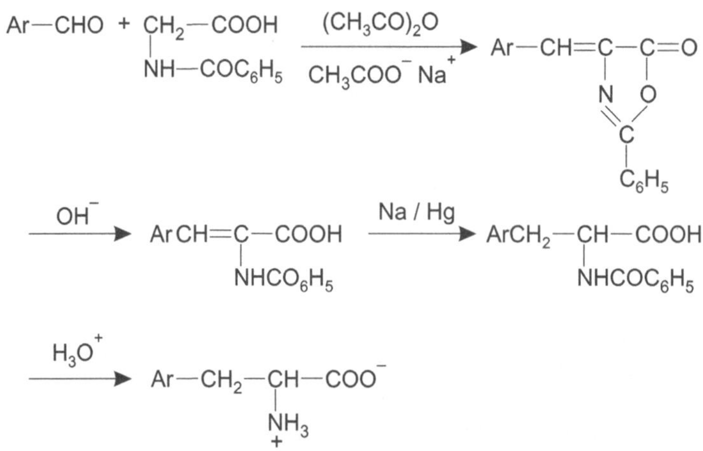

Under the above conditions only aromatic aldehydes condense with hippuric acid. However, it has been found that aliphatic aldehydes may condense with hippuric acid to form azlactones if lead acetate is used instead of sodium acetate.

The azlactone synthesis has been used successfully for preparing aromatic amino acids like phenylalanine, tyrosine, tryptophan and thyroxine. For example, tyrosine may be obtained by the following route:

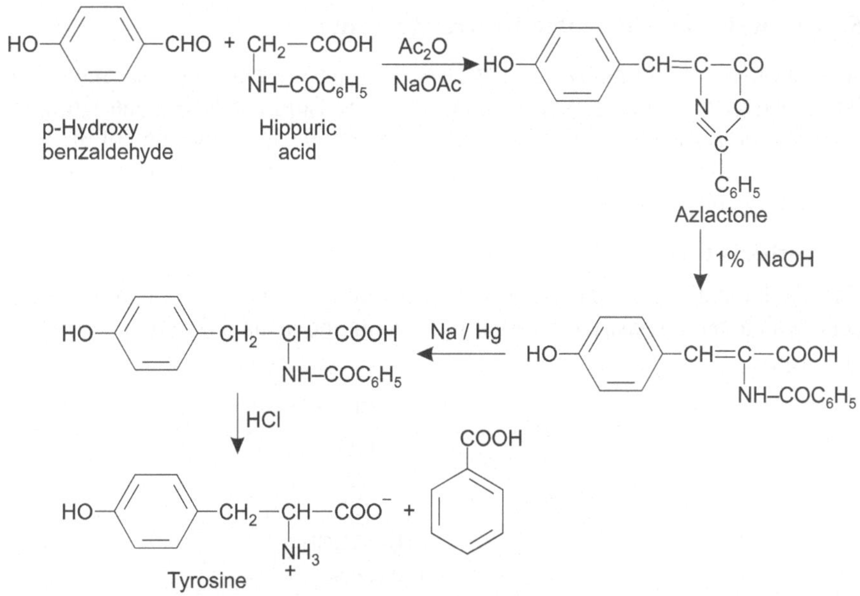

Aromatic aldehydes may also be condensed with 2,5-diketopiperazine in a similar manner. Reductive cleavage of the product by heating with hydriodic acid and red phosphorus gives an α–amino acid. For example,

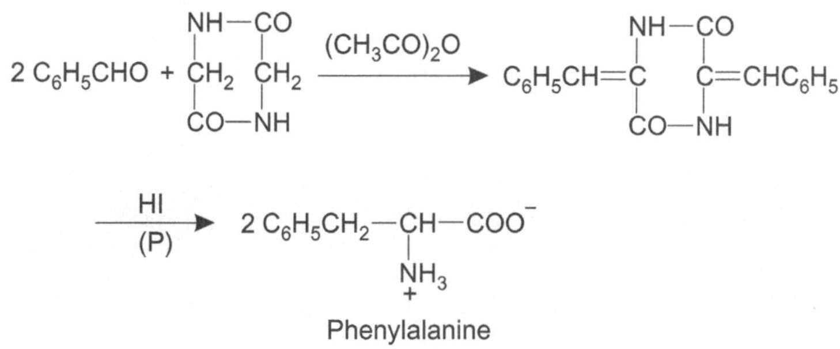

Condensation of the aldehydes with glycine or glycine esters provides only β–hydroxy acids as the first step will occur according to aldol type of condensation.

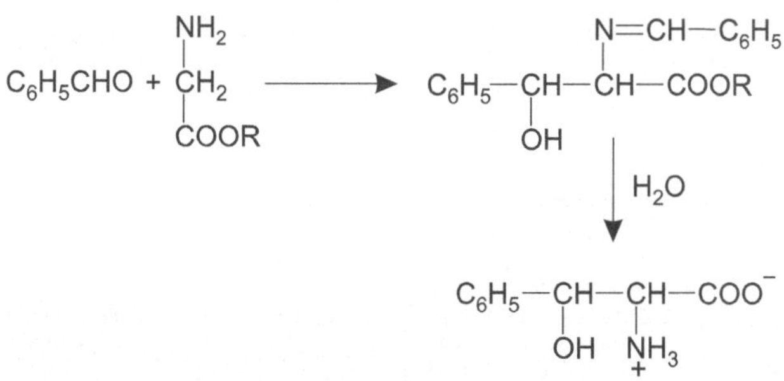

6. Synthesis via Molecular Rearrangements

A large variety of α–amino acids may be synthesised via molecular rearrangements like Beckmann rearrangement, Schmidt reaction, Curtius reaction and Hofmann degradation. Basically in all these reactions intramolecular migration of the alkyl group takes place to the more electron deficient nitrogen atom giving intermediates which ultimately yield α–amino acids.

A. Synthesis through the Beckmann Rearrangement

The Beckmann rearrangement of cyclopentanone oxime gives ω-aminovaleric acid, which can be easily converted to basic amino acids like proline, ornithine and arginine.

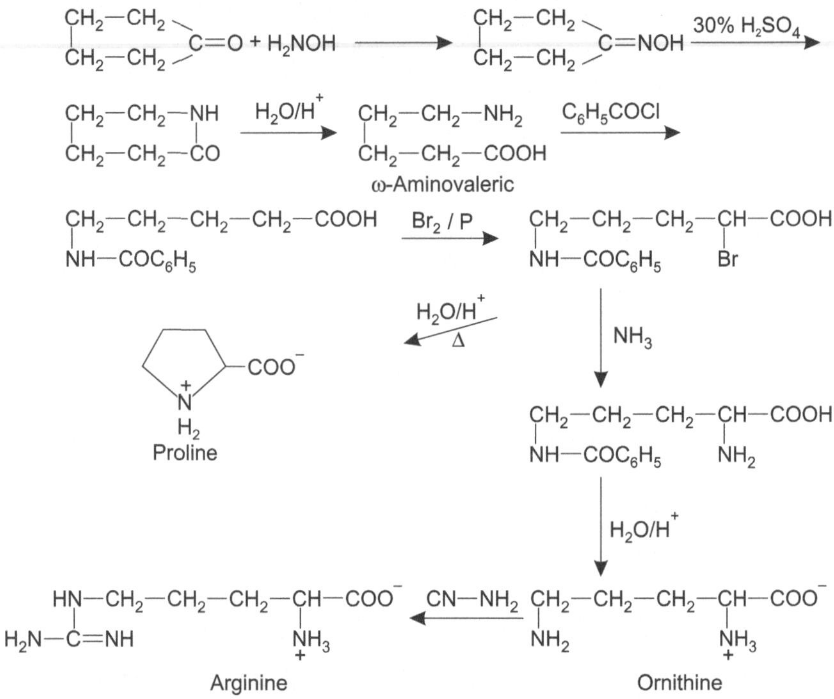

Mechanism of Beckmann rearrangement is as follows:

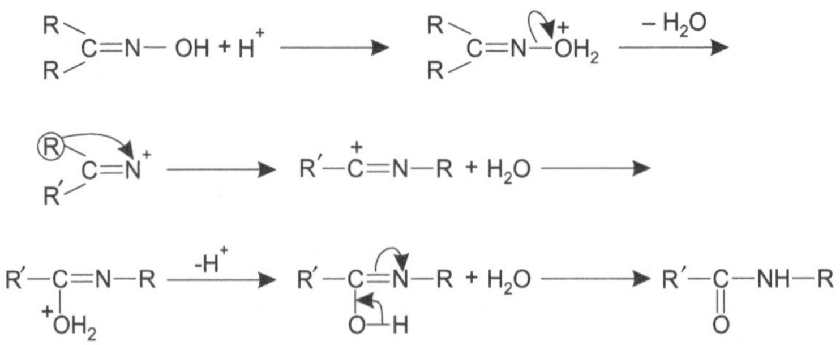

By using cyclohexanone instead of cyclopentanone, we get lysine.

B. By Schmidt Reaction

The Schmidt reaction is used in a variety of ways for the synthesis of amino acids. Simple amino acids may be obtained if this reaction is applied to alkylated malonic acids, *e.g.*,

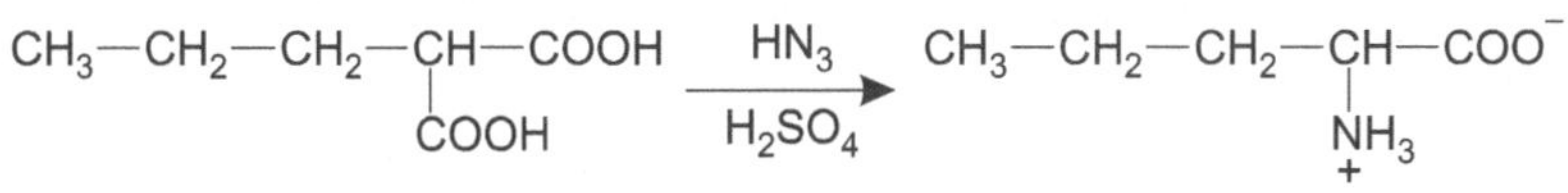

By applying Schmidt reaction to cyclic ketones or keto acids, we get the desired amino acids, *e.g.*,

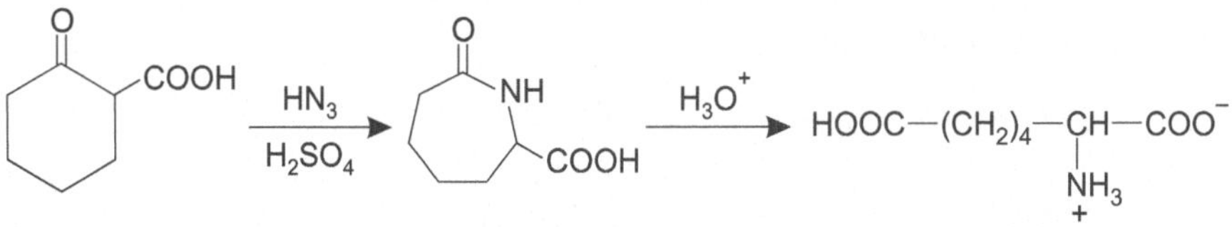

A second application of the Schmidt reaction yields lysine. The carboxyl group that is not α–to the amino group reacts. Actually simple α–amino acids do not react at all in the Schmidt reaction.

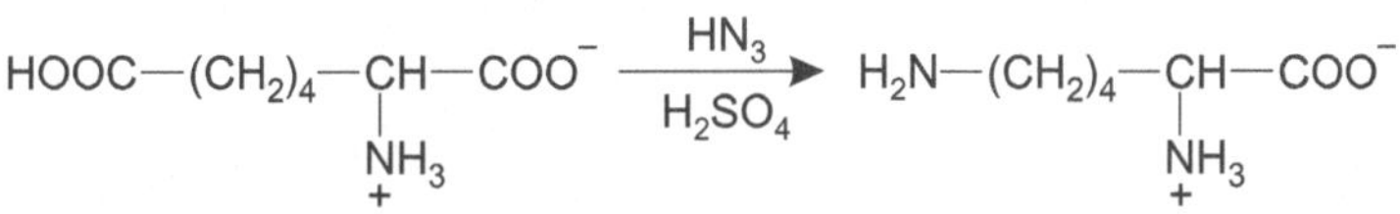

Lysine is also obtained when the Schmidt reaction is applied to cyclohexanone.

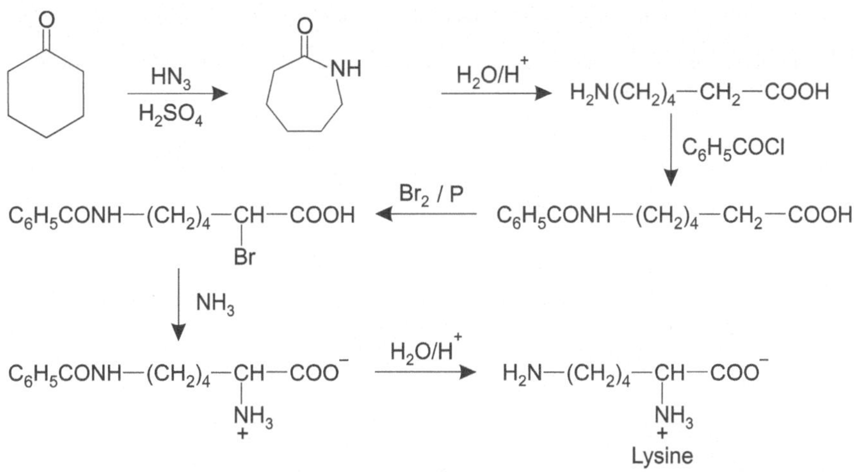

The probable mechanism for the conversion of ketones to amides involves intramolecular migration of the alkyl group to electron deficient nitrogen atom.

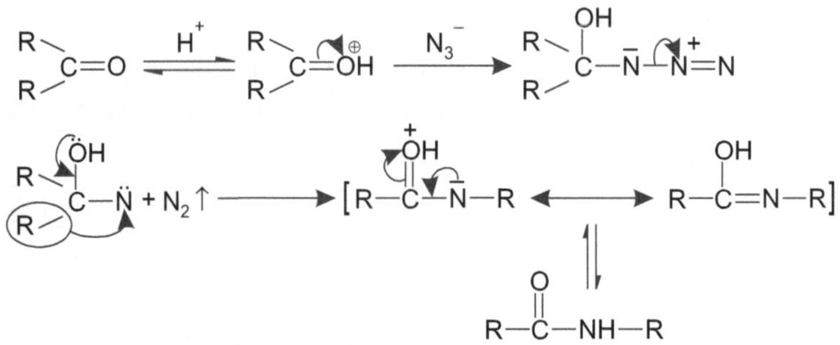

C. By Curtius Reaction

α–Amino acids may be synthesised by the application of Curtius reaction to alkylated malonic half esters.

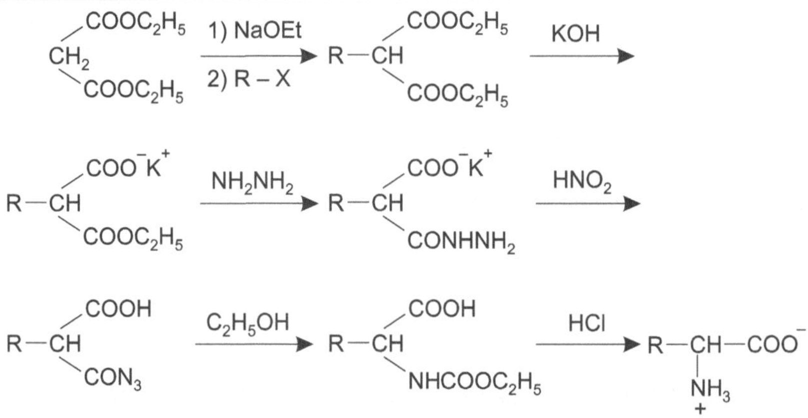

Glycine, alanine, valine and phenylalanine can be obtained very easily by this method.

We can also use ethyl cyanoacetate in place of malonic ester as malonic half ester is quite difficult to prepare.

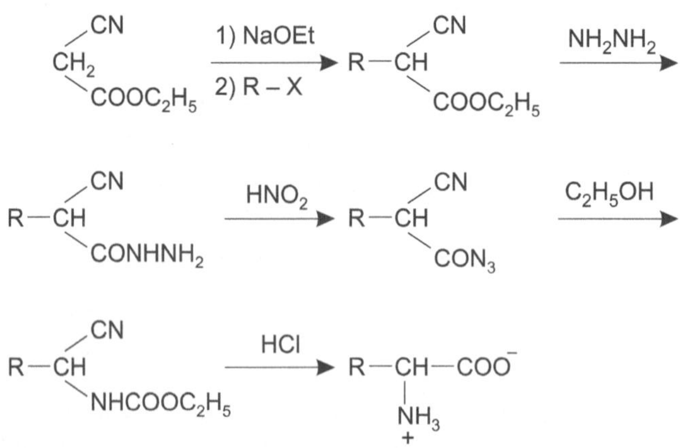

D. By Hofmann Degradation

Ester amides can also be converted to α–amino acids by Hofmann degradation.

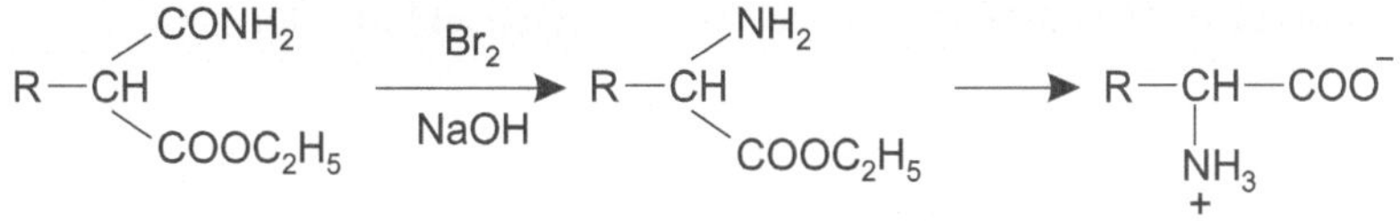

This is a typical conversion of amide to an amine. The formal end-point in the reaction is the isocyanate. Under the reaction conditions, addition of water takes place to yield the unstable carbamic acid, which decarboxylates to give the amine.

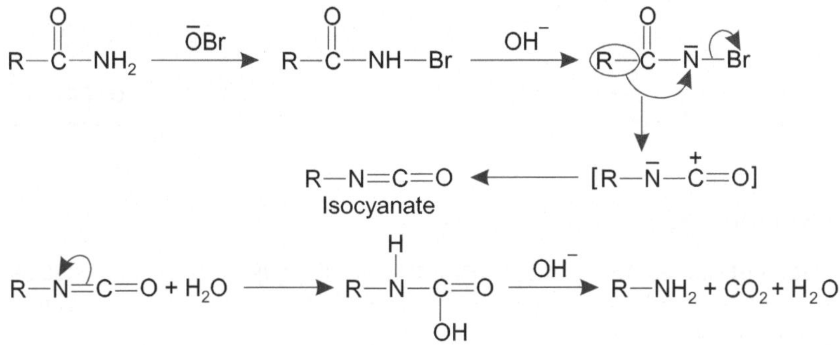

The Curtius and Schmidt reactions are actually closely related to that of Hofmann because all of these involve the formation of the same isocyanate as the intermediate. In both these reactions, N_2 is the leaving group from the same azide intermediate. Here again the migration of the alkyl group R takes place in a concerted process, *i.e.*, it is intramolecular. The azide may be obtained either by nitrosation of an acid-hydrazide (Curtius reaction) or by the reaction of hydrazoic acid on a carboxylic acid—The Schmidt reaction.

7. Reductive Amination

The reductive amination of α–keto acids yields α–amino acids. This may be achieved by carrying out reduction catalytically or with sodium and ethanol in the presence of ammonia. The mechanism of the reduction is not certain but it probably occurs via the imino-acid. The oximes and phenylhydrazones of α–keto acids may also be reduced to α–amino acids in a similar manner.

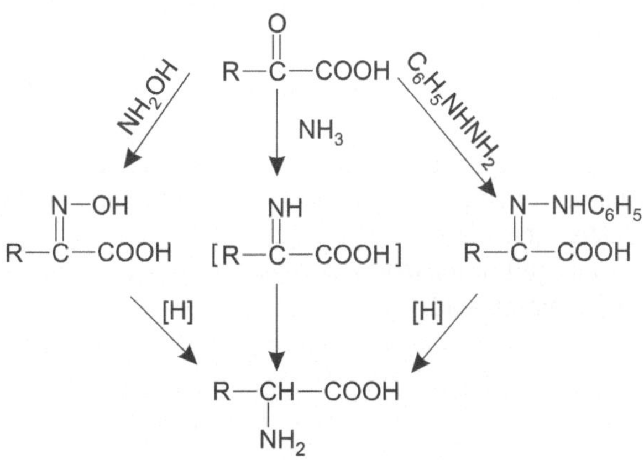

α–Oximino acids are also obtained by the treatment of 2-alkyl or 2-acyl acetoacetic esters with butyl nitrite in sulphuric acid, whereas action of a diazonium salt in the presence of sodium ethoxide affords the corresponding phenyl-hydrazones. Reduction of these can be carried out to get any α–amino acid.

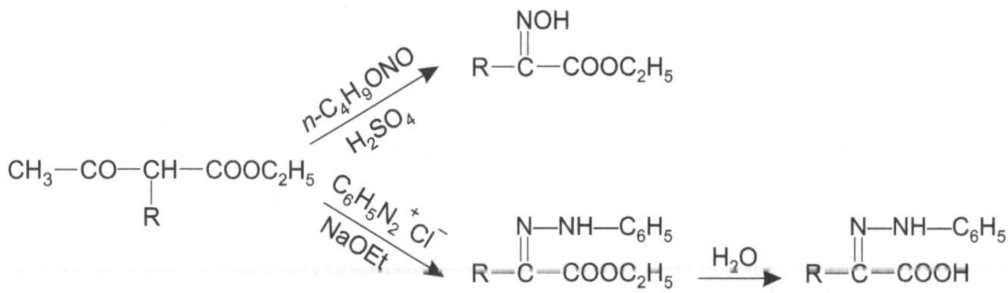

The usefulness of these methods depends upon the availability of α–keto acids. Thus, alanine, phenylalanine, leucine, isoleucine and valine may be prepared in this way.

8. Oxidation of Amino Alcohols

Oxidation of the 2-amino alcohols affords a simple route to α–amino acids but is of less importance because of the limited number of amino alcohols which are readily available.

The conversion of the alcoholic group to a carboxyl residue can be carried out by the usual oxidising agents like alkaline potassium permanganate or acid dichromate, *etc.*, but the oxidation has to be preceded by the protection of the labile amino group. Instead of the usual masking with a phthaloyl or benzoyl residue, protonation by addition of a strong acid is quite sufficient for this process.

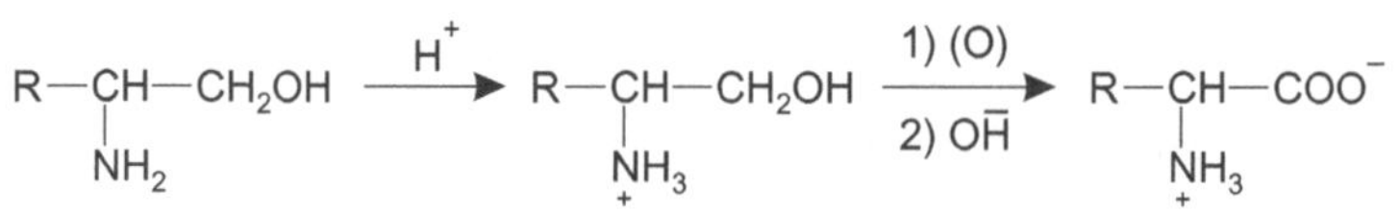

2.7 AMINO ACIDS AS DIPOLAR IONS

Amino acids have relatively high melting points. They usually decompose above 200°C. They have high solubility in water and low solubility in non-polar solvents. They have large dipole moments and these solutions in water have high dielectric constants. All these properties show that amino acids exist as dipolar ions.

As amino acids contain both an acidic group (—COOH) and a basic group (—NH_2), they are amphoteric. The acidic group in an amino acid is a substituted ammonium ion and the basic group is the carboxylate ion. Thus, in the crystalline state, amino acids exist as dipolar ions, also called zwitterions, in which the amino group is protonated and the carboxyl group exists as a carboxylate anion. In aqueous solution an equilibrium exists between the dipolar ion and the anionic and cationic form of an amino acid.

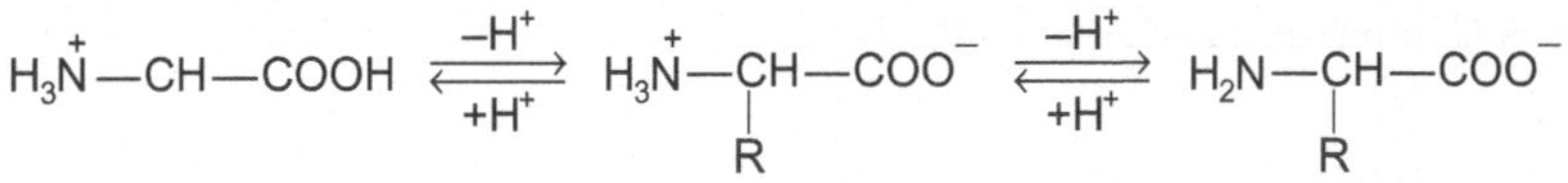

Cationic form (predominant in strongly acidic solution) | Dipolar ion | Anionic form (predominant in strongly basic solution)

The predominant form of the amino acid present in a solution depends on the pH of the solution. In strongly acidic solution, all amino acids are present as cations and in strongly basic solutions they are present as anions. The pK_a for the carboxyl group of cationic form and pK_a of the dipolar ion form will depend upon the pH of the solution.

To correlate the relationship between pH of a solution to the pK_a of a weak acid, let us consider the dissociation of an acid AH^+.

According to the Bronsted Theory of acids and bases, we know that an acid dissociates to give a proton and a conjugate base. For the dissociation of acid AH^+, we have

$$AH^+ \rightleftharpoons B + H^+ \qquad \text{... (1)}$$

where B is the conjugate base of acid AH^+.

The extent of dissociation is determined from its K_a by applying the Law of Mass Action. Thus, K_a for the above equation (1) is given by

$$K_a = \frac{[H^+]\,[B]}{[AH^+]} \qquad \text{... (2)}$$

where K_a is the dissociation constant for the acid AH^+. The term in brackets refer to the concentrations of the species in moles per litre, which we consider equal to the activities of the species.

From the above equation (2), we have

$$[H^+] = \frac{K_a\,[AH^+]}{[B]}$$

By taking common logarithm of both sides

$$\log\,[H^+] = \log K_a + \log \frac{[AH^+]}{[B]}$$

$$\therefore \qquad pH = pK_a + \log \frac{[B]}{[AH^+]} \qquad \text{...(3)}$$

$$pH = pK_a + \log \frac{[\text{Conjugate base}]}{[\text{Acid}]}$$

This is Henderson-Hasselbach equation

or $$pK_a = pH + \log \frac{[Acid]}{[Conjugate\ base]}$$

Thus, pH of a weak acid is related to the pK_a of the acid and the ratio of the concentration of the conjugate base and concentration of the acid.

When the term, $\log \frac{[Conjugate\ base]}{[Acid]}$ is equal to zero, the Henderson-Hasselbach equation becomes

$$pH = pK_a \qquad ...(4)$$

i.e., equation (4) holds good when the molar concentrations of the acid and its conjugate base are equal.

In amino acids, there are at least two ionisable groups —COOH and $\overset{+}{N}H_3$. There are, therefore, at least two pH values for a solution in which equation (4) holds true. For those amino acids which have a third ionisable group on the side chain, there is a third value of the pH where pH = pK_a. That is true for acidic and basic amino acids.

2.8 AMINO ACIDS AS ACIDS AND BASES

We have seen above that the exact structure of an amino acid is a function of the pH, which is related to its pK_a. In acidic solutions, the amino acid is completely protonated and exists as the conjugate acid. For example, glycine exists as glycine hydrochloride in acidic medium.

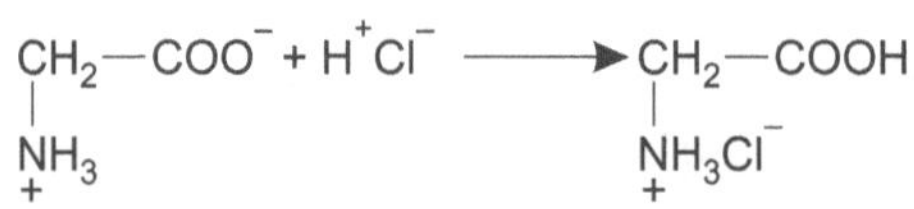

To understand the relationship between pK_as and structure, let us study the titration of any amino acid, say alanine.

We start with 1 equivalent of the cationic form of alanine and titrate it with standardised sodium hydroxide solution. We shall proceed by steps and see the structures of species each time after $\frac{1}{2}$ equivalent of the base has been added.

I. Before the actual titration begins, all the species in solution are in the cationic form and there is a net charge of +1. There are two ionisable groups —COOH and $-\overset{+}{N}H_3$. Carboxyl group is more acidic of the two and will give a proton first.

Step 1: No addition of base

Structures in solution

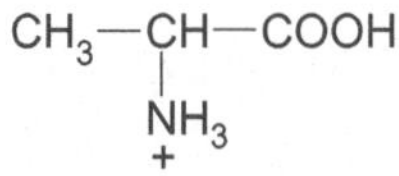

Cationic form
(1 equivalent)

Net charge = + 1

II. After the addition of $\frac{1}{2}$ equivalent of base, *i.e.*, when the hydrochloride has been half neutralised we will have an equal amount of cationic form, 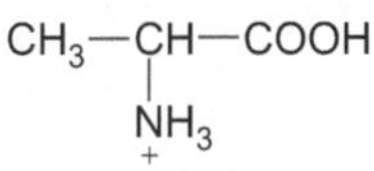 and its conjugate base, 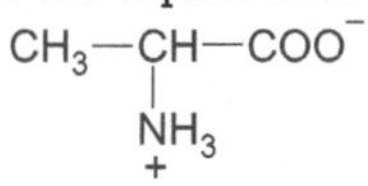

The pH of the solution at this point is equal to pK_{a_1} = 2.34

The net charge of all the species is $+\frac{1}{2}$.

Step 2: After addition of $\frac{1}{2}$ equivalent of base

Structures in solution

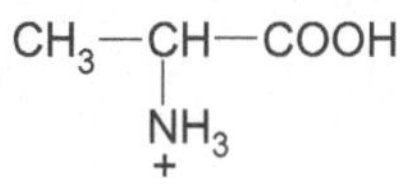

Cationic form
(1 equivalent)

Net charge = $+\frac{1}{2}$

$pH = pK_{a_1} = 2.34$

This dissociation constant refers to the ionisation of —COOH group, which is more acidic of the two acidic groups.

III. After one equivalent of the base has been added, all the —COOH groups have become ionised. The main species in solution is the zwitterionic form of alanine with a net charge of zero. The pH of the solution at this point is simply the pH of the solution of the amino acid in water.

Step 3: After addition of 1 equivalent to base

Structure in solution

$$CH_3{-}\underset{\overset{|}{\overset{+}{N}H_3}}{CH}{-}COO^-$$

Net charge = 0

IV. Addition of a further $\frac{1}{2}$ equivalent of base corresponds to half neutralisation of the zwitterionic form.

At this point we have equal amounts of $CH_3{-}\underset{\overset{|}{\overset{+}{N}H_3}}{CH}{-}COO^-$ acid and its conjugate base, $CH_3{-}\underset{\overset{|}{NH_2}}{CH}{-}COO^-$

The pH of the solution at this point is equal to pK_{a_2} and for alanine this value is 9.87. The net charge of all the species in solution is $-\frac{1}{2}$.

Step 4: After addition of $1\frac{1}{2}$ equivalents of base

Structures in solution

$$CH_3{-}\underset{\overset{|}{\overset{+}{N}H_3}}{CH}{-}COO^- + CH_3{-}\underset{\overset{|}{NH_2}}{CH}{-}COO^-$$

Zwitterion form (½ equivalent) | Anionic form (½ equivalent)

Net charge = $-\frac{1}{2}$

V. When two equivalents of NaOH have been added, all the $-\overset{+}{N}H_3$ groups are neutralised. All the species exist in the anionic form with a net charge of –1.

Step 5: After addition of 2 equivalents of base

Structures in solution

$$CH_3{-}\underset{\overset{|}{NH_2}}{CH}{-}COO^-$$

Anionic Form

Net charge = – 1

Further addition of base causes no change in the composition of the solution because no ionisable group is present.

The titration curve for alanine hydrochloride is shown in Fig. 2.2

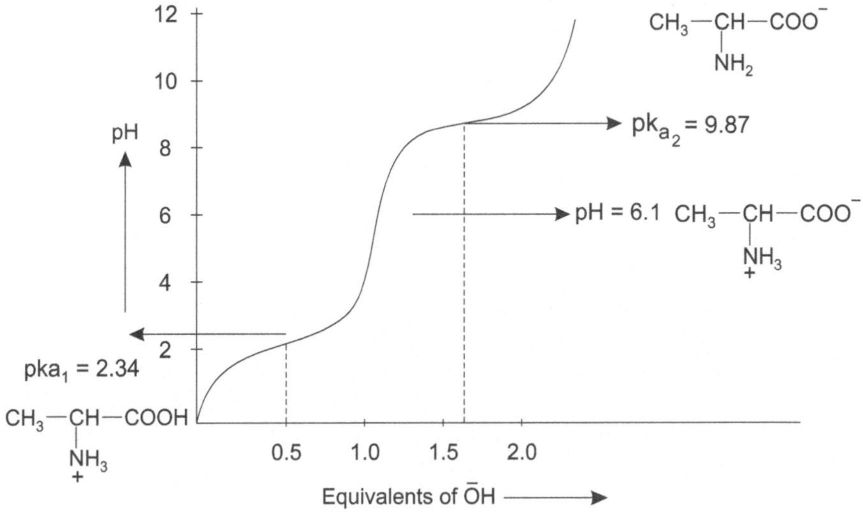

Fig. 2.2 Titration curve for Alanine.

Most of the amino acids show similar pK_{a_1} and pK_{a_2} values. These are listed in Table 2.2.

Acidic amino acids, aspartic acid and glutamic acid, have an additional carboxyl group with pK_a of 3.87 and 4.28 respectively.

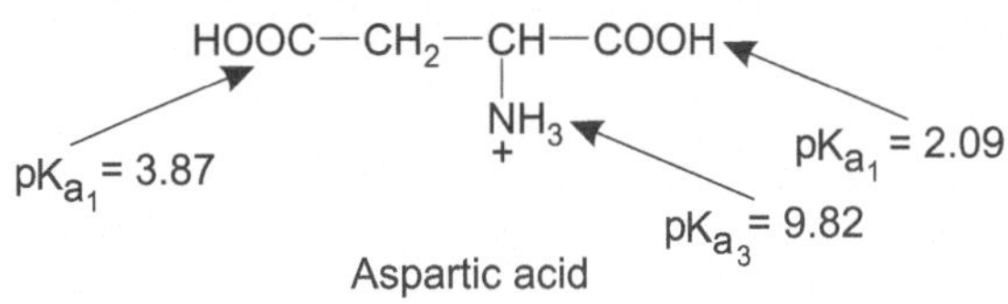

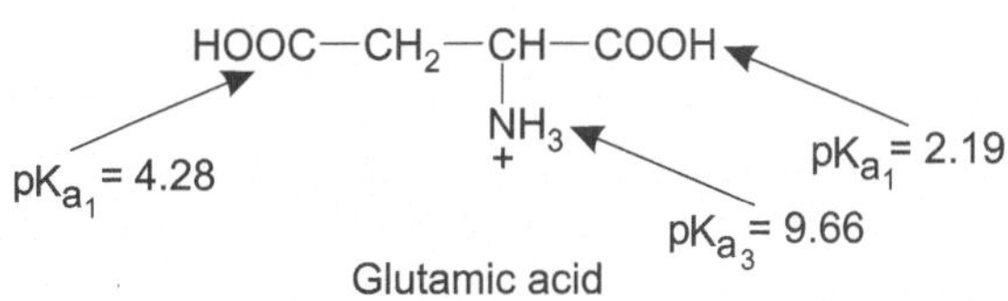

Lysine has two amino groups with pK_a of 8.95 and 10.53. The more basic group is probably the one more remote from the carboxyl group. It has the following zwitterionic structure.

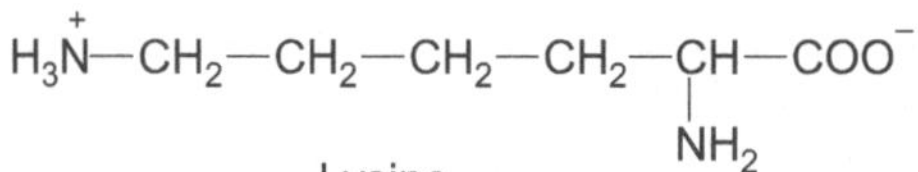

Lysine

Arginine on the other hand contains the strong basic guanidino group with pK_a = 13.2. Therefore, it exists in the following zwitterionic form.

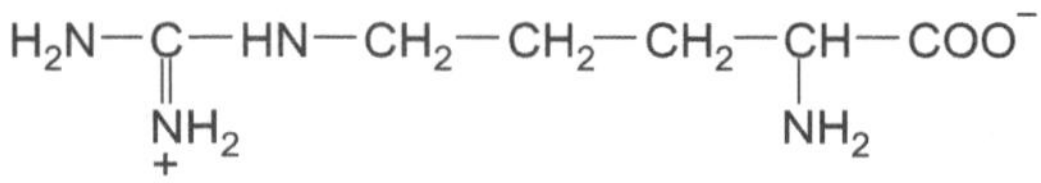

Arginine

2.9 ISOELECTRIC POINT

During the course of the titration of an amino acid, a point is reached at which all structures in solution exist as zwitterions. The pH of the solution at this stage is known as the **isoelectric point.** At the isoelectric point amino acids are electrically neutral and have a zero net charge. At this point the concentration of the dipolar ion is at its maximum and the concentrations of the anions and cations are equal. Each amino acid has a particular isoelectric point.

Strictly speaking, the isoelectric point should refer to the hydrogen ion concentration of the solution in which a particular amino acid does not migrate toward either the anode or the cathode under the influence of an electric field. When a solution of an amino acid is placed in an electric field, the migration toward anode or cathode depends upon the acidity or basicity of the solution. In quite alkaline solution, there is a net migration of the amino acid toward the anode and in acidic solution, there is net migration of the amino acid toward the cathode.

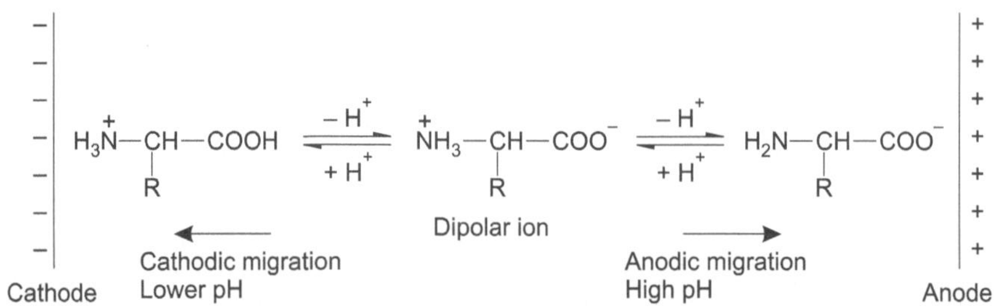

In the titration of the amino acids we can reach a point at which there is a zero net charge without applying an electric field. This may not necessarily be the same as the pH at which there is no migration in an electric field. Sorensen used the term isoionic to refer to the pH at which there is no net charge. Thus, the term isoionic is more appropriate than isoelectric. As there is no difference between the two values for simple amino acids, it has become customary to use the term isoelectric for isoionic. However, the distinction between the two is very important in case of proteins.

The isoelectric point for an amino acid is denoted by pI, *i.e.*, the pH at which an amino acid is isoionic. It can be calculated from its pK_a values. We have seen that for monoamino carboxylic acids there are two ionisable groups. In a generalised way, the two equilibria can be written as follows:

$$X^+ \overset{K_{a_1}}{\rightleftharpoons} X^{\pm} + H^+ \quad ...(1)$$

$$X^{\pm} \overset{K_{a_2}}{\rightleftharpoons} X^- + H^+ \quad ...(2)$$

where X^+ is the cationic form of any amino acid

$X^{\pm}$ is the zwitterionic form.

X^- is the anionic form.

The corresponding equilibrium constants for (1) and (2) are given by:

$$K_{a_1} = \frac{[X^{\pm}][H^+]}{[X^+]} \quad ...(3)$$

$$K_{a_2} = \frac{[X^-][H^+]}{[X^{\pm}]} \quad ...(4)$$

From equation (3),

$$[X^+] = \frac{[X^{\pm}][H^+]}{K_{a_1}}$$

From equation (4),

$$[X^-] = \frac{[X^{\pm}]K_{a_2}}{[H^+]}$$

At the isoelectric point when there is a zero net charge, the concentration of the acid and its conjugate base must be equal

i.e., $[X^+] = [X^-]$

Hence,

$$\frac{[X^{\pm}][H^+]}{K_{a_1}} = \frac{[X^{\pm}]K_{a_2}}{[H^+]}$$

$$[H^+]^2 = K_{a_1} \times K_{a_2}$$

Taking common logarithm on both sides, we get

$$2\log[H^+] = \log K_{a_1} + \log K_{a_2}$$

$$2pH = pK_{a_1} + pK_{a_2}$$

$$pH = \frac{pK_{a_1} + pK_{a_2}}{2}$$

At the isoelectric point pH is denoted by pI. Therefore,

$$pI = \frac{pK_{a_1} + pK_{a_2}}{2}$$

The pK_a and pI values for different amino acids are given in **Table 2.2.**

As shown above the isoelectric point of an amino acid can be calculated from its titration for simple amino acids with no additional acidic or basic group, the isoelectric point lies exactly half way between the two pK_a values. Let us consider the example of alanine. In strongly acidic solution (*e.g.*, at pH 0), it is present mainly in the cationic form. The pK_{a_1} for the carboxyl group of the cationic form is 2.34. This value is quite low than the pK_a of corresponding carboxylic acid (propanoic acid, pK_a = 4.89). This indicates that cationic form of alanine is the stronger acid because it is a positively charged species. It should lose a proton more readily.

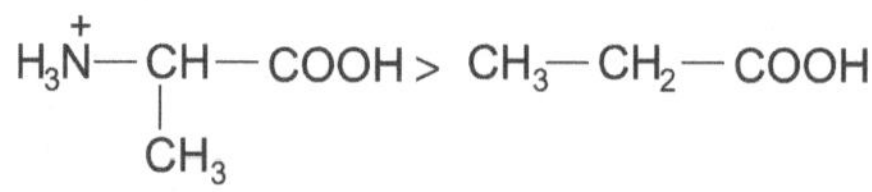

Cationic form of alanine pK_{a_1} = 2.34

Propanoic acid pK_{a_1} = 4.89

The pK_{a_2} for the dipolar ion of alanine is 9.87. Therefore, the isoelectric point, pI for alanine is obtained by taking the average of the two pK_a values

$$pI = \frac{2.34 + 9.87}{2}$$

$$= 6.1$$

When the amino acid contains an extra amino group, the situation is going to be different. For example, lysine has an extra amino group on its 2nd carbon. In strongly acidic solution, lysine will be present as a dication because both amino groups will be protonated. The first proton to be lost as the pH is raised is a proton of the carboxyl group (pK_{a_1} = 2.2).

Lysine exists as a zwitterion with zero net charge between pK_{a_2} and pK_{a_3} as shown below:

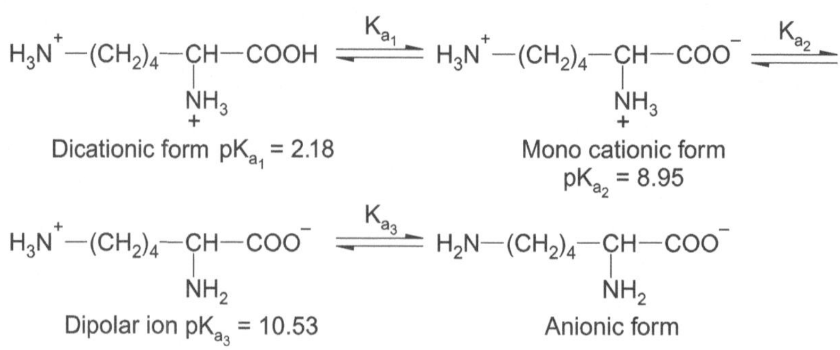

The isoelectric point for lysine is, therefore, the average of pK_{a_2} and pK_{a_3}

$$pI = \frac{8.95 + 10.53}{2} = 9.7$$

For acidic amino acids, the zwitterion structure exists between K_{a_1} and K_{a_2}. Therefore, pI is calculated as the average of the first two pK_a values.

For aspartic acid, we have

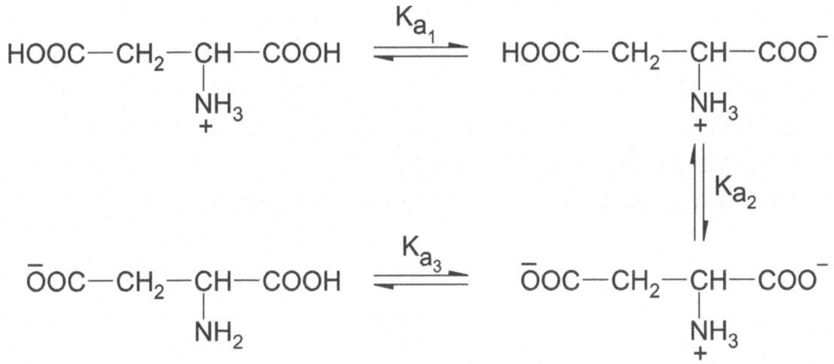

pI is, therefore, given as

$$= \frac{pK_{a_1} + pK_{a_2}}{2}$$

$$= \frac{2.09 + 3.87}{2}$$

$$= 3.0$$

TABLE 2.2 ***The pK_a and pI values for amino acids***

S. No.	*Amino acid*	*pK_{a_1}*	*pK_{a_2}*	*pK_{a_3}*	*pI*
1.	Glycine	2.35	9.78		6.1
2.	Alanine	2.34	9.87		6.1
3.	Valine	2.32	9.62		6.0
4.	Leucine	2.36	9.60		6.0
5.	Isoleucine	2.36	9.66		6.0
6.	Proline	2.00	10.60		6.3
7.	Phenylalanine	2.58	9.24		5.9
8.	Tryptophan	2.38	9.39		5.9
9.	Methionine	2.28	9.21		5.7
10.	Serine	2.21	9.15		5.7
11.	Threonine	2.63	10.43		6.5
12.	Cysteine	1.96	8.18	10.28	5.1
13.	Aspartic acid	2.09	3.87	9.82	3.0
14.	Glutamic acid	2.19	4.28	9.66	3.2
15.	Lysine	2.18	8.95	10.53	9.7
16.	Arginine	2.02	9.04	12.48	10.8
17.	Tyrosine	2.20	9.10	10.10	5.7
18.	Histidine	1.77	6.10	9.18	7.6

2.10 SEPARATION OF AMINO ACIDS

A. Electrophoresis

Electrophroesis is the migration of charged molecules such as amino acids, proteins, etc., when placed in an electric field. The process is used for the separation of different amino acids from a mixture.

The separation is based on the difference on net charges of the components in the mixture. The mixture is applied on to an inert support, which may be paper, cellulose acetate or some polymeric gel. The ends of the support are immersed into a chamber containing a buffer solution of known pH. A direct current is then passed through the solution. Amino acids with a net positive charge migrate towards the cathode whereas those that are negatively charged will move towards the anode. Amino acids that exist as zwitterion will remain immobile (Fig. 2.3).

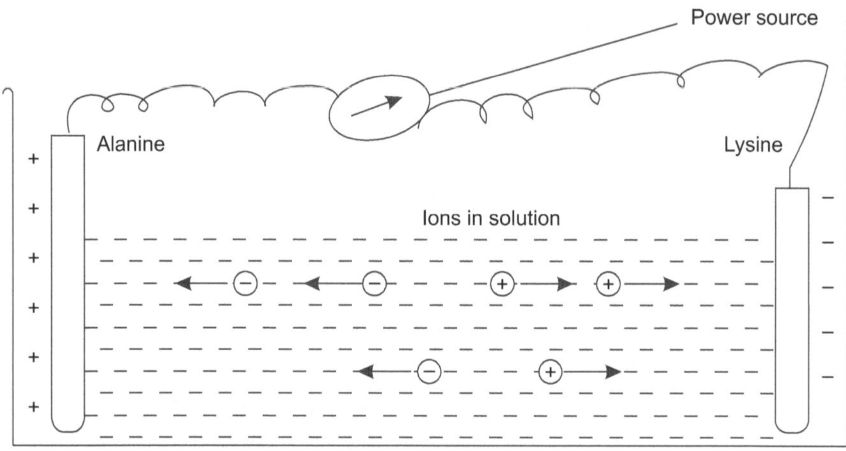

Fig. 2.3 Principle of electrophoresis.

Detection of the amino acids can be carried out by ninhydrin. The paper is sprayed with a dilute solution of ninhydrin and heated at 100°C. Purple coloured spots appear on the paper which can be compared with R_f values of known compounds.

B. Ion-Exchange Chromatography

This is a common method used for the quantitative analysis of amino-acid mixtures. The technique is similar to column chromatography. Here the column consists of a resin which may be strongly acidic cation or a strongly basic anion exchanger. One commonly used resin is a copolymer of styrene and divinylbenzene with negatively charged sulphonic acid groups on some of the benzene rings. When an acid solution of a mixture of amino acids is added, the most basic acids are most tightly bound and the most acidic acids are most weakly bound. Elution is carried out with a series of buffer solutions. The amino acids separated because they flow through the column at different rates.

If a mixture of lysine and glutamate in a solution with pH 6 is loaded onto the above column, glutamate travels down the column rapidly because its negatively charged side chain is repelled by the negatively charged sulphonic acid groups of the resin. The positively charged side chains of the lysine are retained on the column. Such a resin is called a cation-exchange resin. In this Na^+ of the SO_3^- groups are exchanged by the positively charged amino acids.

Resins with positively charged groups are called anion exchange resins because they impede the flow of anions. A common anion exchange resin has $R—\overset{+}{N}(CH_3)_3\ Cl^-$ groups in place of $SO_3^-\ Na^+$ groups.

When a mixture of amino acids possessing different charges passes through a charged column, the amino acids possessing opposite charges get associated with the column while those with the same charge or no charge pass through the column. Thus, separation of a mixture of amino acids is achieved.

This method has now been automated. The instrument that automates ion-exhange chromatography is known as amino acid analyser. When a solution of amino acid mixture passes through the column of an amino acid analyser containing a cation exchange resin, the amino acids move through the column at different rates depending on their overall charge. Fractions of the effluent are collected and analysed for their amino acid content using ninhydrin for formation of the complex with different amino acids. In this way, the identity and the relative amount of each amino acid can be determined.

2.11 CHEMICAL REACTIONS OF AMINO ACIDS

The reactions of amino acids are in general the ones we would expect of compounds containing $—NH_2$ and —COOH groups. In addition, other groups that may be present undergo their own characteristic reactions.

A. Reactions due to Amino Group

1. Alkylation

Alkylation of amino acids is carried out by treating the amino acids with alkyl iodides. Mono alkyl amino acids are obtained by this method.

$$R—\underset{NH_2}{\underset{|}{C}H}—COOH + R'I \longrightarrow R—\underset{NH-R'}{\underset{|}{C}H}—COOH$$

In the presence of excess alkyl halide, these form internal quaternary ammonium salts and are called betaines. The trimethyl derivative of glycine, $(CH_3)_3—\overset{+}{N}—CH_2—COO^-$, known as betaine, occurs naturally.

Amino acids may also be arylated by 2,4-dinitrochlorobenzene or 2,4-dinitrofluorobenzene (DNF B) known as Sanger's reagent. These derivatives are useful for the analysis of peptides. Barger, however, used 2,4,5-trinitrotoluene for arylation of amino acids. The reaction is carried out by warming an alcoholic solution of the amino acid with trinitrotoluene. Nitrous acid is eliminated.

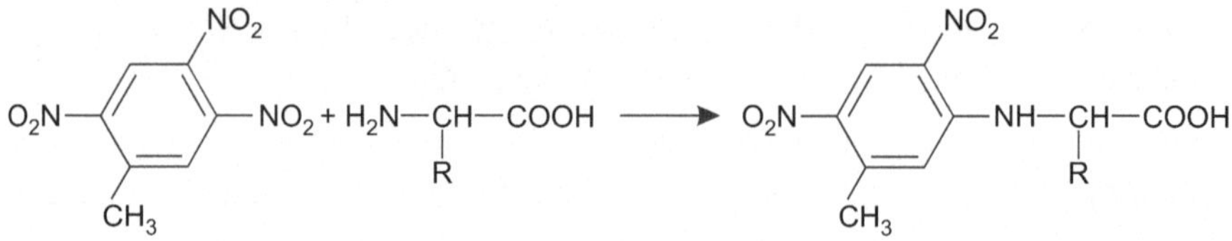

Sanger's reagent is more reactive. It readily gives dinitrophenyl (DNP) derivatives of amino acids.

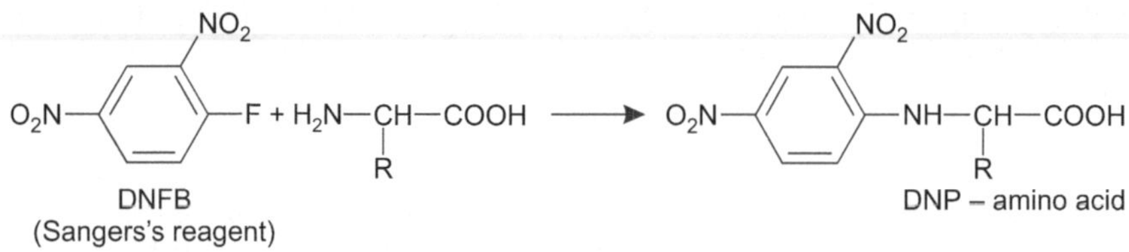

2. Acylation

Amino acids may be acylated in a general way by using acyl halides in aqueous alkali. For acetylation, we can use acetyl chloride.

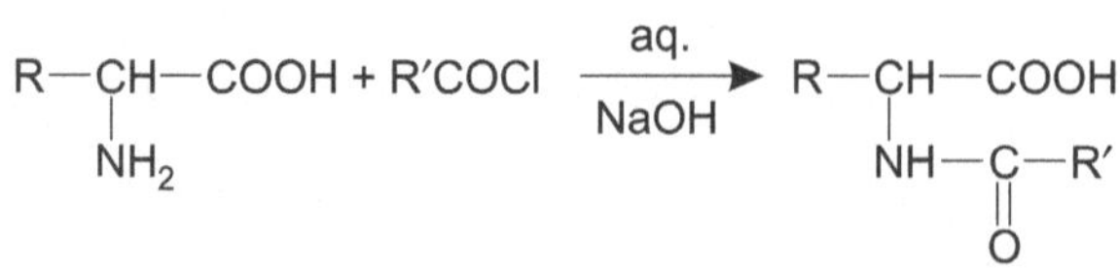

Amides may also be prepared by reaction with acetic anhydride. For example,

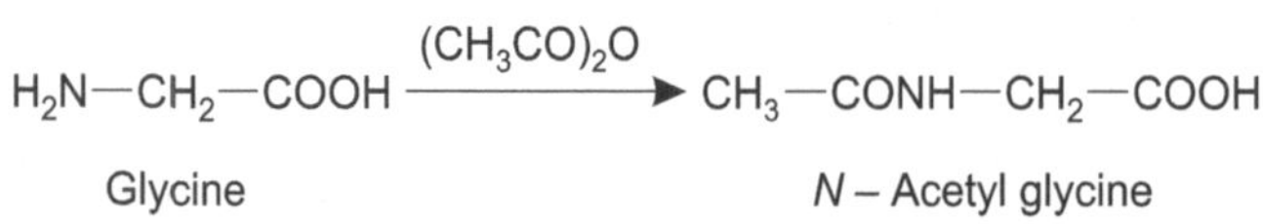

Similarly, benzoylation may be carried out by using benzoyl chloride under Schotten—Baumann conditions (aq. sodium hydroxide at room temperature). However, it is necessary to acidify the aqueous solution to obtain the acidic product.

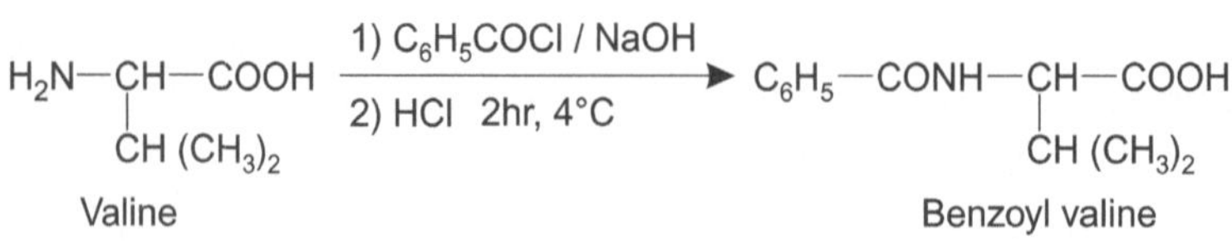

Histidine may be acylated in the following method:

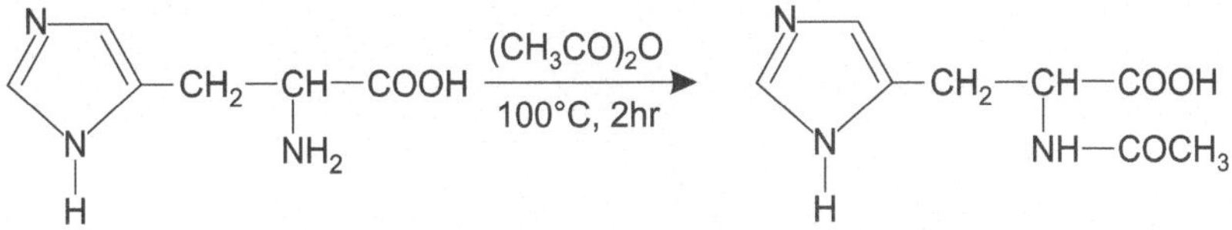

N – Acetylhistidine

The acylated derivatives of amino acids are quite important as they possess sharp and characteristic melting points. The lone pair of electrons on the amino group is withdrawn by the acyl group attached to the nitrogen atom. The benzoyl derivatives are especially useful for characterisation.

Condensation of amino acids with phthalic anhydride gives phthaloyl derivatives, which are of value in peptide synthesis.

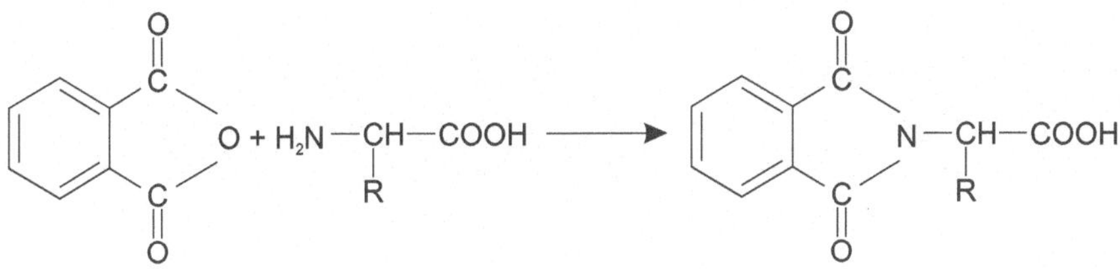

Sulphonamides result by the reaction with sulphonyl chloride. For example,

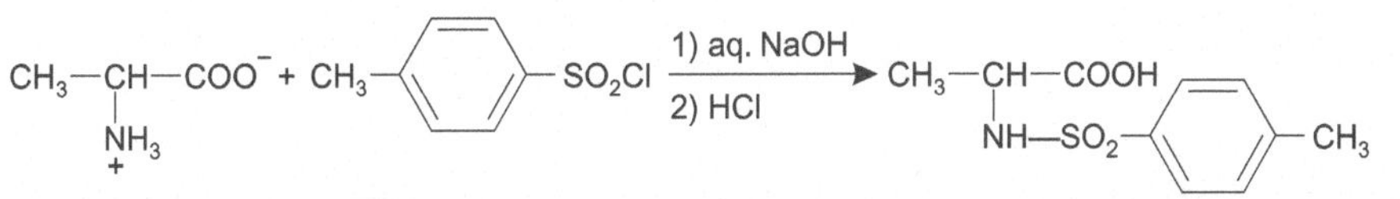

3. Nitrous Acid Reaction

Nitrous acid reacts with amino acids to liberate one mole of nitrogen gas per mole of amino acid and gives the corresponding hydroxy acids.

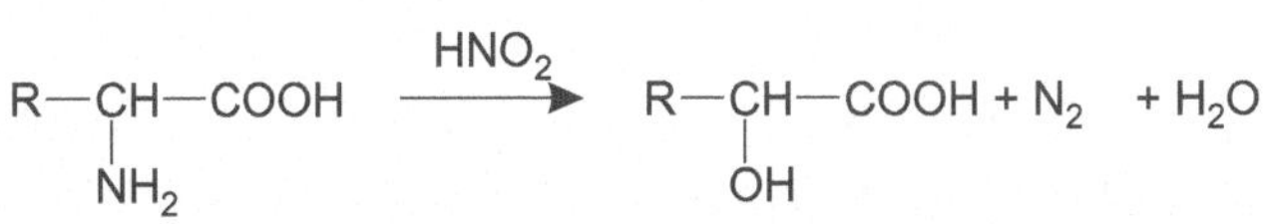

The reaction takes place with α–amino acids at room temperature in less than ten minutes and is of great utility for identification and assay of purity of amino acids. The ε-amino groups of lysine and arginine react in thirty minutes. The method and apparatus was developed by van Slyke and is known after his name.

4. Formol Reaction

The H-atom of the amino group is substituted by methylol group by treating with formaldehyde at room temperature.

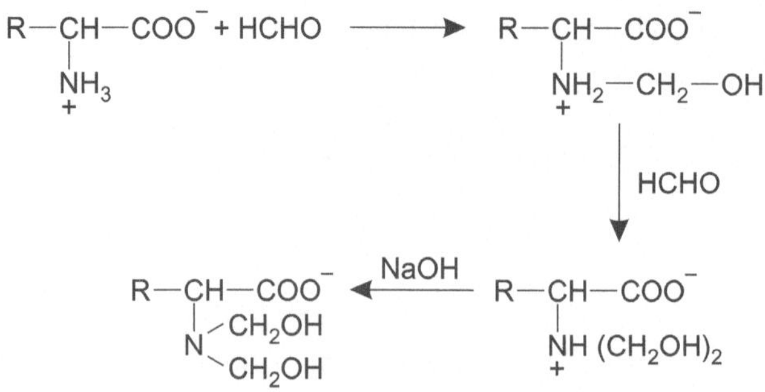

The formol reaction has the same analytical utility as the nitrous acid reaction. Aldehydes other than formaldehyde react in a similar manner.

5. Reaction with Nitrosyl Chloride

Nitrosyl chloride (or bromide) reacts with amino acids to form chloro (or bromo) acids, *e.g.*,

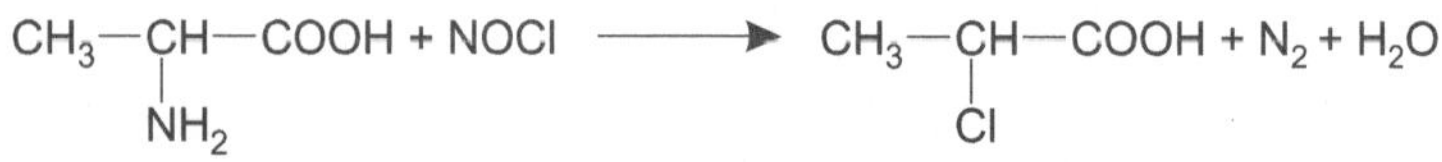

Alanine α-Chloroacetic acid

6. Reaction with Hydriodic Acid

When amino acid is heated with hydriodic acid at 200°C, the amino group is eliminated with the formation of a fatty acid.

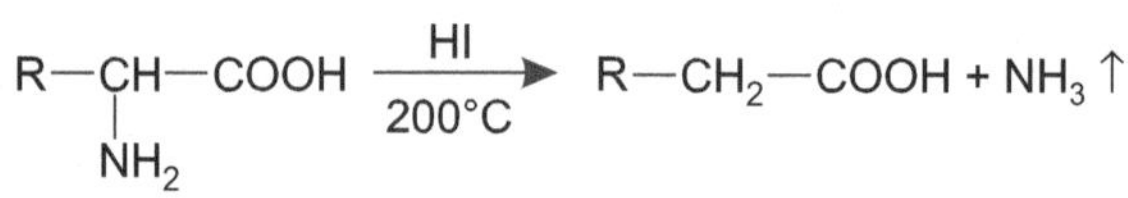

For example, alanine gives propionic acid when heated with HI at 200°C.

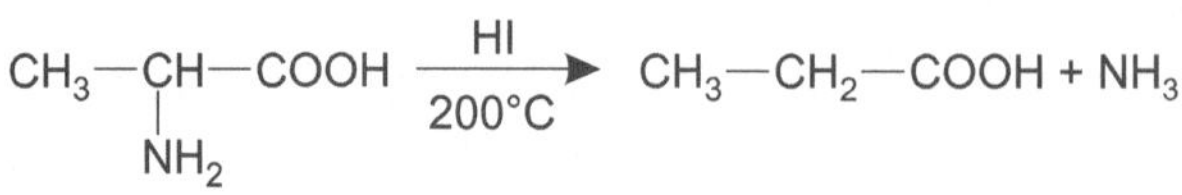

7. Oxidation to Keto Acid

The deamination of amino acids to keto acids is of considerable biological significance. Chemically an amino acid can be converted to a keto acid by treating with hydrogen peroxide. In the biological systems, such oxidations are brought about by L-amino acid oxidases and D-amino acid oxidases. The keto acids, which are obtained by deamination of the amino group of the amino acids are identical with some of the keto acids, which result from carbohydrate metabolism.

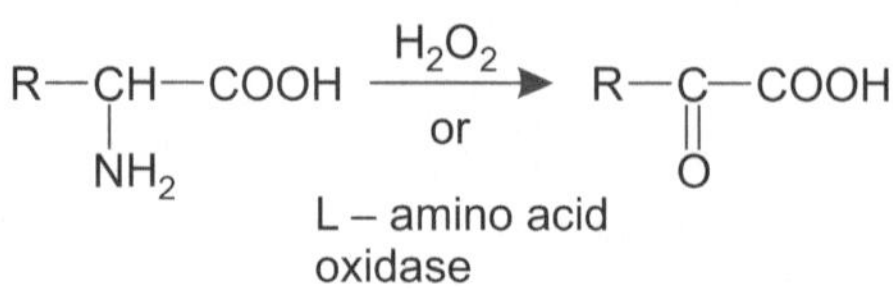

B. Reactions due to the Carboxyl Group

1. Esterification

The carboxyl group of an amino acid may be esterified in the usual manner. The methyl and the ethyl esters are obtained by refluxing the suspension of the amino acid in appropriate alcohol with anhydrous hydrogen chloride. The ester is separated as the crystalline hydrochloride salt.

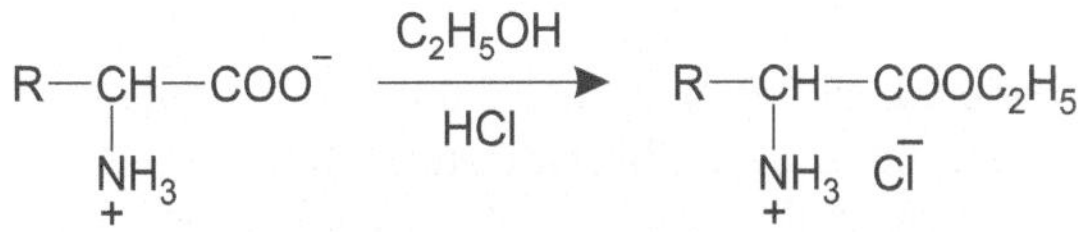

The free ester may be obtained by the action of aqueous sodium carbonate on the ester salt.

Benzyl esters are often prepared by using benzene sulphonic acid as the catalyst. The water produced in the reaction is removed by azeotropic distillation. For example, glycine gives the following benzyl ester derivative.

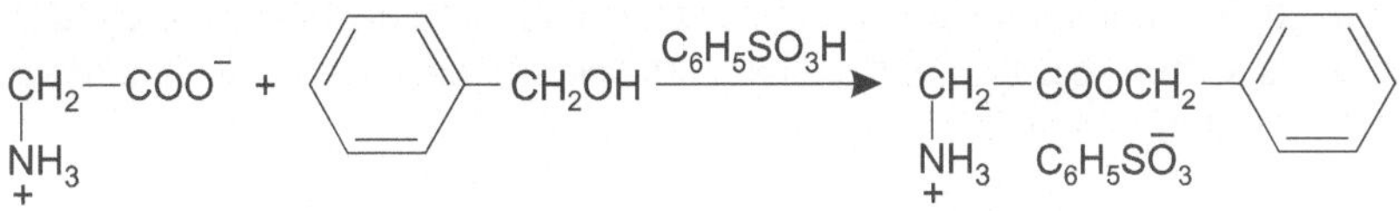

Glycine benzyl ester benzene sulphonate

The esters do not possess the dipolar ion structure found in the unsubstituted amino acids and are, therefore, more volatile. So, it is possible to distil the esters without causing extensive decomposition.

Methyl, ethyl and benzyl esters are employed extensively as intermediate in the synthesis of peptides. However, benzyl esters are more useful because they can be converted back to acids by non-hydrolytic methods. For example, glycine benzyl ester reacts with hydrogen in the presence of palladium to give glycine salt from which the free amino acid can be obtained.

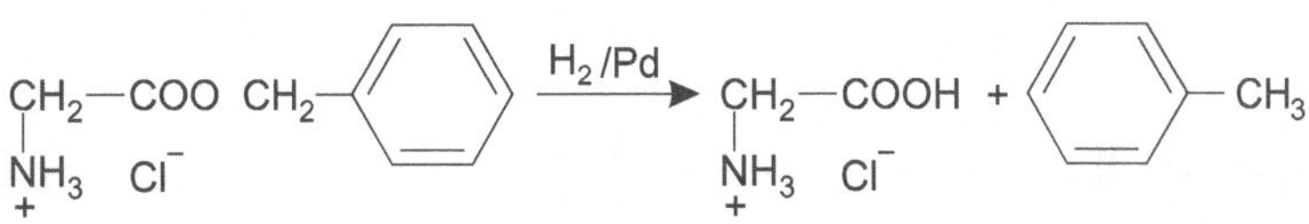

2. Acyl Halide Formation

It is not possible to convert carboxyl group of an amino acid directly to an aminoacyl halide because of the reactivity of the amino group in the presence of halogenating agents. The aminoacyl halides are desirable intermediates in the synthesis of peptides. The amino group is protected first and then treated with phosphorus pentachloride or refluxed with thionyl chloride. Benzoyl group is frequently used as the protecting group.

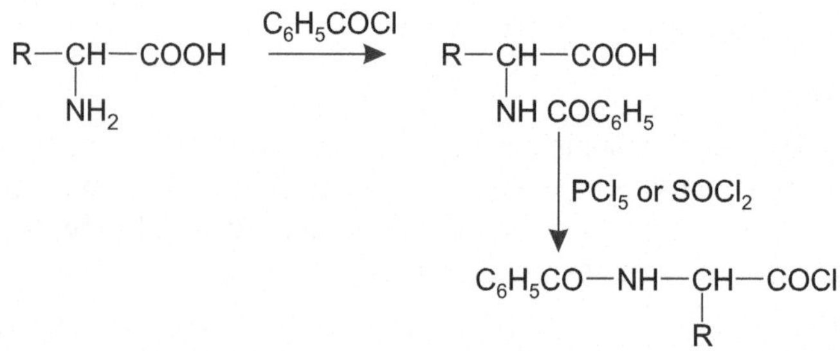

Deprotection of the benzoyl group is carried out in the usual manner. When amino acids are suspended in acetyl chloride and then treated with phosphorus pentachloride, they form hydrochlorides of the acid chloride.

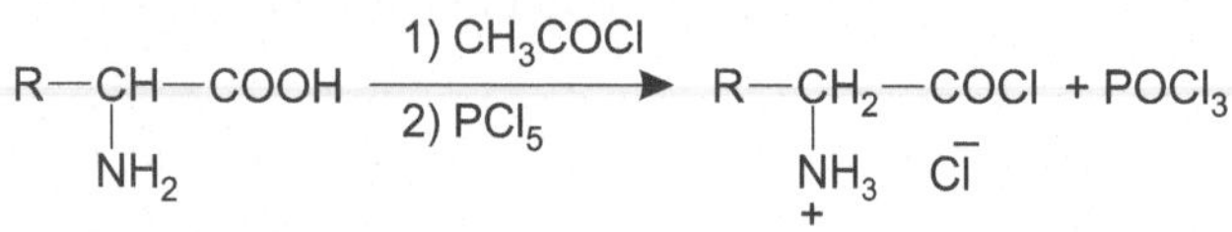

3. Decarboxylation

Decarboxylation of amino acids may be achieved by dry distillation or by heating with bases such as barium oxide. There are specific enzymes which catalyse the decarboxylation of specific amino acids.

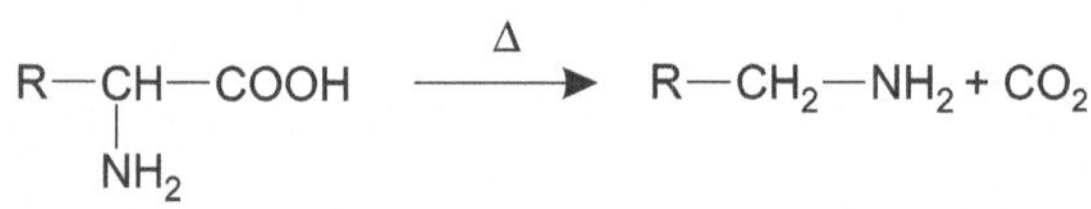

Decarboxylation of amino acids is quite significant biologically. The decarboxylation of histidine to histamine has great physiological significance.

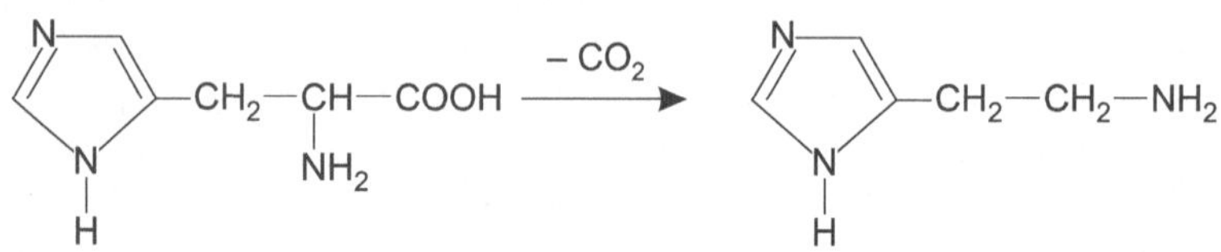

Histidine Histamine

Histamine is a direct causative agent in allergy, shock and similar states and is believed to arise by decarboxylation.

4. Cyclodehydration

Two molecules of an amino acid may react with the loss of water to form an amino acid anhydride, *i.e.*, disubstituted piperazine. The reaction may be brought about by heating the amino acid in high boiling solvents such as ethylene glycol. The esters of amino acids react more readily with the splitting out of two molecules of alcohol.

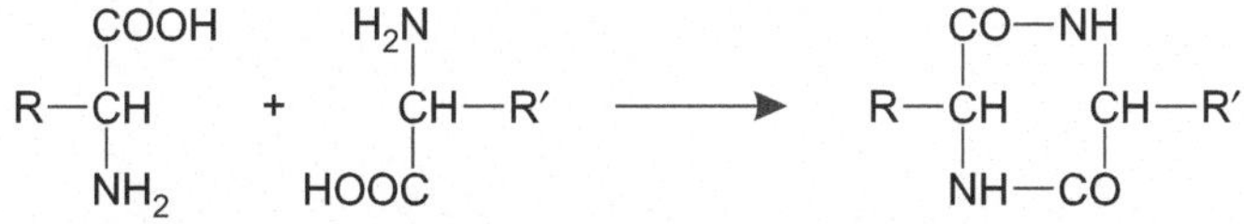

The diketopiperazines are of considerable interest because they have frequently been found in the protein hydrolysates.

5. The Dakin-West Reaction

Amino acids on heating with acetic anhydride and pyridine are converted into methyl α–acetamido ketones. The reaction is often referred to as the Dakin-West reaction.

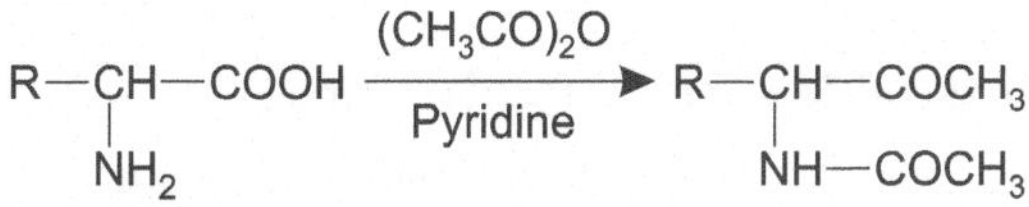

6. Reduction

Amino acids form amino alcohols on reduction with $LiAlH_4$. The optical activity of the alcohols is retained.

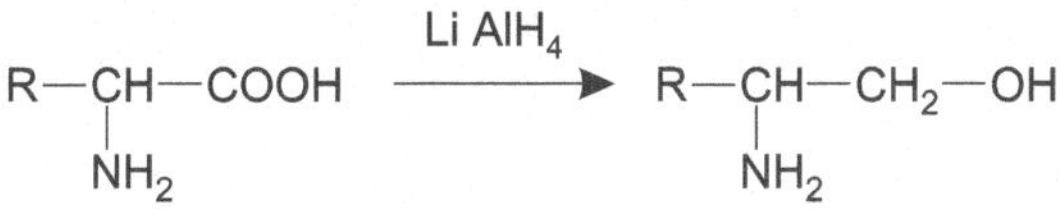

C. Reactions due to both the Amino and Carboxyl Groups

1. Oxidation

The oxidation of amino acids to the aldehyde, in which the carbon containing the amino group is used up, may be carried out by sodium hypochlorite.

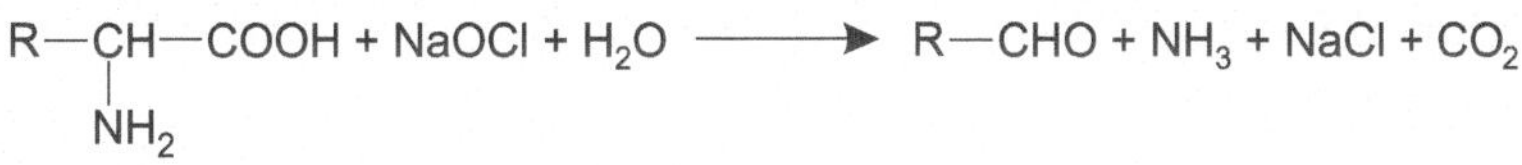

Instead of NaOCl, Chloramine-T may be employed.

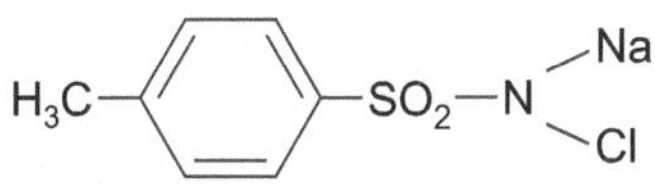

Chloramine – T

Periodic acid or its sodium salt may be used for cleaving β–hydroxy amino acids.

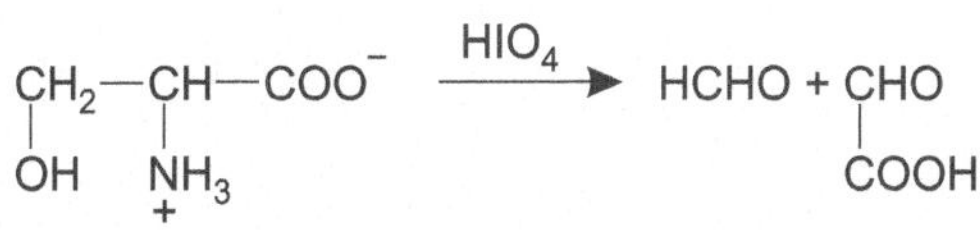

When two moles of chloramine-T are used as the oxidising agent, a nitrile instead of an aldehyde is obtained.

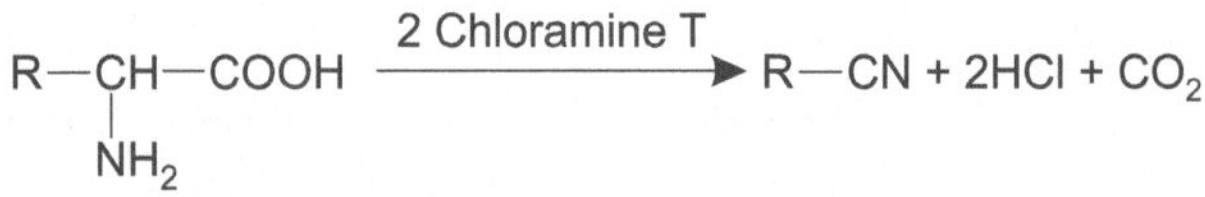

2. Reaction with Organic Acids

Strong organic acids convert the amino acids to cationic form such as

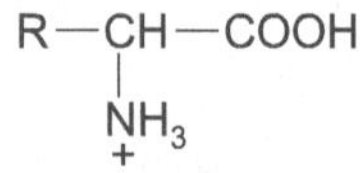

With picric acid (2,4,6-trinitrophenol) an insoluble salt precipitates out because of the affinity of the acid anion for the amino acid cation.

3. Reaction with Metal Ions and Complex Salts

Amino acids form salts with metal ions. Heavy metal ions such as Cu^{2+}, Ag^{+}, Hg^{2+} form insoluble complexes involving the amino group. The copper salt of glycine, which has deep blue colour, is formed by heating copper oxide with an aqueous solution of glycine. These metal ions also precipitate proteins. The structure of the complex of copper with glycine is given below, which is obtained by mixing glycine with an aqueous solution of $CuSO_4$.

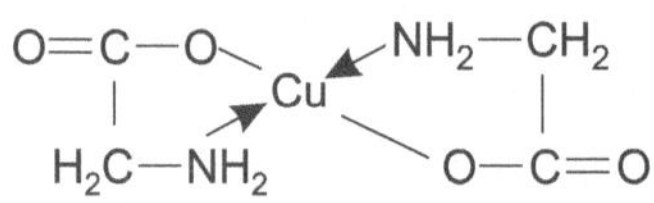

The heavy metal can be removed by decomposition of the complex with hydrogen sulphide and the amino acid is recovered.

Some inorganic complex salts have also been used for selective precipitation of amino acids from protein hydrolysates.

4. Action of Heat

α–Amino Acids

As given above under cyclodehydration, α–amino acids on heating form 2,5-diketopiperazines.

For example, glycine forms 2,5-diketopiperazine

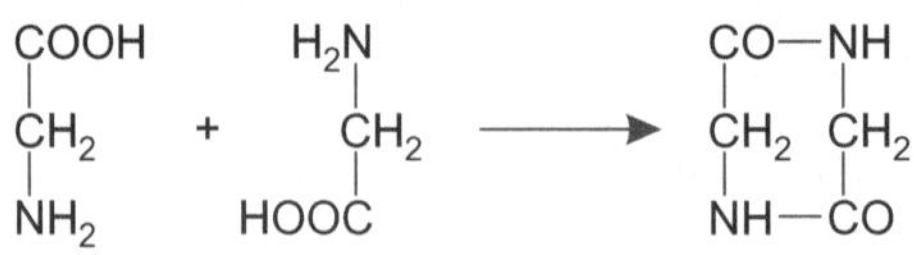

2,5 – Diketopiperazine

β–Amino acids

On heating, β–amino acids eliminate a molecule of ammonia to give α,β–unsaturated acids.

$$R-\underset{\displaystyle NH_2}{CH}-CH_2-COOH \xrightarrow{\Delta} R-CH=CH-COOH$$

β–amino acid

γ–amino acids and δ–amino acids

γ- and δ-amino acids on heating form cyclic amides called lactams.

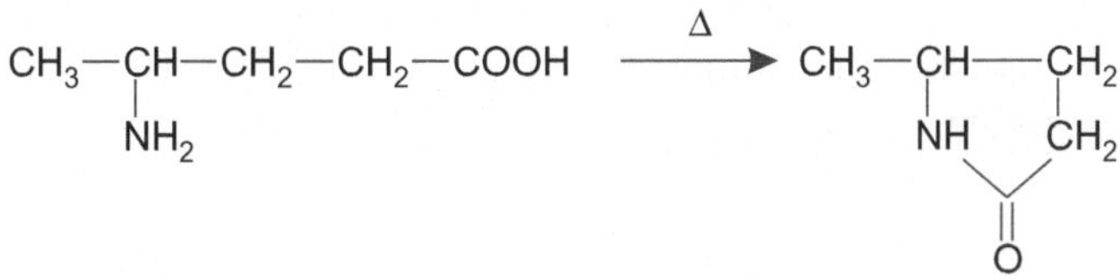

Lactam

5. The Ninhydrin Reaction

All α–amino acids give characteristic violet coloured product when heated with ninhydrin solution. This reaction is the basis of detection of α–amino acids in chromatography techniques. The chromatogram is sprayed with ninhydrin solution in acetone. On heating slowly violet coloured spots develop.

On treating with ninhydrin, which is triketohydrindene hydrate, α–amino acids get converted into an amino derivative which condenses with another mole of ninhydrin to finally give the coloured product. The sequence of reactions is given below:

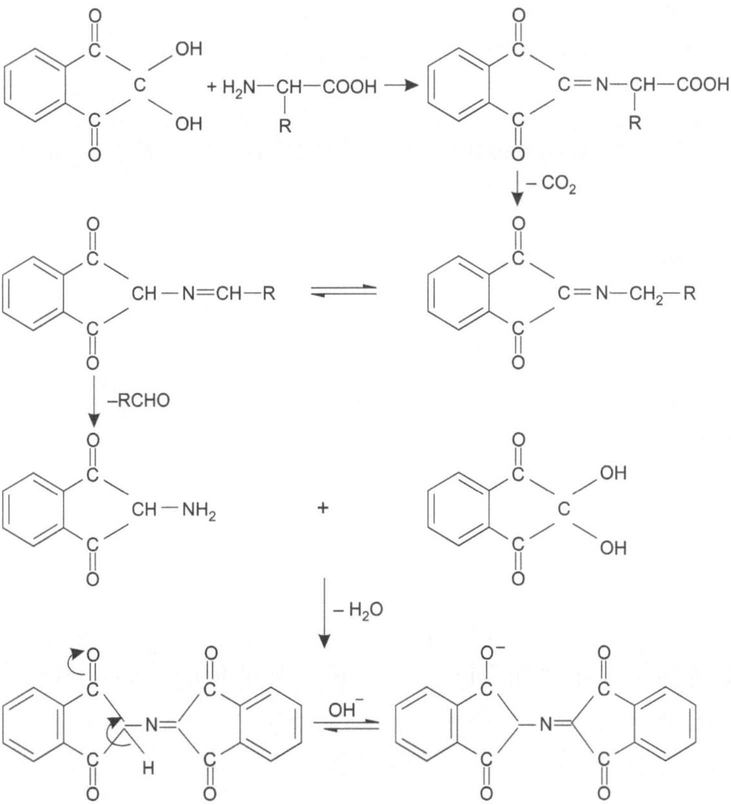

2.12 POLYPEPTIDES

A. Introduction

We have already seen that amino acids may form amide bonds between two molecules, *i.e.*, carboxyl group of one amino acid condenses with the amino group of another. The amide linkage between the amino acids is called a peptide bond and the resulting compounds are called peptides. When two amino acids condense they give a dipeptide. For example, the peptide formed from two molecules of glycine is glycylglycine, a dipeptide. Like glycine, it is also amphoteric and exists as zwitterion.

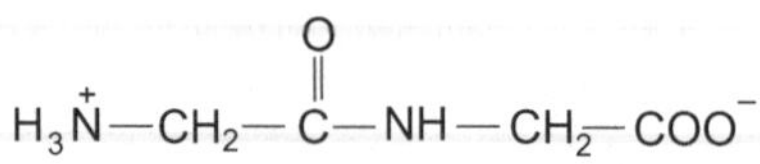

Glycylglycine
(a dipeptide)

Similarly, a tripeptide contains three amino acid units and a tetrapeptide four, and so on. When the number of amino acid units increases beyond ten we get a polypeptide. By convention peptides of molecular weight up to 10,000 are known as polypeptides and above that as proteins. As mentioned earlier, proteins are special types of polypeptides composed of 20–22 different specific amino acids. The term oligo peptide is used for peptides containing about four to ten amino acid residues.

The end of the peptide with the free amino group (– $\overset{+}{N}H_3$) is called the *N*-terminal end or the *N*-terminus and the end with the free carboxyl group (–COO^-) is called the *C*-terminal end or the *C*-terminus. According to convention, the *N*-terminal amino acid residue is written at the left end and the *C*-terminal amino acid residue at the right end.

B. Nomenclature of Peptides

Peptides are named starting with the *N*-terminal amino acid residue. The names of amino acid residues involved in peptide linkages (all except the last) are given –yl suffix of acyl groups. For example, the following dipeptide is named as alanyl-glycine. The alanine residue has the –yl suffix because it has acylated the nitrogen of glycine

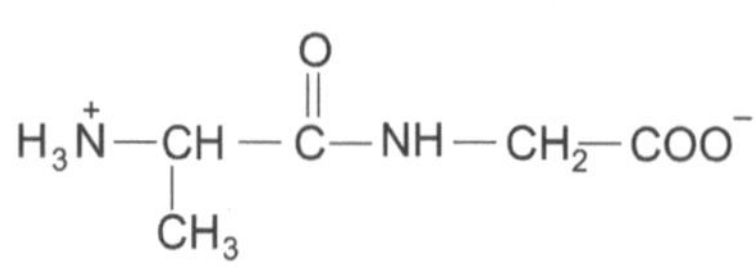

Alanylglycine (Ala–Gly)

The abbreviations given in Table 2.1 are used for different amino acids.

Similarly, a tripeptide (Gly-Ala-Phe) is named as.

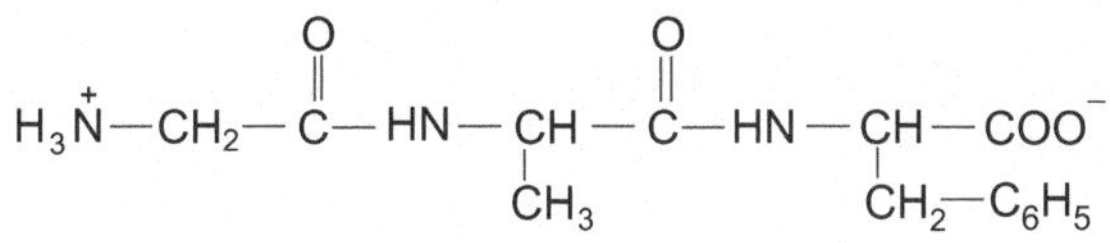

Glycylalanylphenylalanine

2.13 SYNTHESIS OF PEPTIDES

A. The Basic Scheme

The synthesis of a peptide requires the formation of a peptide bond between the carboxyl group of one amino acid and the amino group of another one. The reaction involves a nucleophilic substitution by an amino group at a carbonyl group of a carboxylic acid. It is quite likely that reaction may occur between the carboxyl group and the amino group of the same amino acid. For example, the synthesis of simple dipeptide alanylglycine involves the conversion of alanine to its chloride and then allow it to react with glycine. However, we cannot prevent alanyl chloride from reacting with itself. So, our reaction would give not only Ala-Gly but also Ala-Ala. We may also get tripeptides and so on.

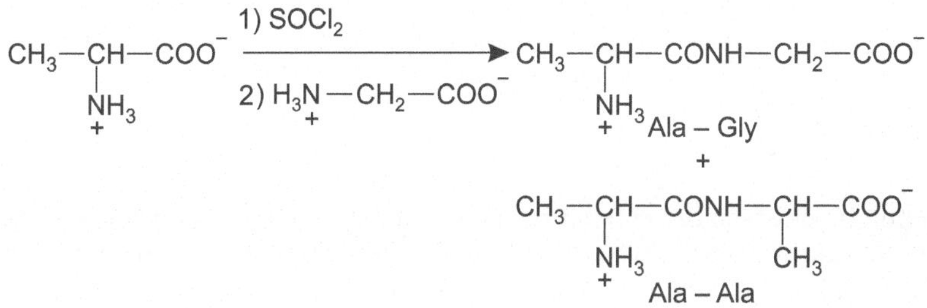

Moreover, there are other nucleophilic groups present in amino acids such as other amino groups, —OH and —SH groups on the side chains of amino acids. These groups are also expected to react with carboxylic acids. Therefore, we must protect these groups so that the reaction can be directed to give the desired product.

The first step in the basic scheme of peptide synthesis is to protect the amino group of the first amino acid before we activate it and allow it to react with the second. The second step involves the activation of the —COOH group of protected amino acid and the formation of the peptide bond with the next amino acid under conditions that do not destroy peptide bonds and other functional groups that may be present in the molecule. After getting the desired peptide the last step is the removal of the protecting group by using a reagent that does not affect the peptide bonds. The steps involved in the peptide synthesis are given below:

1. **Protection of the amino group**

$$Z + \overset{+}{N}H_3\text{—}\underset{\displaystyle R}{\underset{|}{CH}}\text{—}COO^- \longrightarrow Z\text{—}NH\text{—}\underset{\displaystyle R}{\underset{|}{CH}}\text{—}COOH$$

2. **Activation of —COOH group and formation of peptide bond**

$$Z\text{—}NH\text{—}\underset{\displaystyle R}{\underset{|}{CH}}\text{—}COOH \longrightarrow Z\text{—}NH\text{—}\underset{\displaystyle R}{\underset{|}{CH}}\text{—}COCl + H_3\overset{+}{N}\text{—}\underset{\displaystyle R'}{\underset{|}{CH}}\text{—}COO^-$$

$$\longrightarrow Z\text{—}NH\text{—}\underset{\displaystyle R}{\underset{|}{CH}}\text{—}CO\text{—}NH\text{—}\underset{\displaystyle R'}{\underset{|}{CH}}\text{—}COO^-$$

3. **Removal of the protecting group**

$$Z\text{—}NH\text{—}\underset{\displaystyle R}{\underset{|}{CH}}\text{—}CONH\text{—}\underset{\displaystyle R'}{\underset{|}{CH}}\text{—}COO^- \xrightarrow{-Z} H_3\overset{+}{N}\text{—}\underset{\displaystyle R}{\underset{|}{CH}}\text{—}CONH\text{—}\underset{\displaystyle R'}{\underset{|}{CH}}\text{—}COO^-$$

B. Protection of the Amino Group

Numerous methods have been developed to protect an amino group. The protecting group must be carefully chosen because after we have synthesised the desired peptide the protecting group should be removed easily without disturbing the new peptide bonds. Moreover, it should not result in racemisation of the neighbouring chiral atom. We give below some important protecting groups:

1. Benzyloxycarbonyl Group

This method was introduced in 1932 by Bergmann and Zervas of the University of Berlin (later of the Rockefeller Institute). In this method, we do acylation of the amino group by benzylchloroformate earlier known as carbobenzoxy chloride. (Cbz or simply Z group). This is prepared in toluene solution from benzyl alcohol and phosgene.

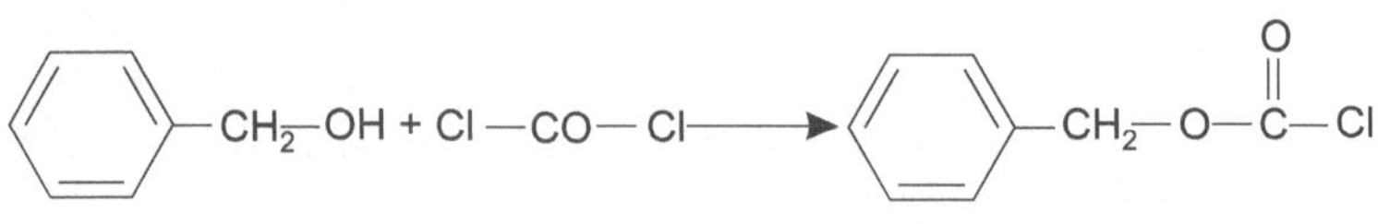

Benzylchloroformate

This may be condensed with the amino acid by the Schotten Baumann reaction. The pH of the reaction mixture is kept high so as to neutralise the HCl so generated in the reaction.

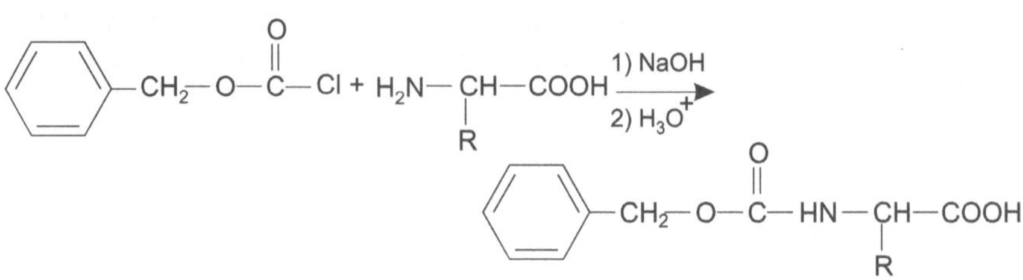

N-Protected amino acid

The benzyloxycarbonyl group can be removed by reagents that do not affect peptide linkages, *i.e.*, by catalytic hydrogenation or by using cold HBr in acetic acid.

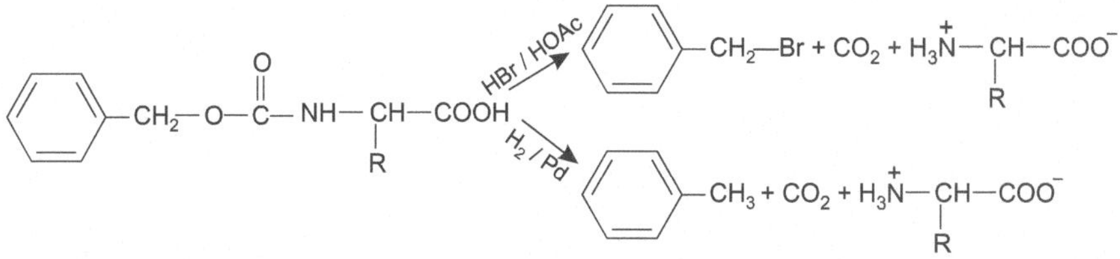

The cleavage of the Z-group by HBr in acetic acid proceeds by S_N2 mechanism as given below:

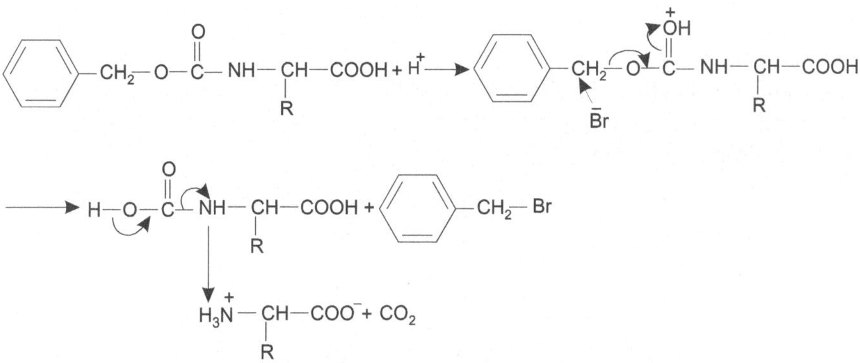

Removal of the Z-group with H_2/Pd depends on the fact that C-O bond is weak and hydrogenolysis takes place at low temperatures.

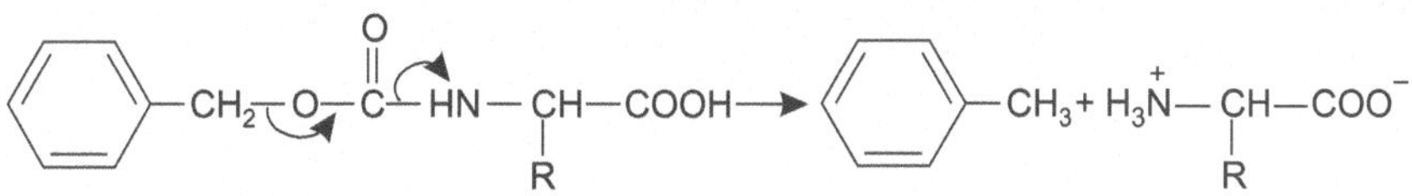

The benzyloxycarbonyl group is normally not affected by mildly basic or other nucleophilic reagents. Therefore, it is quite useful for carrying out peptide synthesis.

2. *t*-Butyloxycarbonyl Group (Boc Group)

This is the most popular group used for protecting α-amino groups. We can introduce this group by using *t*-butylazidoformate or di-tert-butyl carbonate as tert-butylchloroformate is not very stable. When we use azido-formate, azide ion is the leaving group. Formation of the Boc-amino acid is shown below:

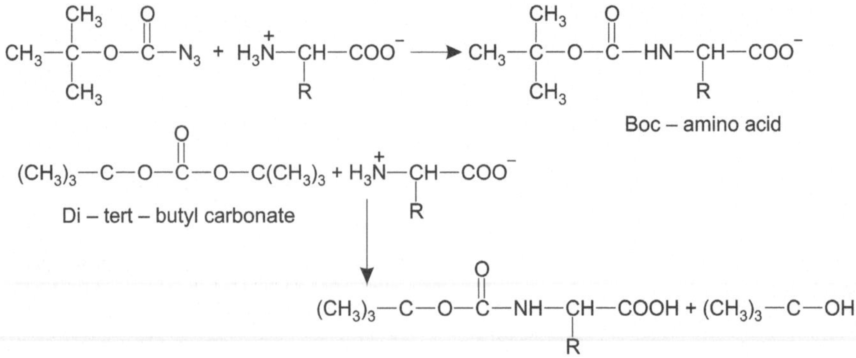

The Boc group is stable to hydrogenation and general reducing conditions and is more labile to acid than the Z-group. Therefore, Boc group can be removed easily under mild acidic conditions such as dry trifluoroacetic acid in dichloromethane or dry hydrogen chloride in ether.

The mechanism of deprotection of the Boc group is similar to the one given for Z-group. It also results initially the more stable tert-butyl cation. In the process it gives the formation of isobutene.

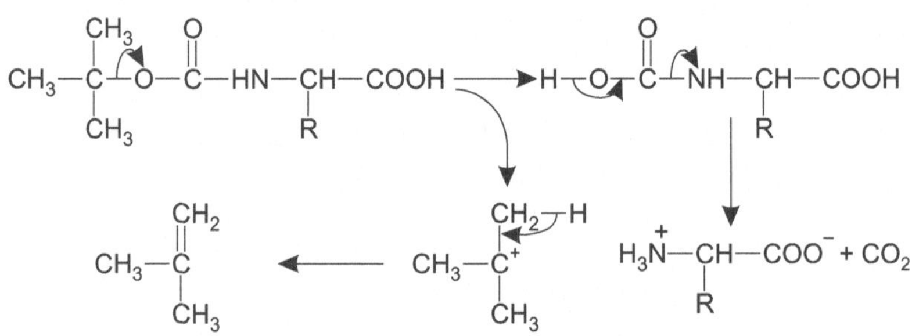

3. 2-(4-Biphenyl)-isopropoxycarbonyl Group-(Bpoc)

This group is introduced *via* the azide. It is more acid labile than the Boc group because of the formation of more stable intermediate carbocation and hence, it is preferred over Boc group.

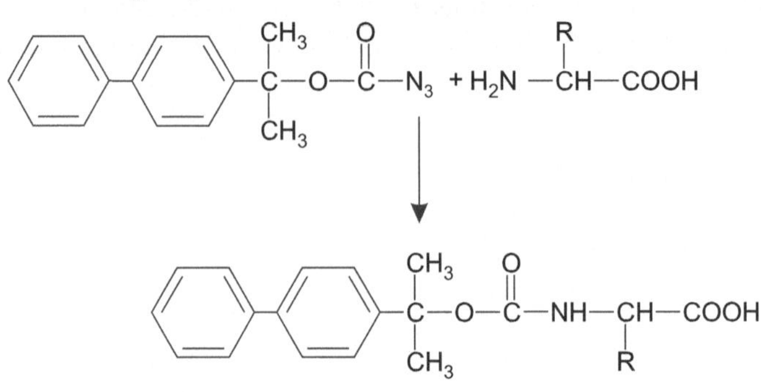

Bpoc Amino acid

Deprotection of the Bpoc group can be carried out by treating the protected amino acid with chloroacetic acid in dichloromethane. The Boc and Z group remain unaffected under these conditions. Therefore, Bpoc group can be used along with these groups and can be selectively removed.

4. Phthaloyl Group

Preparation of *N*-phthaloyl amino acids by direct interaction of phthalic anhydride and amino acids in the fused state or in organic solvents frequently leads to partial or complete racemisation. Therefore, *N*-phthaloyl amino acids are obtained by condensing *N*-ethoxy carbonyl phthalimide with amino acids in alkaline media. The group can be cleaved by treating with hydrazine under refluxing conditions as the *N*-phthaloyl group is sensitive to alkaline conditions.

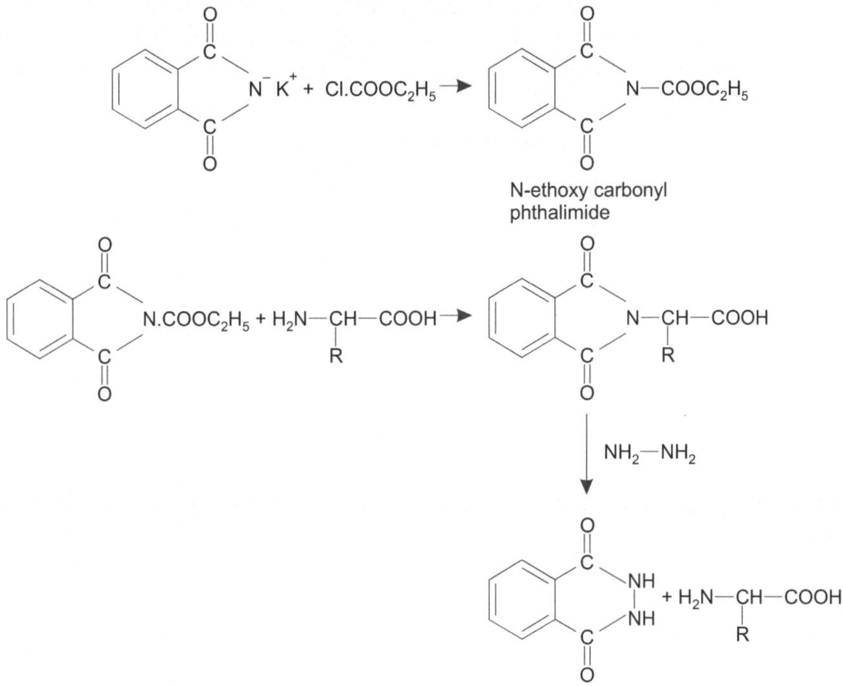

5. Toluene-*p*–Sulphonyl Group (Tosyl)

The tosyl group was used for blocking α-amino groups at an early stage in the development of methods of peptide synthesis. More recently, it has proved useful for protecting the ε-amino group of lysine and arginine. It can be cleaved easily by the reducing action of sodium in liquid ammonia.

C. Protection of the Carboxyl Group

In general, carboxyl groups of amino acids are esterified before the peptide bond is formed. This is necessary because amino acids occur as dipolar ions while esters are simple bases which are soluble in aprotic solvents and formation of the peptide bond may have to be conducted in an aprotic solvent. Peptide bond formation in protic solvents is prone to racemisation . Moreover, it is easier to detach a proton from $\overset{+}{N}H_3-\underset{\underset{R}{|}}{C}-COOR'$ than from $\overset{+}{N}H_3-\underset{\underset{R}{|}}{C}-COO^-$.

Esters of amino acids can be liberated from their hydrochlorides by shaking with a solution of ammonia in chloroform before peptide synthesis or *in situ* during peptide synthesis by the addition of a tertiary base. The former method is preferable because the presence of tertiary base can sometimes lead to racemisation.

The carboxyl group is generally protected by methyl, ethyl or benzyl groups. The ester hydrochlorides are readily obtained by passing dry hydrogen chloride into a suspension of the amino acid in dry alcohol. Amino acids can also be conveniently esterified by first converting into an acid chloride by using thionyl chloride and then an appropriate alcohol is added to get the ester.

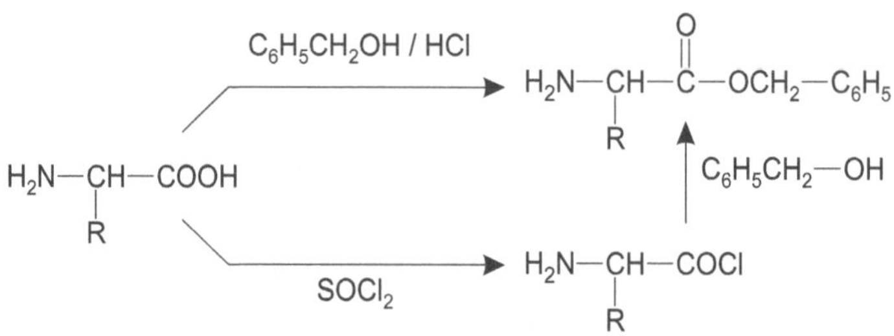

Deprotection of the ester group after peptide synthesis can be carried out by alkaline hydrolysis using mild NaOH. However, alkaline conditions may lead to racemisation.

Benzyl esters are frequently more convenient than methyl or ethyl esters since they can be cleaved to the free acid by hydrogenolysis using palladium-charcoal catalyst. Treatment with HBr in acetic acid also removes the benzyl group to give the free acid.

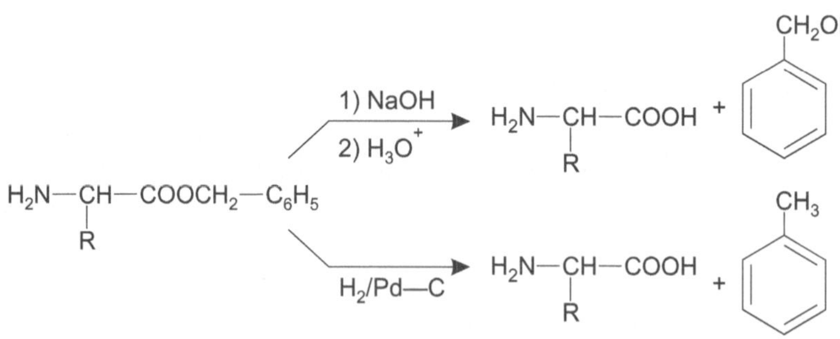

However, *t*-butyl esters, like *t*-butyloxycarbonyl derivatives are resistant to hydrogenolysis but are very sensitive to anhydrous acids. Dry hydrogen chloride in a suitable solvent or trifluoroacetic acid are effective mild reagents for this purpose.

t-Butyl esters are formed when amino acids are allowed to react with isobutylene in the presence of concentrated sulphuric acid.

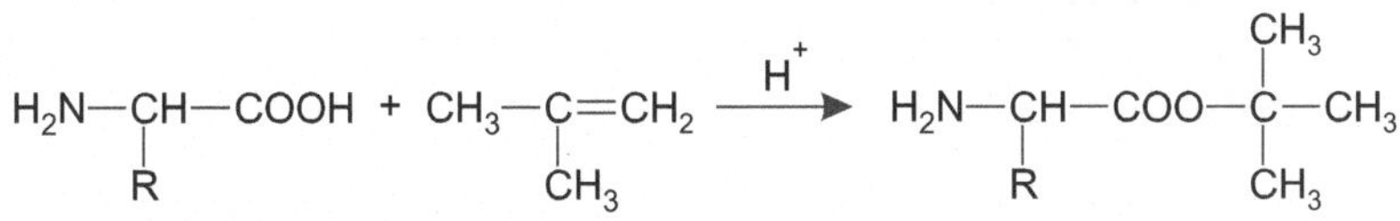

D. Protection of Side-Chains of Amino acids

It is not possible to generalise the protecting groups for the side chains because the choice of protecting group depends upon the nature of the side chain.

The ε-amino group in the side chain of lysine can be protected by any of the methods described for α-amino groups. However, it is necessary to employ separate protecting groups of different stability for the α- and ε-amino groups. The protecting group on the α-position can then be selectively removed to permit the coupling of further amino acid residues. The protecting group on the ε-position is left intact until the end of the synthesis.

Lysine chelates with Cu^{2+} ions by involving α-amino group and $-COO^-$ anion, the ε-amino group can thus be selectively protected. Decomposition of the copper chelate with hydrogen sulphide liberates the α-amino group which can then be protected by using different reagent. For example, ε-amino group can first be protected by Boc group and α-amino group can then be blocked with Z-group. After formation of the peptide bond between –COOH group of lysine and amino group of C-protected amino acid, the Z-group can be removed by hydrogenolysis and the liberated amino group can be coupled with another protected amino acid. Finally, all the protecting groups can be removed with anhydrous hydrogen chloride or trifluoroacetic acid to give the free tripeptide.

The β-carboxyl group of aspartic acid and γ-carboxyl group of glutamic acid are usually protected by benzyl or *t*-butyl groups. It is necessary to choose a protecting group for the *N*-terminus which can be removed without affecting the ester groups in the side chains of aspartic or glutamic acids.

The thiol group in the side-chain of cysteine is a powerful nucleophile and it must, therefore, be protected during peptide synthesis. The *S*-benzyl group is the most commonly used. For example, *S*-benzyl-cysteine is prepared by dissolving cysteine in liquid ammonia, adding sufficient sodium to convert the thiol into the anion and then treating with benzyl chloride. The benzyl group is removed by reduction with excess sodium in liquid ammonia.

The guanidino group in the side-chain of arginine is strongly basic. Mere protonation prevents it from entering into the side reactions during peptide synthesis.

The alcoholic group in the side chain is usually protected either as benzyl or *t*-butyl ethers.

2.14 METHODS OF PEPTIDE SYNTHESIS (COUPLING METHODS)

1. Acid Chloride Method

For formation of a peptide bond it is necessary that the hydroxyl group present in the carboxylic acid function is converted into a good leaving group. The most obvious and simple approach to activate a –COOH group is to convert it to an acyl chloride. In this method, the protected amino acid is converted to the corresponding acyl chloride followed by reaction with amino acids or esters under Schotten-Baumann conditions or treatment with amino esters in organic solvents to give the *N*-protected peptide. Deprotection under hydrolytic conditions gives the free peptide.

This approach played an important role in the early days of peptide synthesis. The reagents normally used for acyl chloride formation, *e.g.*, PCl_5 or $SOCl_2$ are too vigorous to be used in complex or sensitive substrates and may lead to racemisation.

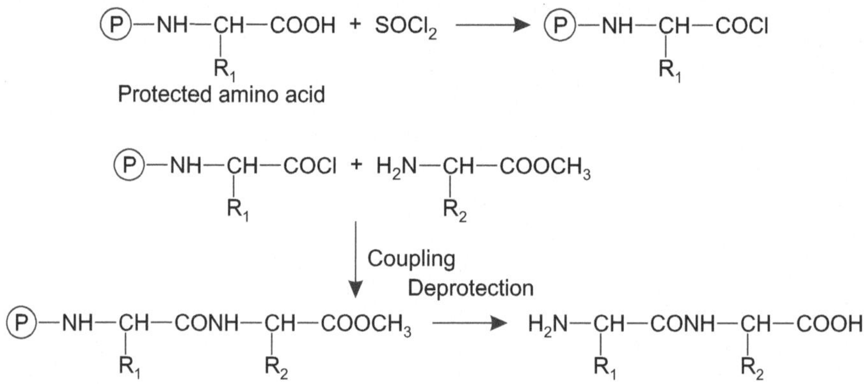

This method has become completely obsolete as many more convenient and efficient methods are available.

2. Curtius Azide Method

Difficulties observed in use of acyl chloride were overcome by Curtius. He used the acid azide for coupling with another amino acid.

In this method, *N*-protected amino acid is converted to its hydrazide which gives the corresponding azide by reaction with nitrous acid. *N*-protected amino acid azide is then reacted with amino acid ester to give the desired peptide. The scheme is given ahead:

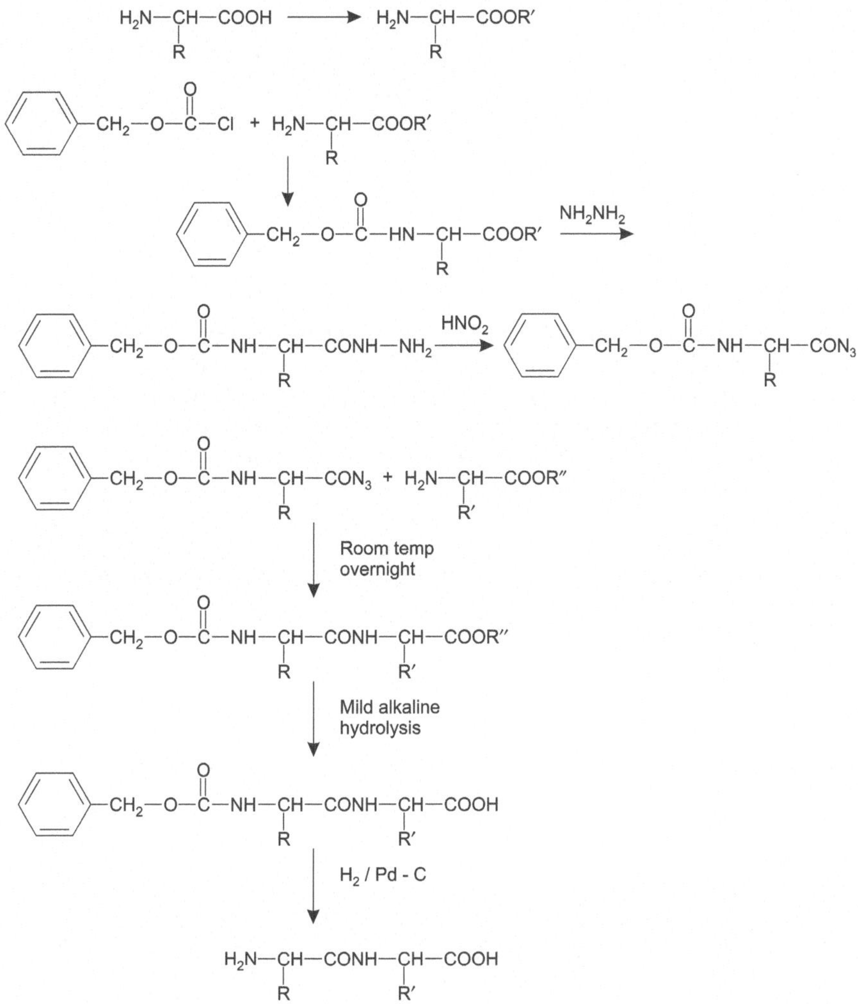

This synthesis is accomplished without the risk of racemisation. However, some disadvantages are there. *Firstly,* the number of steps involved are more. *Secondly,* the acyl azide may undergo the Curtius rearrangement to give the formation of isocyanate. Therefore, the desired product is difficult to purify.

3. DCC Method

In the presence of dicyclohexylcarbodiimide, an acid reacts with an amine to give an amide.

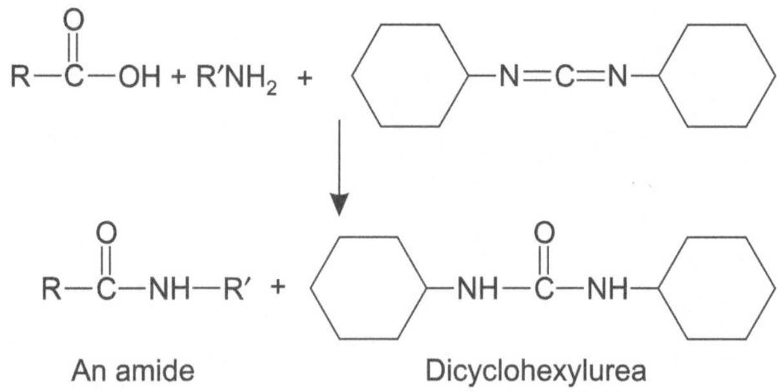

Dicyclohexylcarbodiimide (DCC) as a coupling agent for the synthesis of peptides was developed by Sheehan and Hess in 1955. The non-participating –COOH and $-NH_2$ groups must be protected. The coupling takes place between –COOH and $-NH_2$ just at room temperature. Various inert solvents can be used for the reaction. Methylene chloride or THF are more commonly used. The by-product, dicyclohexylurea, having low solubility in most solvents, separates as a solid and can be filtered off.

DCC can be prepared from carbon disulphide and cyclohexyl amine as given below:

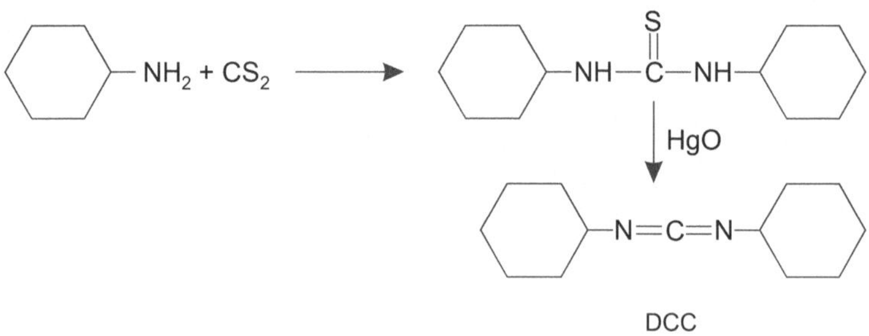

DCC converts the carboxylic acid into an intermediate that has the properties of an acid anhydride.

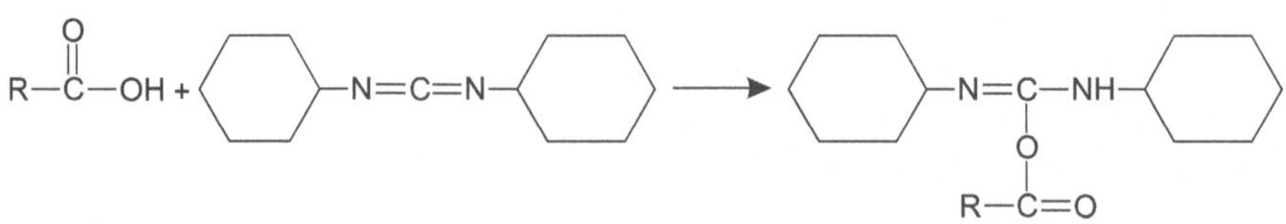

The scheme of reactions for the synthesis of Gly-Ala is given below:

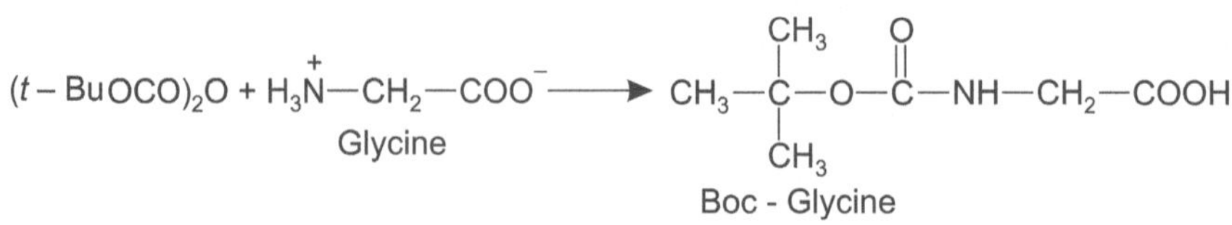

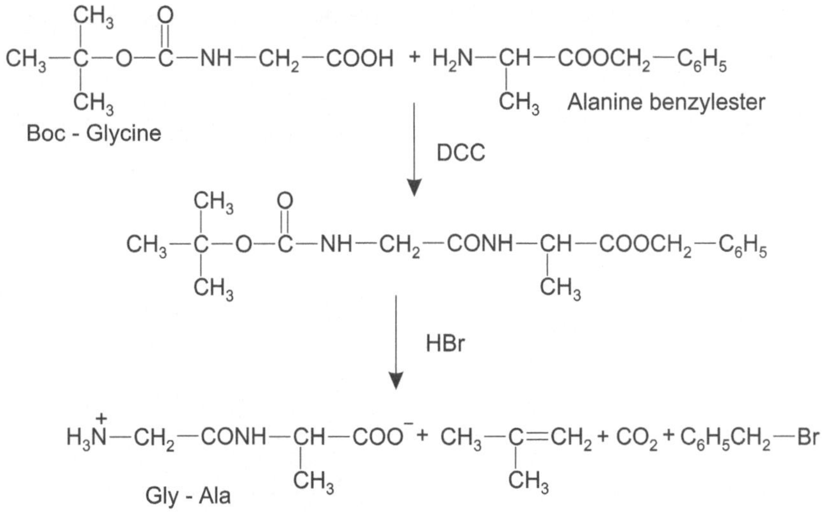

The repetition of the above reactions can give us synthesis of any desired peptide. By following the above approach oxytocin was synthesised by Vincent du Vigneaud in 1955 and synthesis of insulin with 51 amino acids was reported by Sanger in 1963.

4. Mixed Anhydride Method

In this method, the carboxyl group of the protected amino acid is converted to a mixed anhydride using ethyl chloroformate in an organic solvent in the presence of a tertiary base like triethylamine. This method is used in such cases where the separation of the by-product dicyclochexylurea in the DCC method is difficult.

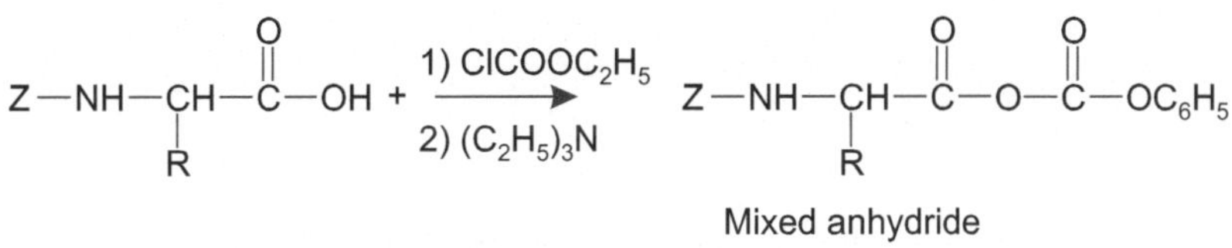

The mixed anhydride can then be used to acylate another amino acid or its ester to give the desired product. The reaction takes place at low temperatures giving high yields and pure products.

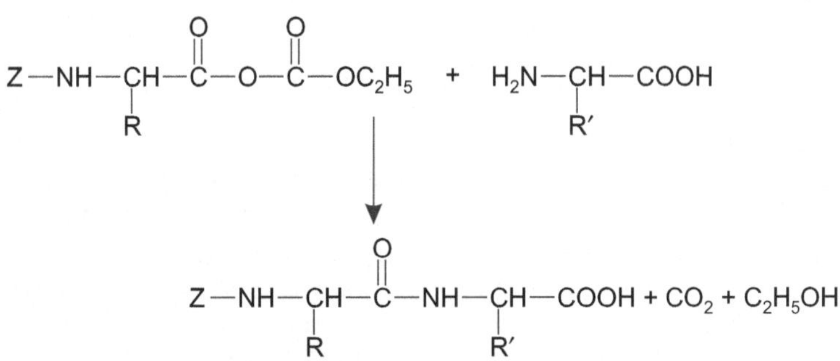

We give below the scheme to be adopted for the synthesis of simple dipeptide Ala-Leu:

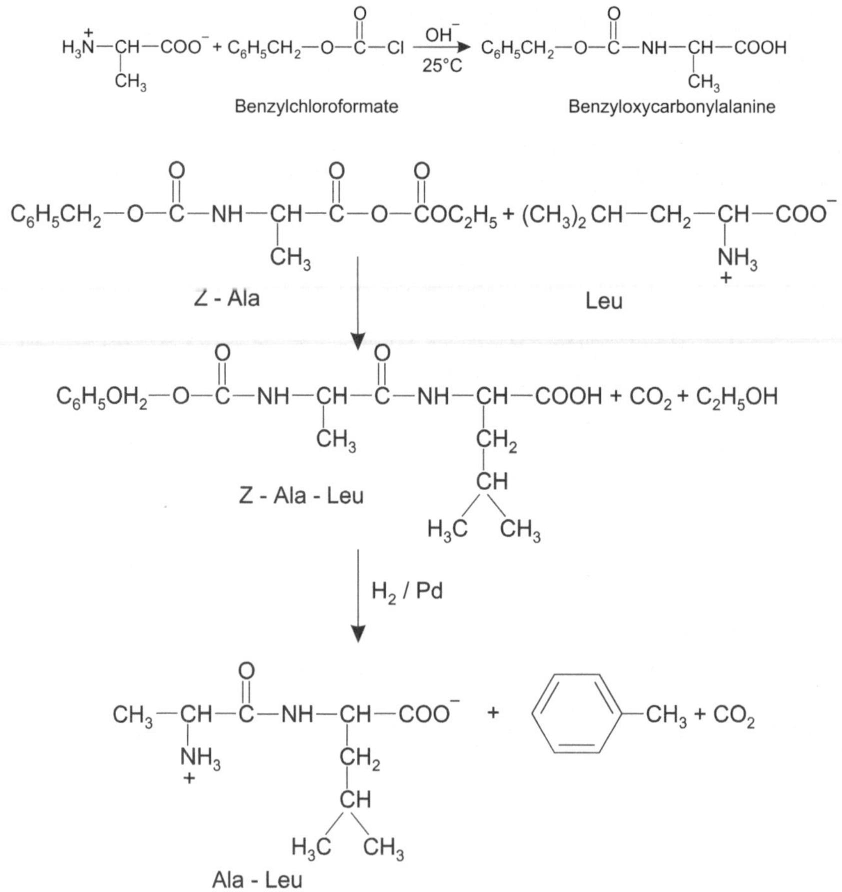

Similarly, we can show all the steps in the synthesis of Gly-Val-Ala using Boc-group as a protecting group.

5. Use of Other Activating Groups

A widely used method for the activation of the carboxyl group and formation of peptide bond involves the use of 'active esters'. The idea is to convert —OH group of —COOH into a good leaving group by the preparation of an active ester of the amino acid. *p*-Nitrophenyl ester is often used. For example, the *p*-nitrophenyl ester of glycine, protected at the amino group by benzyloxycarbonyl group, is prepared. The activated carboxyl group then reacts with another amino acid, say valine, protected at its —COOH group as the methyl ester.

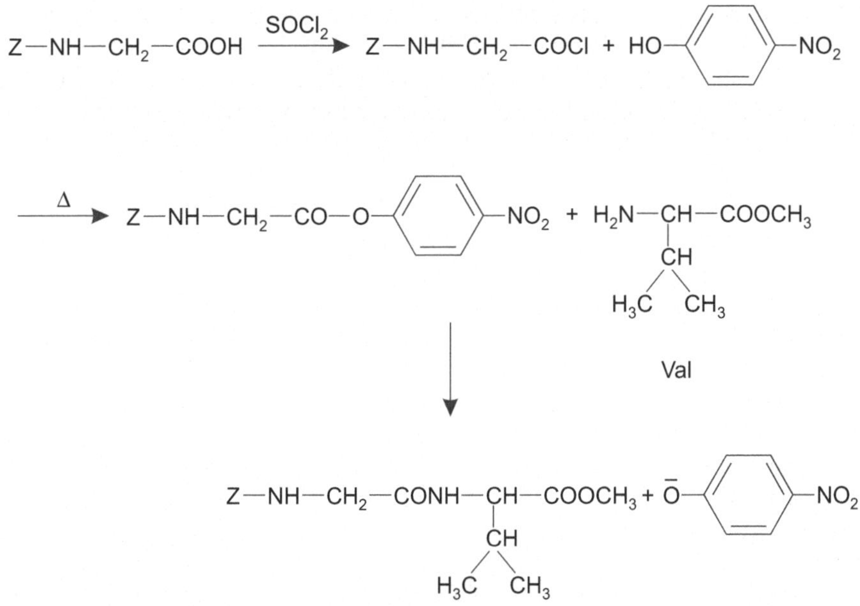

Z - Gly – Val – O – CH_3

The conditions used for the formation of the *p*-nitrophenyl ester are quite harsh and might not be suitable for activating the —COOH group of larger peptides. So other types of esters, formed in the presence of DCC, are used for many peptide synthesis.

Esters of *N*-hydroxy succinimide are often used where *N*-oxysuccinimide anion is a good leaving group.

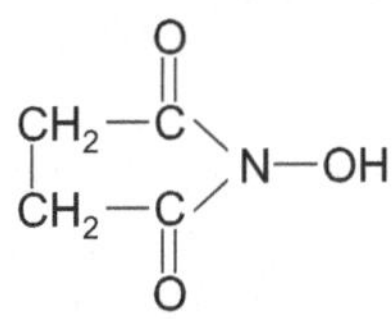

N - Hydroxy succinimide

2.15 SOLID PHASE PEPTIDE SYNTHESIS

A. Introduction

Solid phase peptide synthesis was developed and introduced by R.B. Merrifield of Rockfeller University in order to simplify and accelerate peptide synthesis in a way that would make the synthesis of long peptides practical.

The basic principle of this synthesis is that amino acids can be hooked into a peptide chain of desired sequence while one end of the chain is anchored to an insoluble support. After the desired sequence of amino acids has been linked together on the solid support a reagent can be applied to cleave the finished peptide from the solid support. The reactions have been developed in such a manner so that they go to 100 per cent completion. The great advantages of Merrifield synthesis are the ease of operation and high overall yield. The purification of the resin with its attached polypeptide can be carried out at each stage by simply washing the resin with an appropriate solvent. All the laborious procedures for the purification at the intermediate steps are thus eliminated. Therefore, no mechanical losses are encountered in the intermediate stages. Impurities are simply carried away by the solvent because they are not attached to the insoluble resin. Furthermore, since the method involves the repetitive use of a small number of similar operations, the entire procedure can be automated so that the actual man hours of labour during the synthesis is greatly reduced. All the reactions are carried out in a suitably designed vessel, known as "protein making machine", without any transferring of material from one container to another. Mechanical loss of the material in the transfer operations is thus eliminated. In the automated procedure, each cycle of the Merrifield synthesis requires only four hours and attaches one new amino acid residue.

B. The Basic Scheme of Solid Phase Peptide Synthesis

The solid support is a synthetic polymer that bears reactive groups (X). These groups are made to react with the carboxyl groups of an amino acid in such a way that the amino acid is bound covalently to the polymer support. During this operation the amino group of the amino acid must be protected by a suitable group (Y) so that the amino group will not react with the polymer. The protecting group must be such so that it can be selectively removed without damage to the bond holding the amino acid to the polymer support. After removal of the protecting group a second *N*-protected amino acid is used to couple with the amino group of the first amino acid, thus forming the first peptide chain. By repeating these two steps, deprotection and coupling, the peptide of desired sequence is assembled on the polymer support. At the end a different reagent is applied to cleave the peptide chain from the polymer and also to carry out the deprotection. The basic plan of solid phase peptide synthesis is outlined ahead (Fig. 2.4).

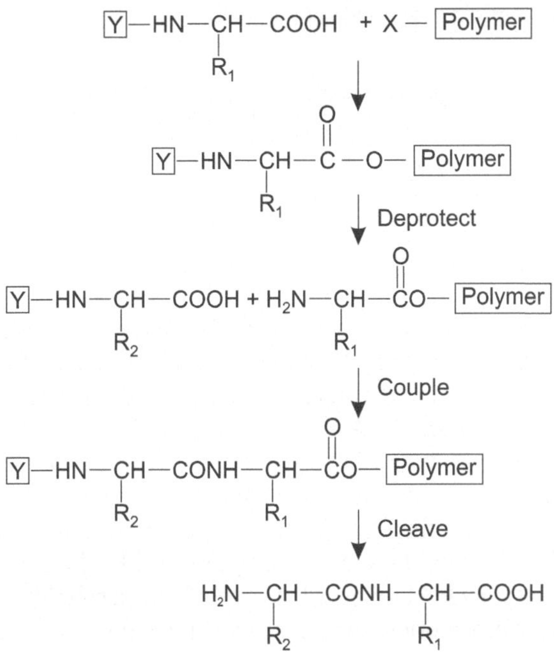

Fig. 2.4 The Basic Plan of Solid Phase Peptide Synthesis.

C. The Merrifield Method

In the Merrifield method, the solid support used is a copolymer of styrene and 2% divinyl benzene. The polymer is composed of long alkyl chains bearing a phenyl ring on every second carbon atom. These chains are cross-linked at every fiftieth carbon by *p*-diethylphenyl residues derived from divinyl benzene. This cross-linking makes the polymer completely insoluble in all ordinary solvents. The polymer also swells in certain organic solvents but is chemically inert during peptide synthesis. The resin is used in the form of fine beads with 20–70 micron in diameter.

The first step in the automated peptide synthesis (Fig. 2.5) involves the attachment of the *N*-protected amino acid residue to the resin support. In order to effect this, the polymer must first be made reactive. That is done by chloromethylating the benzene rings in the chain by a Friedel-Crafts reaction using chloromethyl methyl ether and stannic chloride.

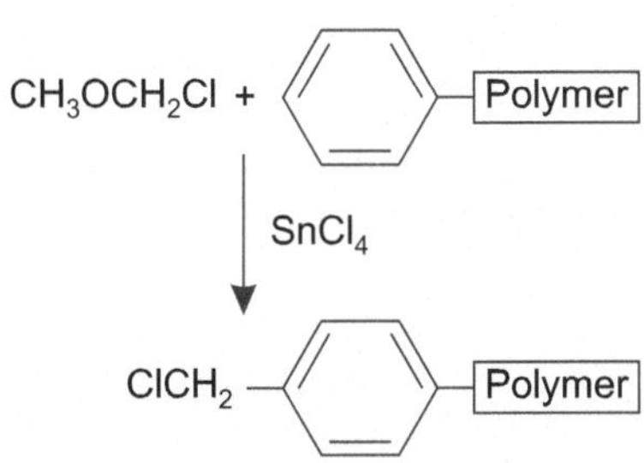

Thus, about one out of 10–100 phenyl groups is chloromethylated giving

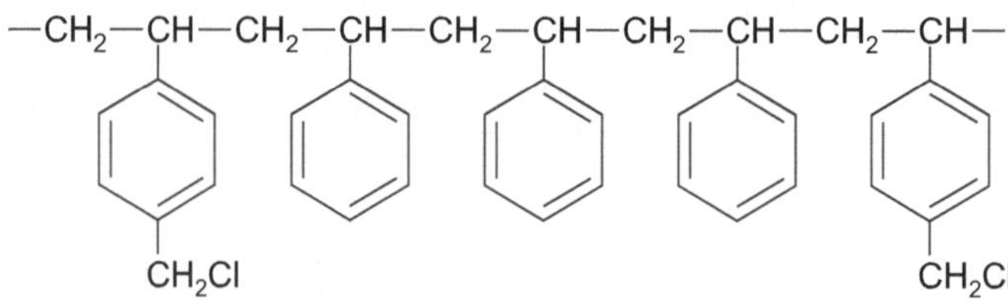

The *C*-terminal amino acid of the desired peptide is now bound to the polymer by shaking a solution of the *N*-protected amino acid salt in an organic solvent such as DMF with the chloromethylated polymer. Nucleophilic displacement of the halogen atom by the carboxylate anion takes place giving rise to the formation of an ester. The protecting group may be *t*-butyloxycarbonyl (Boc group). Excess reagents are removed by filtration and the insoluble polymer bound amino acid ester is washed. The Boc-protecting group is selectively removed by treatment with anhydrous hydrogen chloride in an organic solvent like dioxane or acetic acid. The aminoacyl resin hydrochloride is then treated with a solution of triethylamine to neutralise the hydrochloride, so that the free amino group can couple with next amino acid.

A solution of the next *N*-protected amino acid is then added with a coupling agent and the heterogeneous mixture is shaken until coupling is complete. Usually dicyclohexylcarbodiimide (DCC) is used as the coupling agent, which gets converted into dicyclohexylurea.

The polymer now is bound to *N*-protected dipeptide. Deprotection and neutralisation of the protected dipeptide resin make the amino group ready for coupling with the next amino acid. The three steps—deprotection, neutralisation and coupling are repeated as many times as desired or until the desired peptide chain is linked together on the resin. Throughout all these operations, the reagents and reaction conditions are chosen so that all reactions go to 100 per cent completion. The only purification used at any step is the washing of the resin and its attached peptide with appropriate solvents which can be handled rapidly and effectively.

The finished peptide resin is then suspended in anhydrous trifluoroacetic acid and a slow stream of hydrogen bromide is bubbled through the suspension. This reagent removes the Boc-group by an elimination reaction and cleaves the finished peptide from the resin by a nucleophilic displacement reaction. Certain other protecting groups used for side chains are simultaneously removed from the peptide, but peptide bonds are not harmed.

The complete Merrifield method is outlined below:

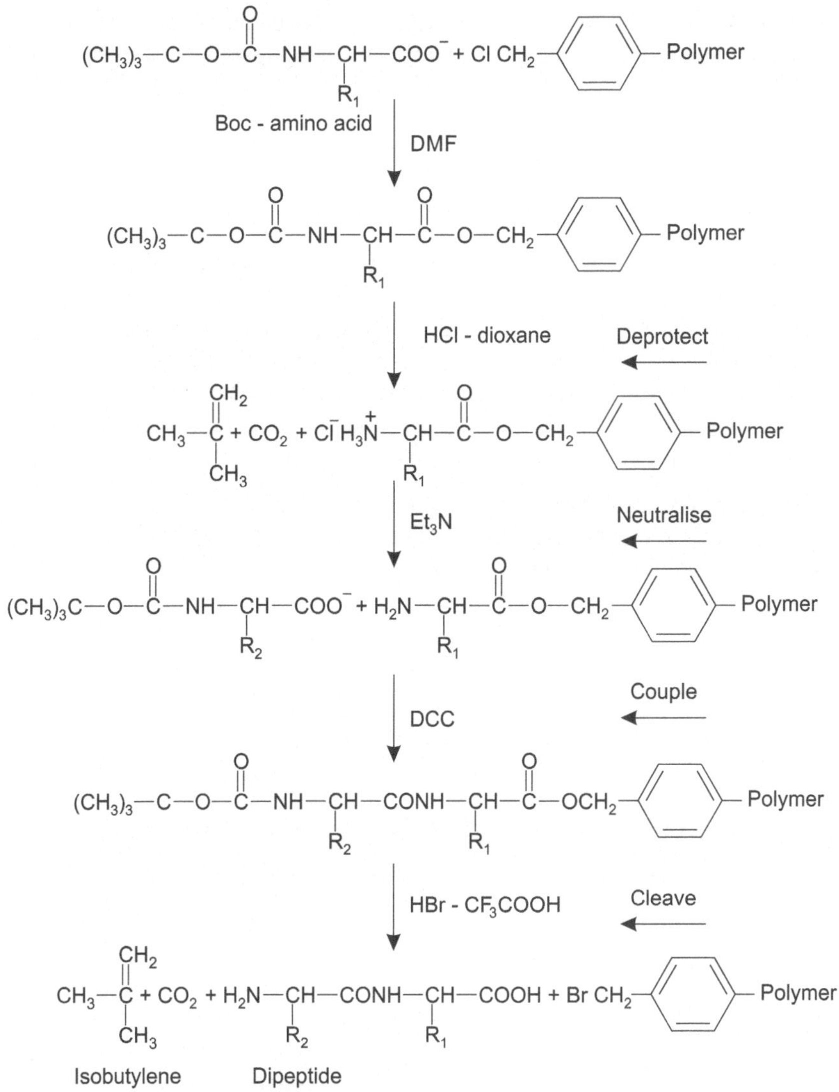

Fig. 2.5 The Complete Merrifield Method.

As Merrifield solid phase peptide synthesis is a stepwise synthesis of peptides from the *C*-terminus, it is free from the danger of racemisation. Only the carboxyl groups of *N*-protected amino acids are activated for coupling, not the carboxyl groups of peptides.

Finally, when Boc-group is used as the protecting group, deprotection and cleavage are carried out in non-hydrolytic solvents, the danger of racemisation is minimised.

The Merrifield technique has been applied successfully to the synthesis of ribonuclease, a protein having 124 amino acid residues. The synthesis involved 369 different chemical reactions and 11,931 automated steps including injection of reagents, filtrations, washings and so on. All these operations were carried out without isolating an intermediate. The synthetic ribonuclease not only had the same physical characteristics of the neutral enzyme, it possessed the biological activity as well.

D. Protection and Deprotection of Amino Groups

Various *N*-protecting groups have been developed by other workers, but *t*-butyloxycarbonyl group used by Merrifield is the standard protecting group in the solid phase peptide synthesis. This group can be removed easily by either anhydrous hydrogen chloride in an organic solvent or anhydrous trifluoroacetic acid in methylene chloride. These reagents do not cleave benzyl esters. Since the *t*-butyloxycarbonyl chloride is too unstable, the Boc-amino acids are usually prepared by the reaction of the amino acid with *t*-butyloxycarbonyl azide.

The amino acid is treated with the Boc-azide at a controlled alkaline pH in a pH-stat or by a mixture of magnesium oxide, sodium bicarbonate and triethylamine. The azide is obtained by the following route:

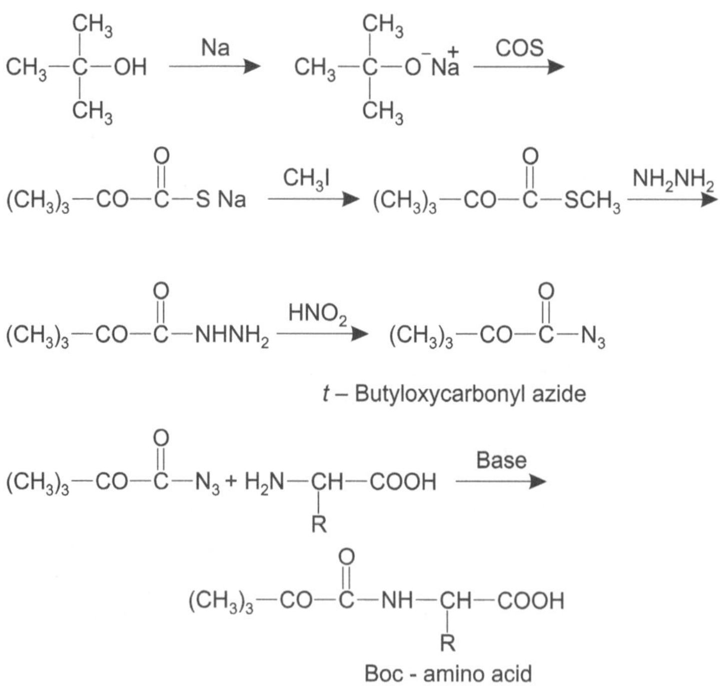

Deprotection of the Boc-amino acid takes place by simple elimination reaction giving rise to the formation of isobutylene and free amino acid with elimination of CO_2.

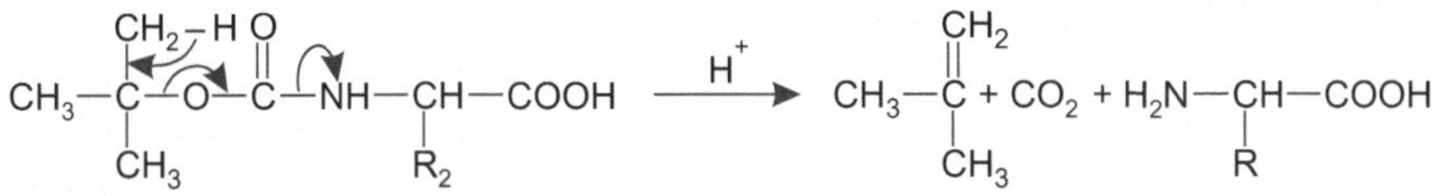

t-Amyloxy carbonyl group (Aoc-group) has replaced to some extent the Boc-group as the NH_2-protecting group because *t*-amyloxycarbonyl chloride is sufficiently more stable than the *t*-butyloxycarbonyl chloride. Therefore, it can be used directly for the synthesis of protected amino acids without recourse to much more expensive azide method.

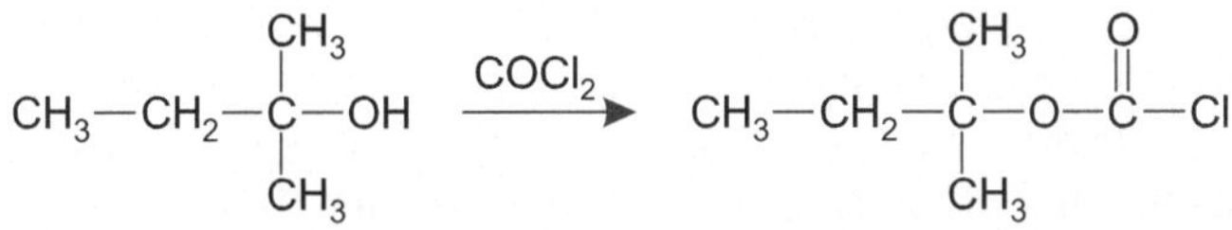

t-Amyloxycarbonyl chloride

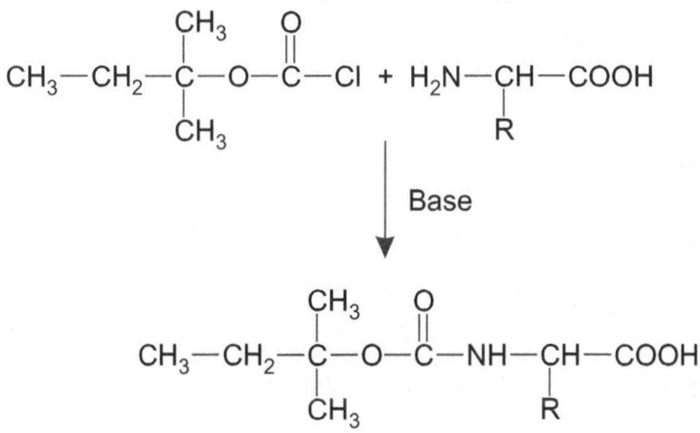

Aoc - amino acid

The most recent advance in $-NH_2$ protecting groups is the introduction of the 2-(4-biphenylyl)-isopropoxy carbonyl group, which has the structure

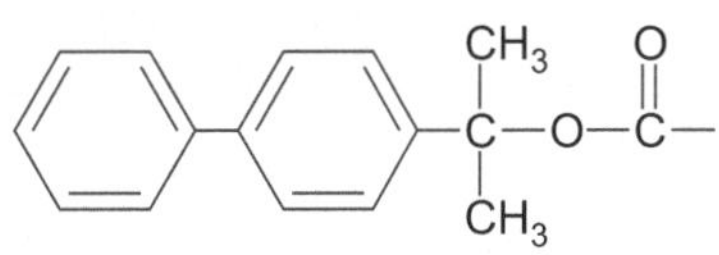

This protecting group is about 2000 times as sensitive as the *t*-butyloxycarbonyl group to acidic cleavage.

2.16 DETERMINATION OF PRIMARY STRUCTURE OF PEPTIDES

A. Introduction

Structure determination of peptides means arriving at the sequence of amino acids in a polypeptide chain. When the polypeptide is a protein, this sequence is referred to as its primary structure. In order to determine the structure of a peptide, we must know:

1. The kind of amino acids that make up the peptide.
2. The number of amino acids of each type.
3. The order in which these amino acids are arranged in the peptide chain.

Therefore, it is necessary to have a qualitative and quantitative analysis of the amino acids present. In the first step, the peptide chain is converted into its amino acid components by complete hydrolysis. In the second step, the hydrolysate that contains the mixture of amino acids from the first step, is analysed to determine the amino acid composition. In order to calculate the number of residues of each amino acid present, we must know the molecular weight of the polypeptide, which is determined by suitable physical methods. Finally, in the third step, the sequence of amino acids is determined.

B. Complete Hydrolysis

Hydrolysis of the peptides may be carried out with acids, bases or with certain enzymes to give a mixture of amino acids. For example, a dipeptide on hydrolysis gives a mixture of two amino acids.

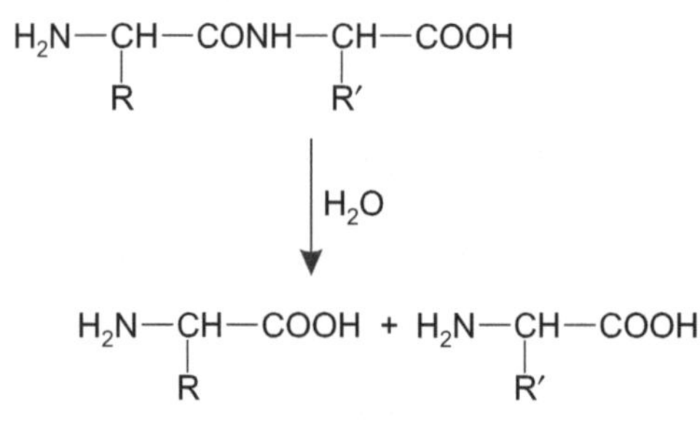

When the hydrolysis is carried out with an acid or with a base, all the peptide bonds are split giving a mixture of amino acids in the hydrolysate, *i.e.*, under the action of an acid or alkali complete hydrolysis takes place. Enzymatic hydrolysis, however, is selective and incomplete, only specific peptide linkages are broken. Enzymatic hydrolysis is usually employed in the third step, *i.e.*, for determining the sequence of amino acids in peptides.

In the acid hydrolysis, the peptide or protein is heated with 6N HCl in a sealed tube at 110°C for 10 to 100 hours, depending on the peptide. The hydrolysate is analysed for the amino acid content. Tryptophan is largely destroyed by acid hydrolysis. Asparagine and glutamine are completely converted into aspartic and glutamic acids respectively. Cysteine and cystine are partially destroyed and better results are obtained by previously oxidising them with performic acid to cysteic acid which is stable. Serine and threonine are slowly destroyed during acid hydrolysis. Finally, valine and isoleucine are slowly liberated because their bulky side chains cause steric hinderance.

Alkaline hydrolysis is effected by heating the peptide at 100°C with 2 to 4 N NaOH for 1 to 10 hours. It is, however, useless for complete analysis of amino acids in a peptide or protein because several amino acids are destroyed by alkaline hydrolysis. Amino acids which are destroyed by alkaline hydrolysis are arginine, cysteine, cystine, threonine and serine. For this reason, acid hydrolysis is usually preferred over alkaline hydrolysis. Tryptophan, however, largely survives alkaline hydrolysis. So, to obtain tryptophan content, a peptide or protein is hydrolysed under alkaline conditions.

C. Analysis of the Hydrolysate

The hydrolysate obtained after complete hydrolysis is examined for its amino acid content. The relative abundance of each amino acid in the hydrolysate is identified and measured. This may be done by any of the suitable methods mentioned earlier. The commonly used methods are paper and thin layer chromatography, electrophoresis and ion-exchange chromatography.

Ion-exchange chromatography is the most popular method for the qualitative analysis of a mixture of amino acids. There are several technical variations, but basically the mixture of amino acids, derived from the hydrolysis of polypeptide or protein, is resolved on a column of a sulphonated polystyrene resin by elution with buffers. More recently the process has been mechanised and the commercial instrument is called an amino acid analyser. Buffer is pumped through the column at a fixed rate, the effluent from the column is automatically mixed with a controlled flow of ninhydrin solution and heated by passage through a reaction coil at 100°C. The developed colour is measured with a visible spectrometer and the absorbance is plotted by a recorder as a function of time. The results have a precision of 100 ± 3 per cent. The amino acids can be identified from their elution positions. By comparison of the chromatogram of an unknown mixture with that of a mixture of known composition, we can arrive at quantitative analysis of the mixture.

D. Cleavage of Disulphide Bridges

After determining the total amino acid content and the kind of amino acids in a polypeptide chain, the next step is to cleave any disulphide bridge that may be present before carrying out sequence studies. The reaction is usually carried out when the peptide or protein contains a cystine irrespective of whether this forms a loop within a chain or links two chains together. This can be achieved by oxidising the substance with performic acid, which converts the two cysteine units into cysteic acid units. This oxidation converts the disulphide bridge to sulphonic acid ($-SO_3H$) groups.

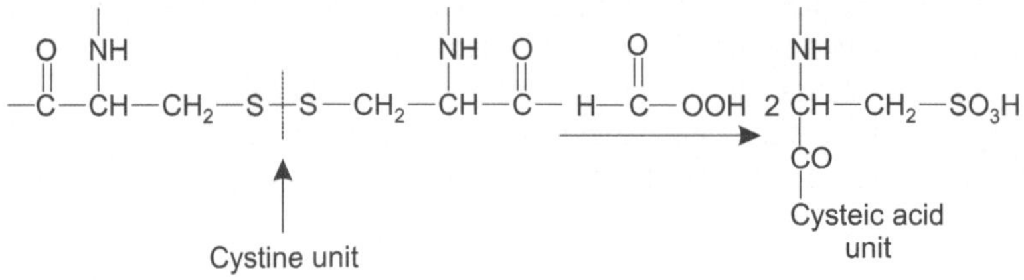

If the compound contains no disulphide bridges this step is not necessary.

Alternatively, the disulphide bridges can be reductively cleaved. Several reagents for reduction have been used, but the best reagent is perhaps 2-mercaptoethanol. When reduction is complete, the liberated thiol groups in the peptide are alkylated with iodoacetic acid. Cysteine is thus converted into *S*-carboxymethyl cysteine.

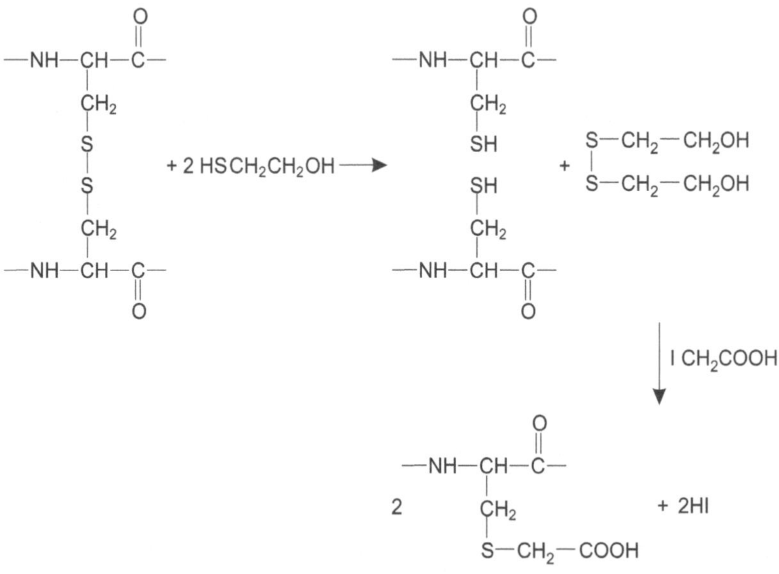

E. Determination of the Amino Acid Sequence

The determination of the correct amino acid sequence of a peptide or protein is carried out in two parts. In the first part, partial degradation of the polypeptide chain is carried out by means of certain enzymes. The fragments so obtained are analysed individually for total amino acid content. The second part involves determination of the amino acids that are present at the *N*-terminus and at the *C*-terminus in the original peptide as well as in the fragments obtained through enzymatic hydrolysis. This is called **End Group Analysis**. The pieces are then fitted together to arrive at the complete sequence of the peptide chain.

The process is then repeated using a different cleavage method usually cyanogen bromide. The second set of peptide fragments are analysed separately and amino acid sequence is determined for each fragment. The complete amino acid sequence is arrived at sometimes by using both sets of data.

F. Selective Cleavage by Chemical Methods

Partial hydrolysis by acids has been widely used and efforts have been made to find conditions which effect selective cleavage of peptides.

Hydrolysis of a peptide with 0.25N acetic acid under reflux preferentially splits the peptide bonds on both sides of the aspartic acid residues. For example, partial hydrolysis of the following pentapeptide:

Val-Glu-Asp-Gly-Leu

will give four fragments with the sequences as given below:

Val-Glu, Asp-Gly-Leu, Val-Glu-Asp, Gly-Leu

However, peptide bonds involving the amino groups of serine and threonine are preferentially cleaved when hydrolysis of the peptide is carried out using 10N hydrochloric acid at 30°C. The cleavage is assisted by the neighbouring hydroxyl group in the side-chain of serine or threonine. The peptide rearranges through oxazolidine tautomer to an O-acyl derivative which is then hydrolysed.

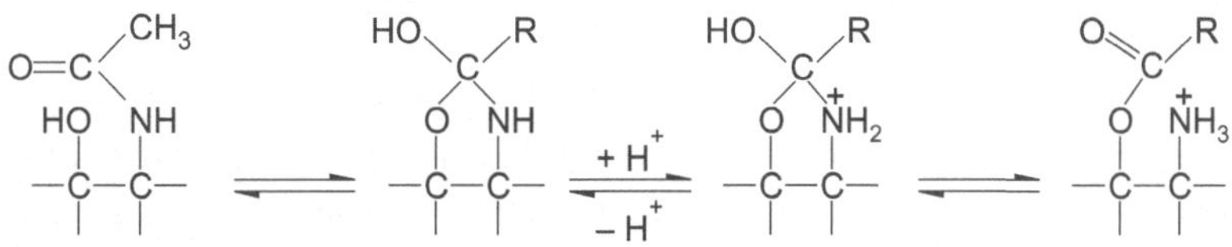

Another useful method for selective cleavage of polypeptide chains employs cyanogen bromide, CNBr. It cleaves dialkyl sulphides to give alkyl thiocyanate and an alkyl bromide.

Cyanogen bromide has been used to cleave the peptide bonds containing the carboxyl group of methionine units; the methionine is converted into a *C*-terminal homoserine lactone unit. The reaction is quite complicated since ring closure and fission of the peptide bond occurs simultaneously.

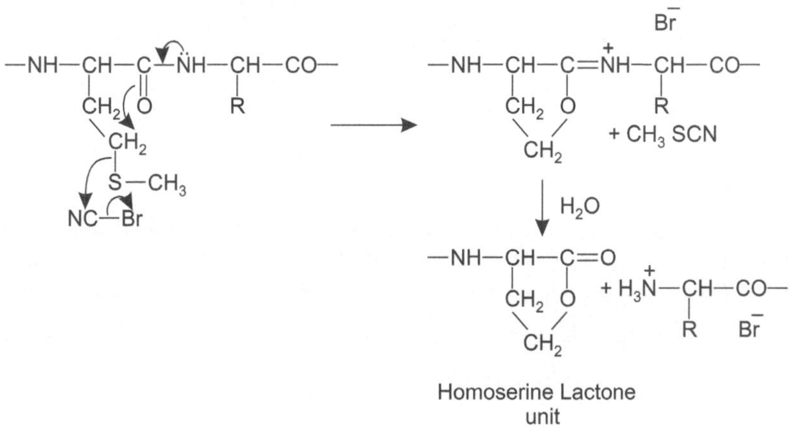

G. Partial Hydrolysis by Enzymes

The partial hydrolysis of peptides is carried out by means of enzymes generally known as proteases. They readily hydrolyse peptide bonds only between specific amino acid residues. The commonly used proteases are trypsin, chymotrypsin and pepsin, all hydrolyse specific peptide linkages to give smaller fragments.

Trypsin, the pancreatic enzyme occurs in the intestine of mammals. This preferentially hydrolyses the peptide bond at the carboxyl position of basic amino acids such as lysine and arginine.

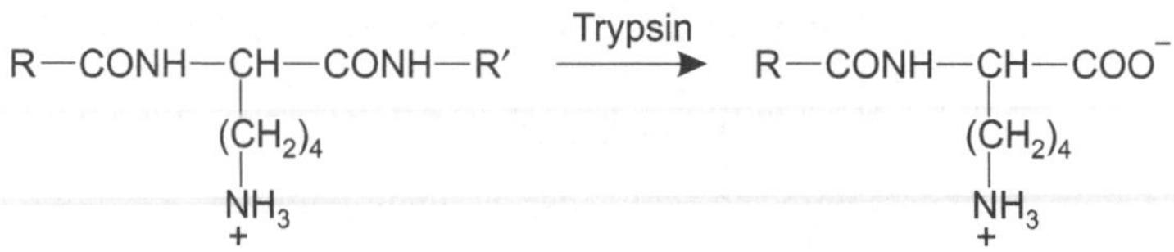

Peptide chain containing lysine

Trypsin is most active in the pH range 7–9, but is fairly rapidly inactivated by self-digestion. Addition of a low concentration of calcium ions markedly inhibits self-digestion. It is possible both to extend and restrict the action of trypsin on proteins. If ε-amino group of lysine is blocked the action of trypsin can be restricted to the hydrolysis of arginyl peptide bonds.

As the action of trypsin is highly specific it has been extensively used for comparing the primary structures of closely related proteins. The method is known as finger printing. It involves hydrolysis of the proteins with trypsin and comparison of the resulting peptides. Separation of these peptides is achieved by electrophoresis on filter paper followed by chromatography at right angles to the direction of electrophoresis. Peptides are spotted by spraying with ninhydrin.

Chymotrypsin, another intestinal enzyme selectively splits the peptide bond at the carboxyl position of phenylalanine, tyrosine and tryptophan (all aromatic amino acids). Like trypsin, chymotrypsin is most active in the pH range 7–9 but is considerably less specific as it also hydrolyses slowly peptide bonds at carboxyl positions of methionine, leucine, isoleucine and valine. Preparations of chymotrypsin may contain traces of trypsin which can be selectively inhibited with naturally occurring proteins like soya-bean trypsin inhibitor. Chymotrypsin is widely used for further degrading tryptic peptides to smaller fragments.

Pepsin, a gastric protease, preferentially cleaves the peptide bond at amino position of phenylalanine, tyrosine and tryptophan. However, pepsin is much less specific. It also splits leucine, aspartic acid and glutamic acid. Pepsin is active in the pH range 1–2.

Other proteolytic enzymes which are useful for degrading proteins include papain (active at pH 5–7) and a bacterial enzyme subtilisin (active at pH 6–8) from Bacillus subtilis. These enzymes have broad specifities and are mainly used for further degradation of large fragments which are produced by tryptic hydrolysis.

The specific action of three proteases on a hepta peptide is illustrated below:

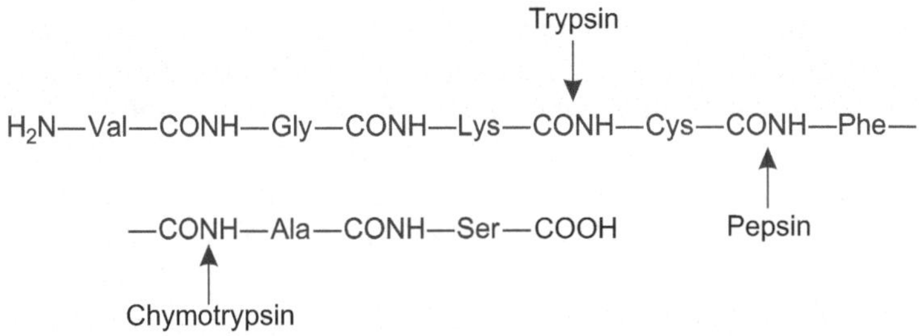

2.17 IDENTIFICATION OF THE N-TERMINAL AMINO ACID (END GROUP ANALYSIS)

There are two main types of methods for identifying the *N*-terminal residue of a peptide or protein.

In the first type, a suitable group is attached to the α-amino group of the *N*-terminal residue so that after hydrolysis with acid or enzymes the labelled residue can be detected by spectrophotometry, fluorescence or radioactive counting and identified by chromatography.

A number of reagents have been developed for condensation of the amino group of the *N*-terminus.

1. Barger's Method

The earlier method which had been most popular for a number of years is that of Barger. As mentioned earlier, trinitrotoluene is condensed with amino end of peptides with the elimination of one of the nitro groups. The peptide is then hydrolysed and the resultant yellow dinitrotolyl derivative of the *N*-terminus amino acid separates out readily, because its solubility properties differ from those of the unsubstituted amino acids obtained by hydrolysis.

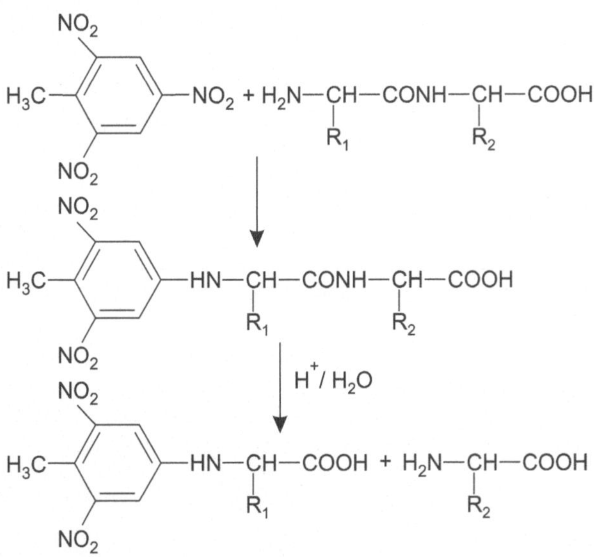

2. The DNP Method (Sanger Method)

A very successful method of identifying the *N*-terminal residue was introduced by F. Sanger of Cambridge University. The method makes use of 2,4-dinitrofluorobenzene (DNFB) which undergoes nucleophilic substitution by the free amino group to give an *N*-dinitrophenyl (DNP) derivative. The DNP derivative of the peptide is obtained by the reaction with 2,4-dinitrofluorobenzene in a buffer at about pH 8. Reaction is likely to occur also with the ε-amino groups of lysine, the phenolic hydroxyl group of tyrosine, the thiol group of cysteine and the imidazole nucleus of histidine.

Complete hydrolysis of the DNP-peptide with acid gives the yellow DNP-derivative of the *N*-terminal residue together with ε-DNP-lysine, *o*-DNP tyrosine and free amino acids, which can be identified by chromatography. Thin layer chromatography is particularly useful for rapid qualitative work.

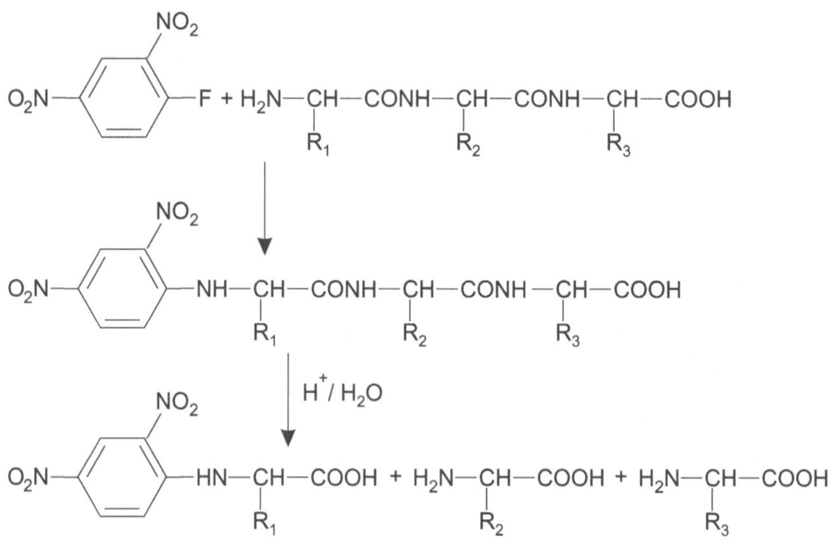

The DNP derivatives of most of the amino acids have been prepared and characterised. The DNP method cannot be used repeatedly since its use requires complete hydrolysis of the DNP derivative.

Partial acid hydrolysis of the DNP-derivative of a polypeptide may give rise to a series of yellow DNP-peptides. If these can be satisfactorily separated (by column chromatography), determination of their amino acid compositions may permit the *N*-terminal sequence to be elucidated. This approach was used by Sanger for elucidating the structure of insulin. This approach can be applied to any peptide. For example, partial acid hydrolysis of

DNP—Phe—Val—Asp—Glu

will give rise to the following

DNP—Phe—Val—Asp
DNP—Phe—Val
DNP—Phe

From these products the original sequence of the tetrapeptide can be reconstructed.

3. The Dansyl Method

A modification of the DNP method is the use of 5-dimethylamino-1-naphthalenesulphonyl chloride (dansyl chloride) since these derivatives are more resistant to acid hydrolysis and give strong fluoroscence in ultraviolet light. Therefore, very minute quantities (10^{-10} to 10^{-9} moles) can be easily detected. The method is about 100 times more sensitive than the DNP-technique. The dansyl amino acids can also be analysed spectroscopically.

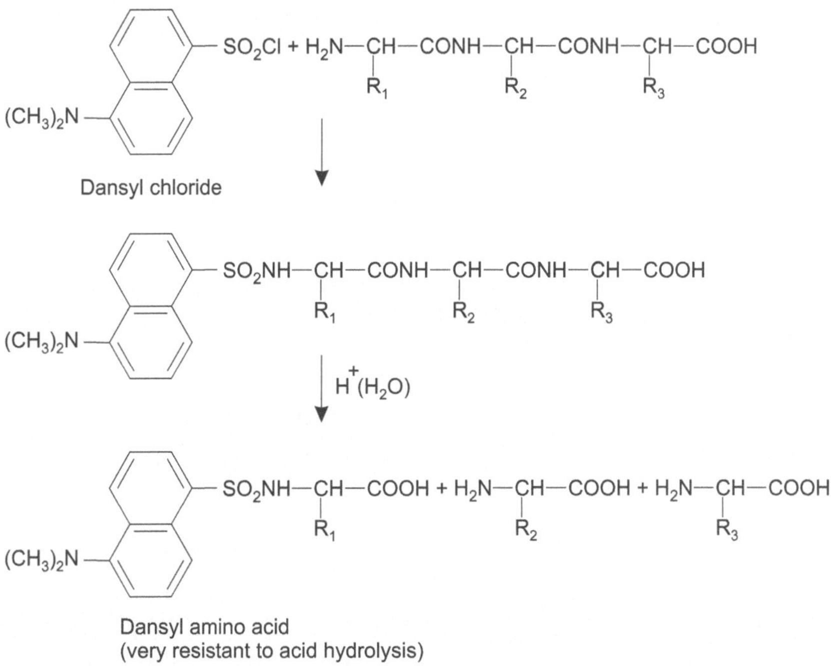

The DNP- and 'dansyl' groups in fact serve as labels to identify the *N*-terminal residue after total hydrolysis although partial hydrolysis may lead to the elucidation of short sequences near the labelled group.

4. The Edman Method

This method comes under the second and more powerful type of methods for identifying the *N*-terminal residue of a polypeptide in which we attach a suitable substituent on the α-amino group followed by selective cleavage of the modified *N*-terminal residue. Repetition of the reaction then provides a method for the stepwise degradation of the peptide and progressive elucidation of its sequence.

The Edman method involves the reaction between phenyl isothiocyanate and the peptide at approximately pH 8. Reaction occurs with the unprotonated α-amino group. The solution is kept at constant pH by the addition of a base from pH-stat. The resultant *N*-phenylthiocarbamoyl derivative is freed from excess of phenylisothiocyanate by extraction with an organic solvent. The labelled peptide is then treated with anhydrous HCl in an organic solvent. These conditions do not hydrolyse the amide linkages, the labelled amino acid undergoes a cyclisation reaction giving rise to the formation of 3-phenyl-2-thiohydantoin and the peptide containing one residue less than the original. The hydantoin so produced can be identified by comparing it with standard amino acid derivatives. The degraded peptide can be isolated and subjected to another cycle of the Edman degradation to identify the *N*-terminal amino acid. The reactions involved are given below:

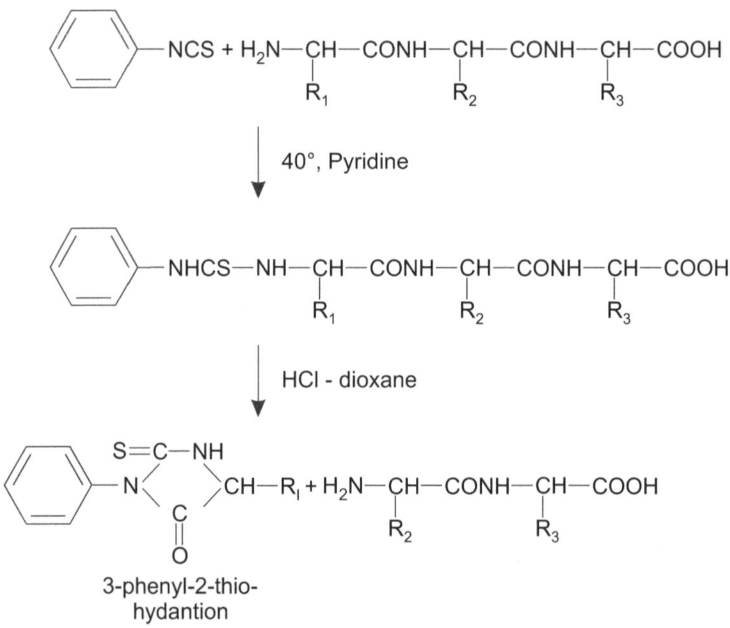

3-phenyl-2-thio-
hydantion

The process has been automated by Edman and Begg. They have described an apparatus which will carry out automatically most of the steps involved in the *N*-terminal sequence determination. The product from the *N*-terminal residue for each cycle is collected separately in a fraction collector and all are identified chromatographically at the end of the degradation. Fifteen cycles can be completed in 24 hours and the yields are about 98 per cent for each cycle. Edman was able to identify 60 amino acids in a protein containing 153 amino acids in its chain.

A number of other acid catalysts have been used for this degradation, but perhaps the best is trifluoroacetic acid. With this reagent cyclisation and degradation proceed rapidly at room temperature with little risk to residual peptide. Use of dilute hydrochloric acid sometimes above room temperature, can cause hydrolysis of other peptide bonds so that subsequent steps liberate more than one 3-phenyl-2-thiohydantoin.

5. Condensation with Isocyanate

A recent method of stepwise degradation, which is related to Edman procedure, involves carbamoylation of the terminal α-amino group with cyanate at pH-7 to 8. The reaction probably occurs between a cyanic acid molecule and the unprotonated α-amino group. The *N*-carbamoylpeptide on treatment with acid gives a hydantoin corresponding to the *N*-terminal residue and a peptide containing one residue less than the original. The hydantoin after isolation can be hydrolysed to the corresponding amino acid which is identified by chromatography.

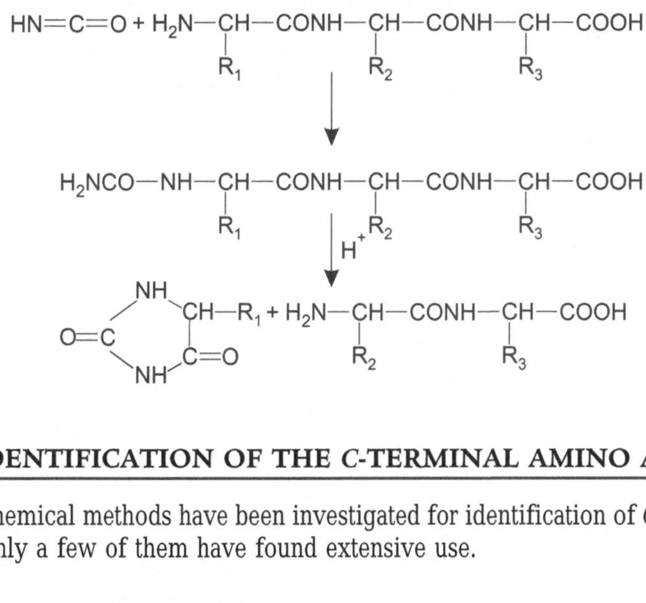

2.18 IDENTIFICATION OF THE *C*-TERMINAL AMINO ACID

Several chemical methods have been investigated for identification of *C*-terminal groups, only a few of them have found extensive use.

1. Reaction with Benzylamine

Benzylamine has been used on the *C*-terminus in many cases. The peptide is converted to an ester and heated in suspension of benzylamine to yield a benzylamide. On hydrolysis the *C*-terminus amino acid can be easily characterised as its benzylamide as given below:

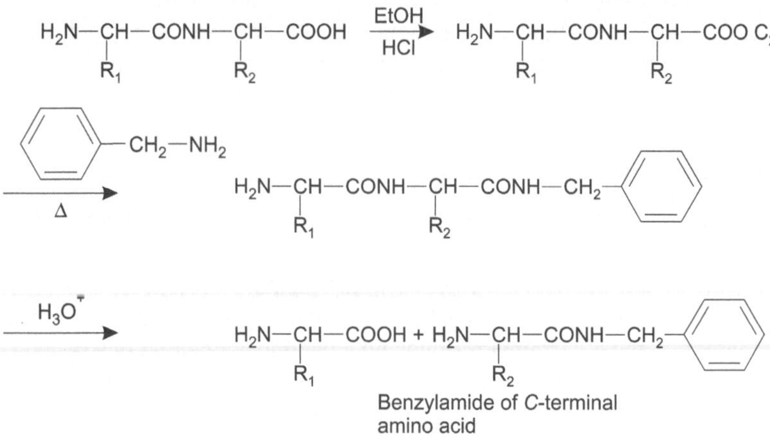

2. Thiohydantoin Formation

This method is analogous to the Edman's method for *N*-terminus amino acids. The *C*-terminal amino acid is converted into a thiohydantoin by treating with ammonium isothiocyanate in presence of acetic anhydride. The thiohydantoin is liberated by free cold alkali and identified. The free amino group, however, must be protected first.

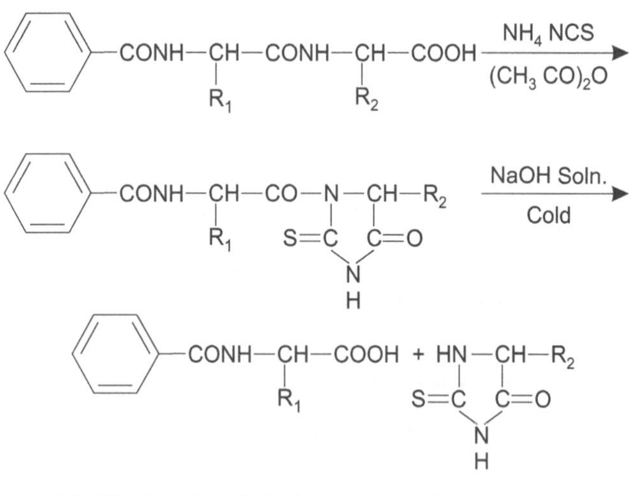

3. Reaction with Hydrazine (Hydrazinolysis)

This is the most widely used method for the identification of *C*-terminal groups. The peptide is heated with anhydrous hydrazine at 100°C. This cleaves the peptide bonds and forms the hydrazide of all the amino acids except the *C*-terminal residue. The mixture of products is subjected to column chromatography using cation-exchange resin. The strongly basic hydrazides are retained on the column, whereas the free *C*-terminal amino acid is eluted and can be easily identified.

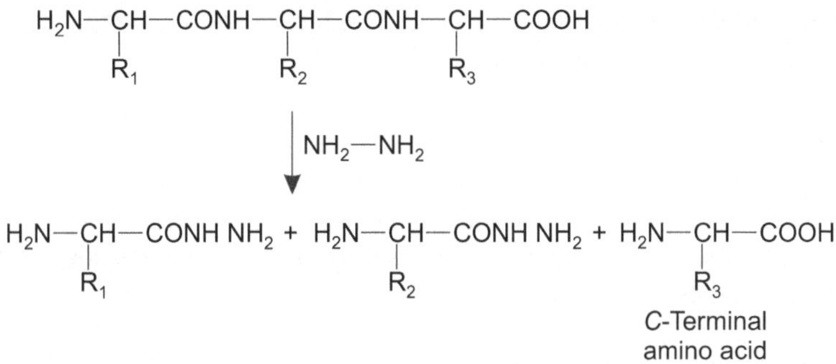

2.19 USE OF ENZYMES

The *N*-terminal residue may be determined enzymatically by treating the peptide chain or a protein with aminopeptidase. Aminopeptidases hydrolyse specifically the *N*-terminal amino acids from peptides and proteins. For example, the tripeptide,

$$H_2N—Gly—Val—Cys—COOH$$

on attack by an aminopeptidase will liberate Gly first. So, glycine would be the *N*-terminal amino acid. The process can go further and valine is released next. Use of aminopeptidases is of considerable importance for determining the *N*-terminal amino acids of short peptides. Under normal conditions the first three or four amino acids can be identified in this way.

The aminopeptidase from kidney has been thoroughly investigated. It preferentially cleaves amino acids with long alkyl side chains (*e.g.*, leucine); aromatic amino acids are cleaved rather slowly and peptides terminating in glycine, serine and proline are rather resistant. Another type of aminopeptidase has been discovered recently which is almost specific for cleavage of *N*-terminal lysine and arginine.

The *C*-terminal amino acid may be identified enzymatically by treating the peptide with carboxy-peptidase, which specifically catalyses the hydrolysis of the peptides from the *C*-terminus only. The reaction is shown below:

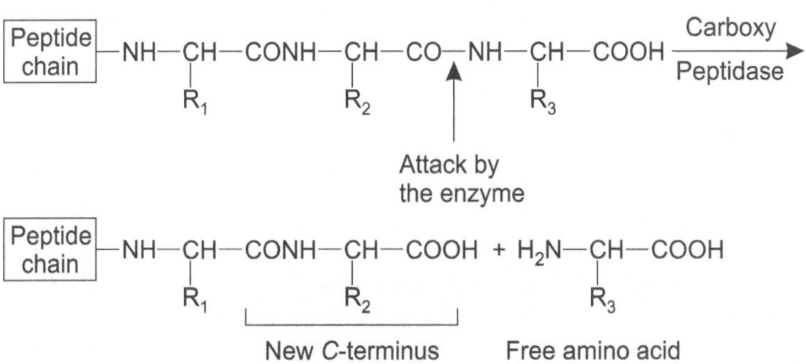

Thus, when the peptide or protein is treated with carboxypeptidase, the first amino acid to appear in solution is the one that occupies the *C*-terminal position. However, after the release of the first amino acid the enzyme continues to function and attack on the next residue and so on. Therefore, under favourable conditions the first three or four amino acid residues can be identified.

The use of two pancreatic enzymes, carboxypeptidases A and B, has been quite successful. These enzymes differ in their reactivity. Carboxypeptidase A has a fairly broad specificity although it preferentially liberates aromatic amino acids. *C*-terminal lysine, arginine and proline are not hydrolysed. Carboxypeptidase B is, however, much more selective and liberates only the basic amino acids, lysine and arginine when these are *C*-terminal.

2.20 TIME COURSE ANALYSIS

The sequence of amino acids is determined by following the increase in concentration of the amino acids liberated at regular time intervals after treating the peptide with aminopeptidase or carboxypeptidase. This method is called **Time Course Analysis**. For example, when the peptide is incubated with carboxypeptidase the *C*-terminal amino acid would be the first to be released. If we plot a graph between the concentration of the amino acid liberated with time, we find that concentration of the *C*-terminus amino acid will be more than the next amino acid at any interval of time, *i.e.*, the *C*-terminus amino acid will be liberated first. For example, the time-course analysis for the tripeptide,

$$H_2N—Gly—Lys—Phe—COOH$$

with carboxypeptidase gives the following information

$$Phe > Lys = Gly$$

The tripeptide on treatment with carboxypeptidase liberates first phenylalanine indicating that it is the *C*-terminal amino acid. Simultaneously, the dipeptide H_2N—Gly—Lys—COOH is generated. As time goes on, the dipeptide is hydrolysed by carboxypeptidase giving equal amounts of lysine and glycine. Therefore, concentration of Phe will always be greater than Lys at any interval of time.

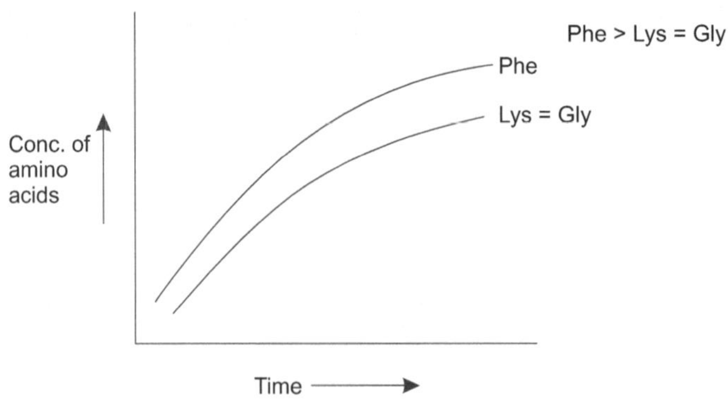

Fig. 2.6.

Thus, conclusion of the time-course analysis of the above tripeptide with carboxypeptidase is that phenylalanine is the *C*-terminal amino acid. However, it does not give us any information about the sequence of other two amino acid residues (Fig. 2.6).

Similarly, time-course analysis of the same peptide with aminopeptidase gives us the following results:

$$\text{Gly} > \text{Lys} = \text{Phe}$$

Glycine is the first amino acid to be released. Its concentration will be more than lysine at any interval of time. It concludes that glycine is the *N*-terminal amino acid. Again, the time-course analysis with aminopeptidase does not specify the position of the other two residues. However, if the results of the two time-course analysis are combined, we arrive at the complete sequence.

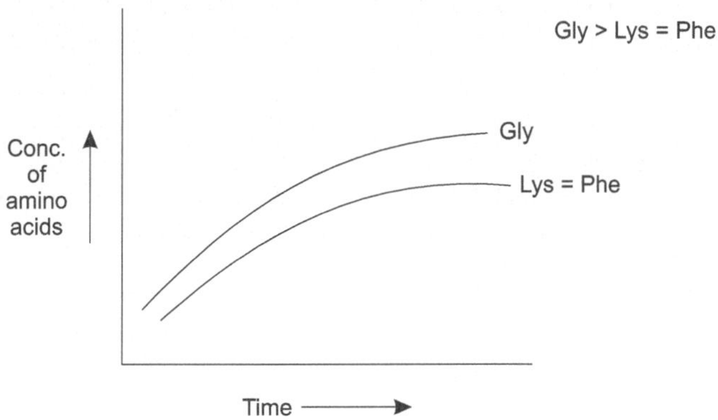

Fig. 2.7.

Results of the time-course analysis can be interpreted more clearly when the amino acids are released at the same rate. If the amino acids are released at different rates, we would have difficulty even in deciding which amino acid was the *C*-terminal amino acid.

For peptides having more number of amino acids the time-course analysis is useful for deciding the sequence of the two acids from both the *N*-terminus and *C*-terminus. For example, a heptapeptide gives the following results in the two time-course analyses:

1. Carboxypeptidase gives

$$\text{Ala} > \text{Val}$$

2. Aminopeptidase gives

$$\text{Asp} > \text{Leu}$$

From these results, we can interpret that alanine is the *C*-terminal amino acid, followed by valine and aspartic acid is the *N*-terminal residue followed by leucine. Thus, sequence of two acids on each of the *N*-terminus and *C*-terminus is decided.

H_2N — Asp — Leu ----------- Val — Ala — COOH

N-Terminus (Asp — Leu) *C*-Terminus (Val — Ala)

2.21 IMPORTANCE OF PROTEINS

Proteins are the primary constituents of all living matter as they are found in all living cells. They are the most abundant organic molecules in animals. Proteins perform several important biological functions. Some important functions are listed below:

1. **Proteins serve as structural material:** Hair, horns, hoofs, feathers, fur and tough outer layers of skin are all composed largely of a protein called keratin. Keratin is a structural protein. Collagen, another structural protein is a major component of bones, muscles and tendons.
2. **Proteins act as enzymes:** These are biocatalysts which regulate almost all life processes, catalyse the chemical reactions that occur in living organisms. We know these bring about digestion of food and carry out the biosynthesis of various cell components. We have already encountered carboxypeptidase, pepsin, trypsin and chymoptrypsin-enzymes that catalyse the hydrolysis of peptide chains.
3. **A few proteins function as hormones:** Insulin, secreted by pancreas, is a polypeptide which regulates glucose metabolism. Pituitary gland secretes a growth hormone, which is a protein. β-Endorphin, a polypeptide with morphine like activity that appears to be a natural pain reliever.
4. **Proteins are responsible for many physiological functions** such as the transport and storage of oxygen in the body. As haemoglobins they transfer oxygen to remote corners and the protein, myoglobin stores oxygen in the muscles. Proteins are also responsible for contraction of muscles. Cytochrome C and cytochrome P-450 also have a transport function. These compounds carry electrons in the oxidative phosphorylation cycle. Lipoproteins of the blood plasma transport lipids between the intestine, liver and adipose.
5. **Protective proteins act as antibodies giving protection against a disease:** The blood proteins, thrombin and fibrinogen are resopnsible for blood clotting and thus prevent loss of blood at the time of injury. These materials form insoluble complexes with foreign substances that invade the blood-stream.
6. **Proteins are also associated with hereditary characters:** Nucleoproteins form chromatin material. During cell division, chromatin condenses to form chromosomes. These are carriers of hereditary characters.
7. **Some proteins have other protective functions:** Snake venoms and plant toxins protect their owners from other species. In addition to snake venom, the toxic proteins of cotton seed and caster bean are highly poisonous to vertebrates.
8. **Cell secretions like mucus are glycoproteins:** These provide slippery texture and help in protection.

9. **Spiders and silkworm secrete a thick solution of protein fibroin,** which solidifies into thread of high tensile strength and is used for weaving web or cocoon. We get silk from the cocoon.

10. **Proteins are stored as reserve food:** Some proteins are used for storage. Ovalbumin is employed by nature as a food reservoir in egg white. Casein plays a similar role in mammalian milk. Similarly, proteins are strored as reserve food in various seeds, pulses, rice, maize and peas.

2.22 CLASSIFICATION OF PROTEINS

We have seen that the proteins are the most important, most varied and most abundant organic compounds in the body of all living organisms.They show such a wide range of structural and functional diversity that it is difficult to classify them on the basis of one single criteria. They may be classified according to their chemical composition or their shape and structure.

I. Classification of Proteins on the basis of their Shape

On the basis of their shape, proteins are divided into two broad classes. These are

A. Fibrous proteins
B. Globular proteins

A. Fibrous Proteins

These are tough and usually insoluble in water and aqueous solutions of acids and bases. Fibrous proteins have long and thread-like structure and tend to lie side by side to form fibres. Fibrous proteins serve as the chief structural material of animal tissues.

Examples of simple fibrous proteins are:

1. Keratins: α-Keratins are found in outer layers of skin, in hair and wool. They consist of α-helices that are twisted together like the strands of a rope.

β-Keratins are found in hard tissues such as nails, horns, hooves and bird feathers. They have a β-pleated sheet structure.

2. Collagens: Collagen proteins are found in white fibrous connective tissue, which constitute tendons. These are also the main components of bone, teeth and cartilage.

3. Elastins: They are found in yellow elastin tissue like ligaments and blood vessels.

4. Fibroin: This protein is present in silk fibres.

B. Globular Proteins

Globular proteins are soluble in water or aqueous solutions of acids, bases or salts. These solutions are colloidal in nature due to the large size of protein molecules. Globular proteins are folded into roughly spherical shapes.

Globular proteins serve a variety of functions which are related to maintenance and regulation of life processes. They do all such functions that require mobility and hence solubility. They usually function as enzymes, hormones or transport proteins. Examples of globular proteins are insulin, haemoglobin, ribonuclease and fibrinogen, which is converted into the insoluble fibrous protein, fibrin and this causes the clotting of blood.

II. Classification of Proteins on the basis of Structure

Based on structure, proteins may be classified as simple, conjugated and derived.

A. Simple Proteins

Simple proteins on hydrolysis yield only amino acids. They do not contain any non-protein part attached to them. Most of the simple proteins are globular in nature. As mentioned above some fibrous proteins are also simple proteins.

Major groups of simple globular proteins are:

1. Albumins: These are soluble in water and aqueous solutions of acids, bases and salts. They can be precipitated by saturating the solution either with a neutral salt such as sodium sulphate in slightly acidic solution or with acidic salts such as ammonium sulphate. On heating, albumins get coagulated. Albumins are widely distributed in nature. Examples are egg white or ovalbumin, blood-serum albumin, soyabean albumin, gliadin (wheat protein), casein in milk, legumelin (protein of pulses), *etc.* These are stored as food reserves.

2. Globulins: These proteins are insoluble in water but readily dissolve in dilute neutral salts such as NaCl. They can be precipitated by half saturation with ammonium sulphate. These coagulate on heating and heat coagulation is enhanced by the addition of dilute acids. They are found both in animal and plant tissues. Examples are ovoglobulin (eggs), lactoglobulin (milk), serum globulin in blood plasma, myosin in muscles are animal globulins. Vegetable globulins include legumin (in peas), tuberin (potatoes), and edestin (wheat and hemp seeds).

3. Protamines: They are basic proteins, highly soluble in water and dilute aqueous solutions of acids, bases and ammonia. They form crystalline salts with mineral acids and are not coagulated by heat. These are the simplest of all naturally occurring proteins and have lowest molecular weights. Protamines occur almost entirely in animals and are main components of sperm cells of certain fishes.

4. Histones: They are also basic proteins of high molecular weight. They are soluble in water, dilute acids but not in ammonium hydroxide. They occur as part of nucleoproteins. Histones are not easily coagulated by heat.

5. Prolamines: These proteins are insoluble in water but are soluble in dilute aqueous solutions of acids and bases. They are found in plants only and are not coagulated by heat. Gliadin from wheat, zein from maize and hordein from barley are common examples of prolamins.

6. Glutelins: They are insoluble in water, alcohol and neutral salt solutions but are soluble in dilute aqueous solutions of acids and bases. They can be easily coagulated by heat. They are found exclusively in seeds of grains.

B. Conjugated Proteins

These proteins are composed of a simple globular protein combined with some non-protein substance. This non-protein substance is known as prosthetic group. Depending upon the nature of the prosthetic group, the conjugated proteins are divided into the following classes:

1. Glycoproteins: In glycoproteins, the simple proteins combine with carbohydrates. For example, mucin of saliva, heparin of bile juice, immunoglobulins of plasma, mucopolysaccharides of cartilage and tendon are glycoproteins.

2. Chromoproteins: In chromoproteins, simple proteins are combined with coloured pigments like flavins, cartotenoids, porphyrins, etc. Haemoglobin, myoglobin and cytochromes are some examples of chromoproteins.

3. Phosphoproteins: These proteins contain phosphoric acid as the prosthetic group. These are soluble in dilute alkalis and are precipitated by the addition of acids. Casein, the milk protein and ovovitellin of eggs are examples of phosphoproteins.

4. Metalloproteins: In these proteins, the protein molecule is bound to some metal ion like iron, copper or zinc. Siderophilin, an important plasma protein has a great affinity for iron.

5. Lipoproteins: In these, simple proteins are combined with lipids and have a variable composition. These proteins are present in the brain, blood plasma, milk and egg yolk etc.

6. Flavoproteins: These are enzymes. In these proteins, the protein moiety (apoenzyme) is permanently attached to flavin compound. Flavoproteins are enzymes of Kreb's cycle and participate in electron transport system of respiratory chain.

7. Nucleoproteins: In nucleoproteins, the protein molecules are combined with nucleic acid. These proteins are protamines or histones. The chromatin material of the nuclei is composed of nucleoproteins.

C. Derived Proteins

These are the degradation products obtained from native proteins either by hydrolysis or by coagulation by heat. These are further classified according to the following types:

1. Coagulated Proteins: These are coagulated or denatured proteins obtained when ordinary proteins are heated. In these proteins, the molecular mass does not change, only the properties like solubility, precipitation, etc., are altered, *e.g.*, coagulated egg white.

2. Meta Proteins: They are obtained by the hydrolysis of complex proteins by the action of digestive enzymes, acids or alkalis. The initial products are called proteoses. These are insoluble in water but soluble in dilute acids and alkalis. These can be salted out by half saturation with ammonium sulphate. Further hydrolysis gives peptones, which are soluble in water and are not coagulated by heat. They cannot be salted out by ammonium sulphate. Finally, hydrolysis gives polypeptides and simple peptides, which also behave like peptones.

2.23 STRUCTURE OF PROTEINS

A. Primary Structure

We have already discussed the primary structure of proteins. The primary structure of a protein refers to linear sequence of amino acids in its polypeptide chain and location of disulphide bridges, if any. It represents the number, nature and sequence of amino acid molecules in the polypeptide chain. All the properties of a protein molecule are determined, directly or indirectly, by the primary structure. Any folding or catalytic activity of the protein molecule depends on the proper primary structure (Fig. 2.8).

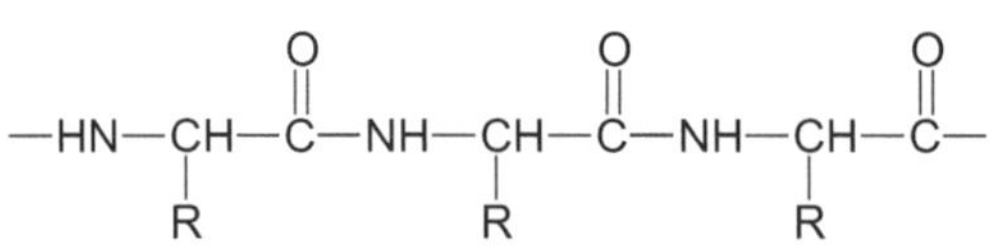

Fig. 2.8 A polypeptide chain showing primary structure.

B. Secondary Structure

The folding of linear polypeptide chains into a specific arrangement, represents the secondary structure of a protein molecule. It results by hydrogen bonding between the carboxyl oxygen and amide hydrogen atoms of the component amino acids of the peptide chain. These hydrogen bonded arrangements, if present, are called secondary structure of the protein. These bonds can occur either within the molecules of one polypeptide chain forming α-helix, or between different polypeptide chains of protein to form a pleated sheet.

1. The α-Helix: The α-helix structure of proteins was first proposed by Linus Pauling and Robert Corey using X-ray analysis of proteins. They were able to understand the secondary structure of proteins for the first time. In the α-helix (a helix that looks like the thread on the right handed screw) conformation, the polypeptide chain assumes the form of a spiral staircase with 3.6 amino acid residues per turn (Fig. 2.9). Although an α -helix may turn in either direction, the α-helices of almost all natural proteins are right handed which means that it rotates clockwise as it moves away from a viewer at either end. The coils of the helix are held together by hydrogen bonds lying parallel to the main axis and all side chains extend outward from the axis of the spiral. Such a structure is flexible and elastic but it gives stability to the peptide chain. The stability of this structure is attributed to a series of regularly spaced hydrogen bonds and van der Waals forces between atoms in the main chains.

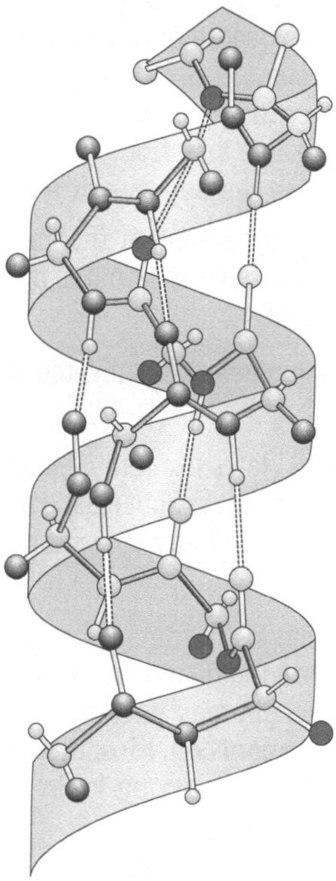

Fig. 2.9 A representation of the α-helical structure of a polypeptide.

The α-helical structure is found in many proteins. It is the predominant structure of the polypeptide chains of fibrous proteins such as myosin, the protein of muscle, and of α-keratin, the protein of hair, unstretched wool and nails.

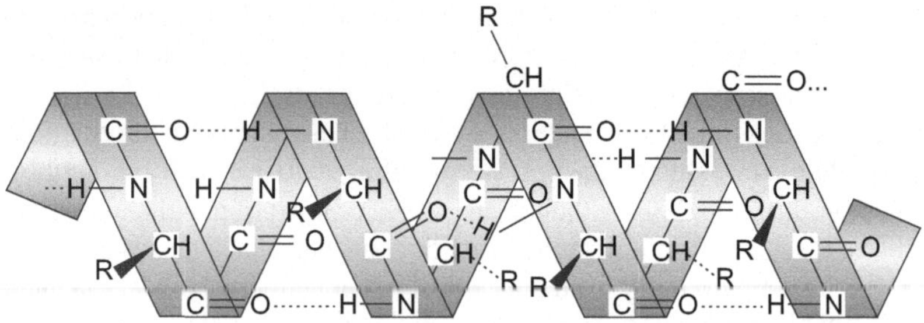

Fig. 2.10 The α-helical arrangement.

2. The β-Pleated Sheet: The second type of secondary structure is the β-pleated sheet. In a β-pleated sheet structure hydrogen bonding occurs between two or more peptide chains and these bonds are at right angles to the main chain. In this arrangement, each carboxyl oxygen on one chain forms a hydrogen bond with an N-H hydrogen on an adjacent chain. The adjacent hydrogen-bonded peptide chains can run in the same direction or in opposite direction. In a parallel β-pleated sheet the adjacent chains run in the same direction, whereas in an antiparallel β-pleated sheet, the adjacent chains run in the opposite direction (Fig. 2.11).

As the substituents (R) on the α-carbons of the amino acids on adjacent chains are close to each other, we can have maximum hydrogen-bonding interactions only if the substituents are small.

For example, silk has large segments of β-pleated sheets because it has a large number of small amino acids.

β-keratin found in feathers and claws is another example of this type of structure. Fibres containing β-pleated structures are not elastic because the polypeptide chains are fully extended. The number of side-by-side strands in a β-pleated sheet ranges from 2 to 15 in a globular protein.

A protein may or may not have the same secondary structure throughout the length. Some parts may be curled into an α-helix, while other parts are lined up in a pleated sheet. Parts of the chain may have no secondary structure at all. Such a structureless part is called a random coil. Most globular proteins contain segments of α-helix, or β-pleated sheet separated by kinks of random coil allowing the molecule to fold into its globular shape.

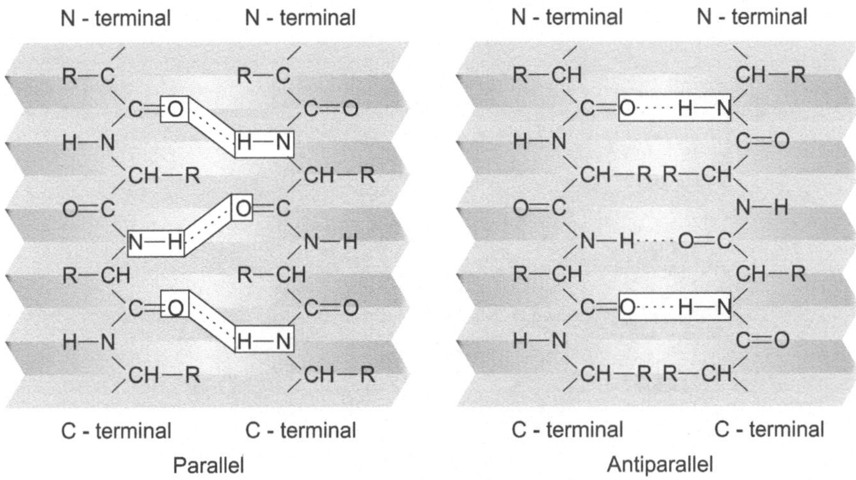

Fig. 2.11 The β-pleated structure.

C. Tertiary Structure

The tertiary structure of a protein is the final three-dimensional shape that arises due to further foldings of its polypeptide chains. These foldings do not occur randomly. They occur in a particular way, a way that is characteristic of a particular protein. Proteins fold in solution in order to maximize their stability. The stabilizing interactions that occur in folding are covalent bonds, hydrogen bonds, electrostatic attractions and van der Waals or hydrophobic interactions. So, a protein tends to fold in a way that maximizes the number of stabilizing interactions.

Disulphide bonds are the only covalent bonds that can form when a protein molecule folds. Other interactions that occur in folding are much weaker but they are so many of them that they become most important interactions in determining how a protein folds. An important feature of most proteins is that the folding takes place in such a way as to expose the maximum number of polar hydrophilic groups to the aqueous environment and enclose a maximum number of non-polar hydrophobic groups within its interior.

The soluble globular proteins tend to be highly folded than fibrous proteins. However, the fibrous proteins also have a tertiary structure. For example, α-keratin, fibrinogen are fibrous proteins and have tertiary structure. Albumin, myoglobin and plasma globulins are globular proteins possessing tertiary structure (Fig. 2.12).

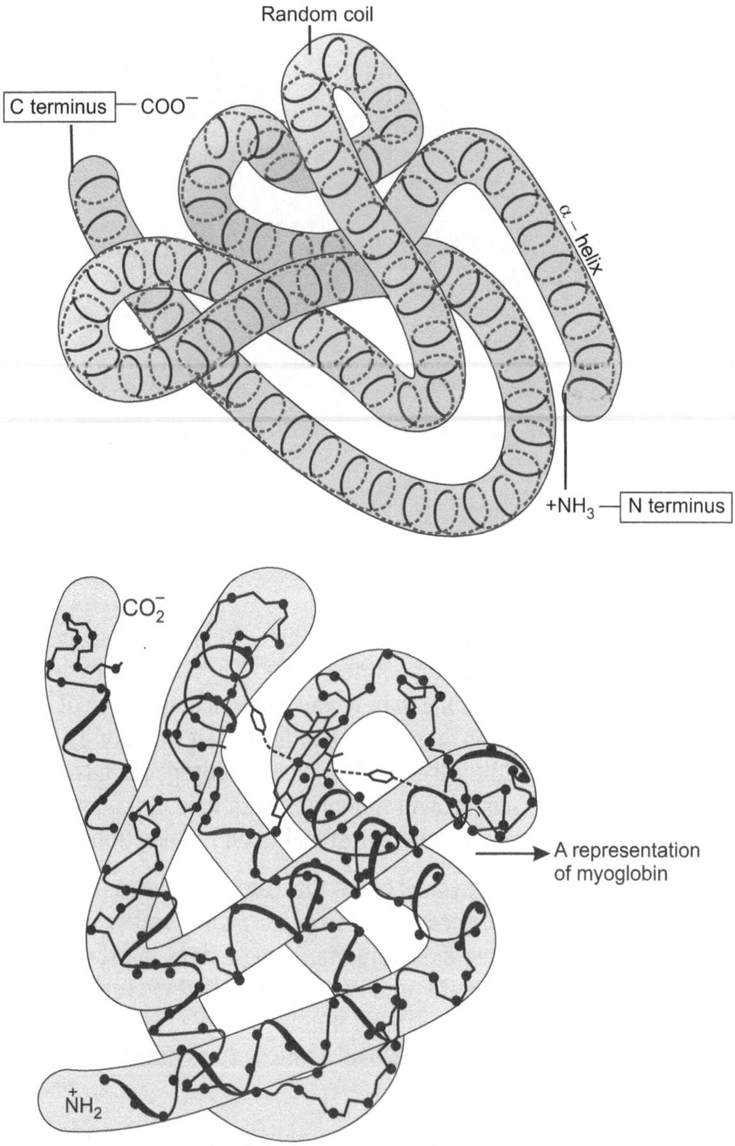

Fig. 2.12 Tertiary structure of a protein.

D. Quaternary Structure

The quaternary structure involves the non-covalent association of two or more peptide chains in the protein molecule. Proteins that have more than one peptide chain are called oligomers. The individual chains are called sub-units. In a quaternary structure, these subunits are held together by the same kinds of interactions that hold the individual protein chains in a particular three dimensional conformation. They are hydrophobic interactions, hydrogen bonding and electrostatic interactions. The quaternary structure of a protein describes the way the sub-units are arranged in space. For example, haemoglobin consists of four peptide chains fitted together to form a globular protein (Fig. 2.13).

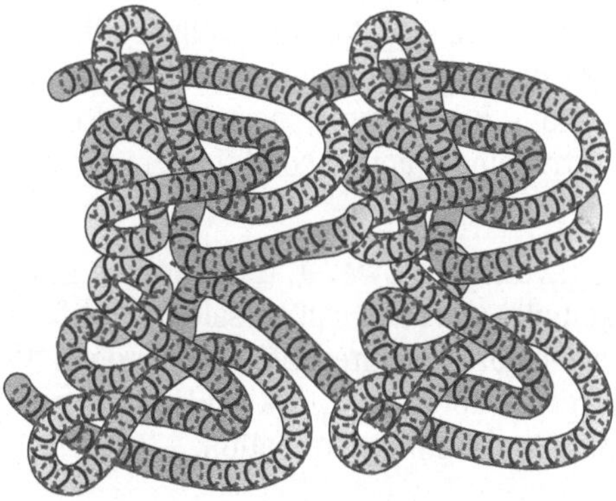

Fig. 2.13 A representation of quaternary structure.

2.24. DENATURATION OF PROTEINS

We have seen earlier that proteins take part in various biological reactions. For a protein to be biologically active, it must have the correct structure at all levels. The sequence of amino acids in the primary structure must be right. The secondary and tertiary structures are very important. Proteins act only when they retain their natural conformations, *e.g.*, in enzymes the active site must have the right conformation. In primary structure, we have covalent bonds. But all other structures are maintained by weak solvation, non-covalent interactions or hydrogen bonding. Proteins molecules may undergo disruption of their overall structure when subjected to heat or small changes in the environment. This change in structure (chemical or conformational) is called denaturation of protein molecules. Secondary and tertiary structures are disrupted and protein molecules lose their biological activity. The process of denaturation, however, does not change the primary structure of proteins. The amino acid sequence remains the same. Denaturation can also be brought about by strong acids or bases, X-rays, UV light, high pressure and heavy metals, etc.

Denaturation is generally an irreversible process. However, in some cases it can be made a reversible process. The cooking of an egg is an example of protein denaturation at high temperature. Egg white contains soluble globular proteins called albumins. When egg white is heated, the albumins unfold and coagulate to produce a solid mass. When cooked egg white is cooled it does not become uncooked. This is irreversible denaturation. We cannot get back the original egg white.

Similarly, when a protein is subjected to an acidic pH, some of the side-chain carboxyl groups are protonated and lose their ionic charge. Conformational changes result leading to denaturation. In basic solutions, amino groups become deprotonated losing their ionic charge which again causes conformation changes and denaturation. The process of curding is an example of irreversible denaturation. Milk turns sour because of bacterial conversion of carbohydrates to lactic acid. When pH increases, the soluble proteins in milk are denatured and precipitate. This process is called curding. However, some proteins are more resistant to acidic and basic conditions. For example, most digestive enzymes such as amylase and trypsin remain active under acidic conditions in the stomach even at pH of about 1.

Denaturation may be reversible, however, if the protein has undergone only mild denaturation. For example, a protein can be salted out of solution by a high salt concentration, which denatures and precipitates the protein. When the precipitated protein is redissolved in a solution with a lower salt concentration, it usually regains its activity and its natural conformation.

❑❑❑

CHAPTER 3

ENZYMES

Learning Objectives

In this chapter, we will study:

- Enzymes – their nomenclature and classification
- Characteristics of Enzymes – Catalytic power, enzyme specificity and stereo-specificity.
- Factors influencing enzyme activity – effect of temperature and pH on enzyme activity ; effect of substrate and enzyme concentration
- Cofactors – coenzymes
- Enzyme inhibitors
- Mechanism of enzyme catalysis – the lock and key model
- Alcoholic Fermentation
- Glycolysis
- Citric acid cycle

3.1 INTRODUCTION

We consume various food ingredients - carbohydrates, proteins and fats any time we take our meals. These are digested and give us energy for performing the various metabolic functions in our body. The secret ingredient in living organisms is catalysis. The catalysts in living cells which facilitate the metabolic reactions are called **enzymes**. The miracle of life is that chemical reactions in the cell occur with great accuracy and at astonishing speed. Without the proper **enzymes** to process the food we eat, it might take us years to digest our breakfast or lunch.

With the exceptions of some RNAs that have catalytic activity, all enzymes are proteins or their derivatives and vary in molecular weight from 10,000 up to 5,00,000. Of all the functions of proteins, catalysis is probably the most important. In the absence of catalysis, most reactions in biological systems would take place too slowly to provide products at an adequate speed for a metabolising organism. A number of enzymes have been isolated and obtained in a crystalline form. Other enzymes are derivatives of proteins formed by combination with some other group such as a metal. Enzymes containing a wide range of metals including iron, copper, zinc, manganese and magnesium are known. In other cases, the protein is combined with a non-proteinous organic molecule. We shall study these under cofactors.

3.2 NOMENCLATURE AND CLASSIFICATION OF ENZYMES

The first enzymes to be discovered were named according to their source or mode of activity. Ptylin, which hydrolyses starches, is found in saliva (from the Greek word *ptyalon*, means 'spittle'). The enzyme pepsin is present in the digestive juices of the stomach (derived from the Greek, *pepsis*, means digestion). The first active enzymes were those present in yeast. Zymases (originated from the Greek *zyme* which means 'a leaven') are found in yeast. They catalyse the fermentation of glucose to ethyl alcohol and carbon dioxide. Many other enzymes have been given uninformative names *e.g.*, trypsin, chymotrypsin (present in pancreatic juices). Papain is present in the fruit papaya.

As more enzymes were discovered, a more systematic nomenclature developed. The trivial or common names of the enzymes are derived by adding the suffix–***ase*** to the names of substrate on which they act or the reaction they catalyse. A substrate is the molecule on which the enzyme exerts its catalytic action. For example, **urease** catalyses the hydrolysis of urea to ammonia and carbon dioxide, **arginase** catalyses the hydrolysis of arginine to ornithine and urea. The enzyme which catalyses the hydrolysis of the disaccharide, maltose, is named as **maltase**. Similarly, sucrase acts on sucrose and **lipases** act on lipids. Hydrolases catalyse hydrolysis reactions and transferases facilitate the transfer of groups from one molecule to another.

However, the above nomenclature has not always been practical. As the number of newly discovered enzymes is increasing rapidly, a more systematic nomenclature has been developed by the International Commission on enzymes which was set up by International Union of Biochemistry (1961). The commission recommended two types of nomenclature ; one systematic and the other working or trivial.

According to the new system of nomenclature, enzymes are divided into six major classes, which are further divided into subclasses according to the type of reaction catalysed.

Each enzyme is assigned a recommended name, usually short and appropriate for everyday use, a systematic name which identifies the reaction it catalyses and a classification number. The classification number is used where accurate identification of an enzyme is required. The coding system is mainly used by research workers and it is beyond the scope of this book.

A. Systematic Names

The systematic names of the enzyme consist of two parts. The first part contains the name of the substrate, while the second part indicates the nature of the reaction and it uses *–ase* as the suffix. For example, alcohol dehydrogenase in which alcohol is the substrate and it is oxidised by the enzyme. However, in case of a bimolecular reaction the first part contains the names of both the substrates separated by a colon. For example, for the biomolecular reaction

$$\text{ATP + Creatine} \rightleftharpoons \text{ADP + Phosphocreatine}$$

the systematic name for the enzyme is ATP : Creatine phosphotransferase. The enzyme catalyses the transfer of a phosphate group from ATP to creatine. The recommended name of the enzyme, that normally used, is creatine kinase. Although, systematic names are quite logical, but they are inconvenient to use. Therefore, the International Commission on enzymes later decided to give more importance to the trivial names.

B. Major Classes of Enzymes

1. **Oxido-reductases:** They catalyse oxidation-reduction reactions. Transfer of electrons takes place from one molecule to another.
2. **Transferases:** Transferases catalyse the transfer of functional groups from one substrate to another

$$AB + C \rightleftharpoons A + BC$$

3. **Hydrolases:** They catalyse hydrolytic reactions, in which a substrate is hydrolysed into two simpler products. During hydrolysis, hydrogen atoms from water enter one of the products while the hydroxyl groups end up in the other product. The reaction is shown in a simplified form as follows:

$$AB + H_2O \rightleftharpoons AH + B.OH$$

4. **Lyases:** They catalyse the reactions involving the removal of a group leaving a double bond or the addition of a group to a double bond. Decarboxylases which regulate the release of CO_2 from respiratory substrates are examples of lyase enzymes. Deaminases catalyse the release of ammonia from amino acids.
5. **Isomerases:** Isomerases control the conversion of one isomer of a compound to another isomer of the same compound, *e.g.*,

$$ABC \rightleftharpoons ACB$$

The interconversion of sugar isomers in glycolysis is catalysed by isomerase enzymes *e.g.*, glucose-6-phosphate is converted to fructose-6-phosphate by hexosephosphate isomerase.

$$\text{Glucose-6-phosphate} \rightleftharpoons \text{Fructose-6-phosphate}$$

6. **Ligases:** This group of enzymes catalyse reactions in which new chemical bonds are formed. ATP provides the energy to make the new chemical bonds. For example, DNA and RNA ligase control the synthesis of macromolecules such as nucleic acids.

Let us examine some of these in detail.

C. Oxido-reductases

These enzymes are of two kinds:

1. **Oxidases:** Carbohydrates, fats and proteins, the main sources of energy, do not get oxidised easily in the presence of atmospheric oxygen. But in the body they are readily oxidised. The enzymes which catalyse these reactions by making use of molecular oxygen are called oxidases. Oxidases catalyse the transfer of hydrogen to molecules of oxygen as shown below:

$$\underset{\text{Substrate}}{AH_2} + \frac{1}{2} O_2 \rightleftharpoons \underset{\text{Oxidised substrate}}{A} + H_2O$$

   Cytochrome oxidase is one of the most important of this group. It catalyses the oxidation of reduced cytochrome.

$$\underset{\text{Reduced cytochrome}}{\text{Cyt. } H_2} + \frac{1}{2} O_2 \rightleftharpoons \text{Cytochrome} + H_2O$$

   Catalase acts on H_2O_2, converting it to H_2O and O_2. It is present in most living cells and its function appears to be removal of H_2O_2 which is toxic to living cells.

   Flavin enzymes have prosthetic groups, which are mono or dinucleotides containing riboflavin. This functions by accepting hydrogen atoms and is subsequently oxidised by molecular oxygen. The enzyme xanthine oxidase, found in liver and milk contains flavin adenine dinucleotide, iron and molybdenum. It catalyses the oxidation of hypoxanthine to xanthine and the latter to uric acid.

   The **phenol oxidases** are copper containing proteins which catalyse the oxidation of phenol derivatives to quinones.

   Ascorbic acid oxidase works with ascorbic acid as the coenzyme.

2. **Dehydrogenases:** Dehydrogenases catalyse the oxidation of substrates by transferring hydrogen to coenzymes such as NAD^+. These dehydrogenases are highly specific in their action. The overall process for dehydrogenation may be represented as follows:

$$\underset{\text{substrate}}{AH_2} + \underset{\text{coenzyme}}{2NAD^+} \rightleftharpoons \underset{\text{oxidised substrate}}{A} + \underset{\text{reduced coenzyme}}{2NADH}$$

For example, alcohol dehydrogenase controls the rate at which ethanol is oxidised to ethanal.

$$\underset{\text{Ethanol}}{CH_3CH_2OH} + 2NAD^+ \rightleftharpoons \underset{\text{Ethanal}}{CH_3CHO} + 2NADH$$

In reactions catalysed by oxido-reductases substrates are oxidised whilst oxygen or coenzymes are reduced.

D. Hydrolases

The earliest enzymes to be discovered were all hydrolases. These are especially important because many of them are associated with the digestion of foods in our body. The food materials we take are composed of large molecules. They must be broken down into simpler components so that they can be absorbed and utilised by the body. The breakdown of large molecules in the digestive tract is carried out by hydrolytic enzymes or hydrolases. They catalyse introduction of the elements of water at a specific bond of the substrate.

Hydrolases are also found universally in plant and animal cells and in microorganisms. Because the concentation of the water is so high ***in vivo***, hydrolytic reactions are genreally irreversible. In addition to the solvents, hydrolases have only a single substrate. They are easy to detect and study. For this reason we know more about them than about other enzymes.

Hydrolases have been further divided into the following groups :

(*i*) **Esterases - Cleaving Ester Linkages:** These enzymes catalyse hydrolysis of organic esters to form alcohol and acids. Simple esterases such as liver esterase catalyse the hydrolysis of esters from lower alcohols and fatty acids. **Lipases** catalyse the hydrolysis of fats *i.e.*, of triglycerides to form glycerol and fatty acids. **Pancreatic lipase** (steapsin) is responsible for the hydrolysis of most of the fat that we take.

Phosphatases hydrolyse esters of phosphoric acid. This is a large and complex group of enzymes, some of which appear to be highly specific. These are important factors in carbohydrate and lipid metabolism.

(*ii*) **Carbohydrases:** These enzymes hydrolyse the glycosidic linkage of simple glycosides, oligosaccharides and polysaccharides. Those which hydrolyse starch and glycogen are called amylases. There are two amylases, α- and β-amylase. α-Amylase hydrolyses starch and glycogen to dextrins, β-amylase hydrolyses starch to maltose. These enzymes are widely distributed in plants. Ptyalin, a salivary amylase (found in saliva) and amylosin, the pancreatic amylase, are thought to be mixtures of α- and β-amylase.

The disaccharide-splitting enzymes maltase, sucrase and lactase are important enzymes of the digestive tract as they hydrolyse common disaccharides. Maltase hydrolyses α-glycosides and emulsin attacks on β-glycosides.

(*iii*) **Nucleases:** Nucleases cause hydrolysis of nucleic acids liberating smaller nucleotides. Two specific nucleases have been obtained in crystalline form. Ribonuclease hydrolyses nucleic acids of ribose type. Deoxyribonuclease causes depolymerisation of DNA.

Nucleosidases hydrolyse nucleosides to sugars and purines or pyrimidines. As nucleotidases remove phosphoric acid from nucleotides to give nucleosides, they are phosphatases and often classified as such.

(*iv*) **Amidases:** Amidases hydrolyse amides. Two amidases are quite important – arginase and urease. Arginase is found in the liver and is an important factor in the synthesis of urea. Urease, the enzyme which hydrolyses urea to carbon dioxide and ammonia is of special interest to the biochemist because it is used in the estimation of urea in blood and urine.

(*v*) **Proteases:** The proteases hydrolyse proteins to simpler molecules. They do not completely hydrolyse proteins to amino acids. The enzymes of the digestive tract are obvious examples. Pepsin found in the gastric juice and trypsin and chymotrypsin present in the pancreatic juice initiate the digestion of protein foods in the body. They attack only at specific peptide bonds.

Rennin, found in the gastric juice of calves is a proteolytic enzyme, which acts on casein of milk. It is used commercially in the manufacture of cheese.

(*vi*) **Plant Proteinases:** Papain is obtained from the unripe fruit of papaya. Bromelin is found in pineapples and ficin is present in the milky sap of the fig tree. Bromelin is of special interest to the house-wife. In making fruit gelatin desserts, fresh pineapples cannot be used because bromelin present will hydrolyse the gelatin forming products that will not gel. Cooked pineapple may be used because cooking ultimately destroys the enzyme.

(*vii*) **Peptidases:** They act on peptide bonds adjacent to a free amino or carboxyl group. If the amino acid containing a free amino group is split off, the enzyme acting is known as aminopeptidase, whereas a carboxypeptidase splits the amino acid having a free carboxyl group.

E. Transferases

These enzymes catalyse the transfer of a group from one substance to another. We consider some examples :

(*i*) **Transglycosidase:** Phosphorylase is an example of this type of enzyme which can split 1,4- glucosidic bond in starch with the help of phosphoric acid

$$\text{Starch} + PO_4^{3-} \xrightarrow{\text{Phosphorylase}} \alpha\text{-D-glucose-1-phosphate}$$

(*ii*) **Transphosphorylase:** The formation of D-glucose-6-phosphate from D-glucose with the help of ATP is catalysed by the enzyme phosphotransferase. One phosphate unit is transferred from ATP to glucose and ADP is obtained.

D-Glucose + ATP ⟶ D-Glucose-6-phosphate + ADP

This enzyme is commonly known as glucokinase.

Phosphoglucomutase catalyses the conversion of glucose-1-phosphate to glucose-6-phosphate

(*iii*) **Transacetylase:** Acetyl-CoA is formed from coenzyme A. The transacetylation is performed by the coenzyme molecule. Acetyl-Co A can transfer the acetyl group to choline forming acetyl-choline and CoA.

(*iv*) **Transaminases:** These enzymes transfer an amino group of an amino acid to an α-keto acid and promote the formation of a new amino acid. An example is the reaction between glutamic acid and oxaloacetic acid to produce α-keto glutaric acid and aspartic acid. The transaminases are important enzymes associated with protein metabolism.

(*v*) **Transmethylase:** With these enzymes the transfer of methyl group takes place from one compound to another.

3.3 CHARACTERISTICS OF ENZYMES

Certain characteristics of enzymes are different from inorganic catalysts. Some of these are :

A. Catalytic Power

Enzymes are the most efficient catalysts known. They can increase the rate of a reaction by a factor of up to 10^{20} over uncatalysed reaction. However, non-enzymatic catalysts enhance the rate of reaction by factors of 10^2 to 10^4 only.

We know that catalysts are effective in small amounts. When analysing the catalytic power of an enzyme, we should know the amount of starting material or substrate that is converted to product in a unit time by a given quantity of the enzyme. This quantity is called the **turnover number** and is defined as the number of moles of substrate converted into product per minute by one mole of enzyme. The turnover numbers of enzymes vary greatly ranging from one hundred to over three million. The turnover numbers of some enzymes is given in Table 3.1.

TABLE 3.1. Turnover numbers of some enzymes

Enzymes	*Turnover Numbers*
Catalase	4×10^7
Carbonic anhydrase	1.0×10^6
Chymotrypsin	1.9×10^2
Lysozyme	0.5
Lactatedehydrogenase	1.0×10^3

Catalysts ordinarily have no effect on the equilibrium of a reversible chemical reaction. They merely speed up the reaction until it reaches equilibrium. Enzymes act in a similar way, they hasten the process in either direction.

Catalysts are usually unchanged in the reaction: This property of ideal catalysis does not hold true by enzymes. As proteins are easily inactivated or denatured by high temperature or very alkaline or acidic conditions. So, enzymes work under optimum conditions of temperature and pH. We shall discuss this in detail at a later stage.

B. Control of Enzyme Action

Biochemical pathways usually involve a number of linked reactions. Each reaction is catalysed by a specific enzyme. For example,

$$A \overset{\text{Enzyme X}}{\rightleftharpoons} B \overset{\text{Enzyme Y}}{\rightleftharpoons} C \overset{\text{Enzyme Z}}{\rightleftharpoons} D$$

In this there are three specific enzymes for each of the three steps involved in the conversion of A to D. In some instances accumulation of one of the products formed near the end of the chain inhibits the action of an enzyme in one of the earlier reactions. In the example shown above, the presence of a high concentration of the product D may slow down the rate at which enzyme X converts A to B. This is called negative feedback inhibition. Negative feedback ensures that reactants are used efficiently and prevents the excess manufacture of end products. Thus, control is exercised on the pathway of the above reactions. Such a control of enzyme action helps to maintain stable environment in living oganisms.

Pathways can be shut down if an organism has no immediate need for their products, which saves energy for the organism.

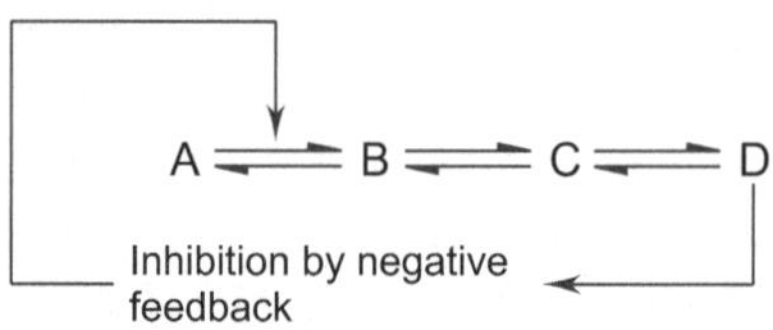

In glycolysis, three reactions are control points. The first is the conversion by glucose to glucose-6-phosphate, which is inhibited by glucose-6-phosphate itself. The second is the production of fructose-1, 6-bisphosphate, inhibited by ATP and lastly reaction of phosphoenolpyruvate to pyruvate is also inhibited by ATP.

It is frequently observed that control is exercised near the start and end of a pathway as well as at points involving the key intermediates as exemplied above.

C. Enzyme Specificity

Enzymes are highly specific in their reactivities. They catalyse only one reaction or a group of closely related reactions. An enzyme may be specific with regard to the type of reaction it catalyses or with regard to the kind of substrate on which it acts. That means some enzymes are reaction specific and others possess substrate specificity. For example, hydrolases carry out the hydrolysis reactions and oxidases catalyse only oxidations. A carbohydrase will attack only on carbohydrates and nucleases attack on nucleic acids. Maltase catalyses the hydrolysis of the disaccharide, maltose.

Urease, however, is a unique enzyme. It catalyses only the hydrolysis of urea. It has no effect on other compounds, even closely related ones such as amides. Such absolute specificity is rather rare among enzymes characterised to date

$$H_2N{-}\overset{\displaystyle O}{\overset{\|}{C}}{-}NH_2 + H_2O \xrightarrow{\text{Urease}} 2\,NH_3 + CO_2$$

Among reaction specific enzymes group specificity or substrate specificity is observed side by side. For example, *α*-glucosidases will hydrolyse all compounds containing an *α*-linked glucose. Invertase catalyses both hydrolysis of sucrose and of raffinose which have the same linkage, *i.e.*, the enzyme is *β*-fructofuranosidase. On the other hand lipases hydrolyse fat molecules irrespective of the nature of fatty acids present in the fat. Peptidases, however, are specific with regard to a particular linkage, *i.e.*, the peptide bond but they also require a special grouping in its vicinity. Carboxypeptidase, for example, cleaves only those terminal amino acids in a peptide chain having a free carboxyl group. Aminopeptidases on the other hand cleave the *N*-terminal amino acid *i.e.*, an amino acid having a free amino group in its vicinity.

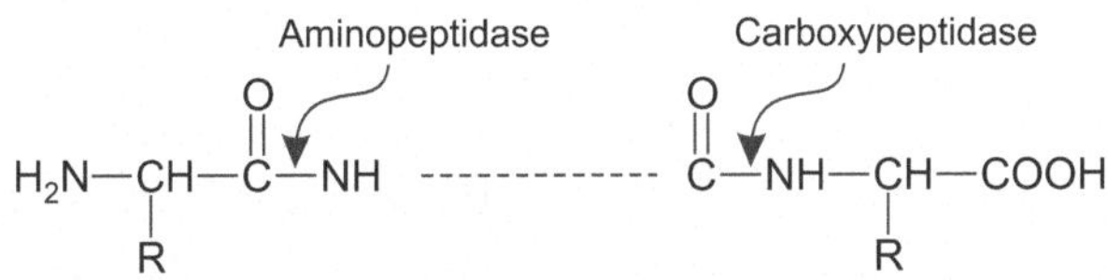

Substrate specificity of peptidases are as follows :

Pepsin : It attacks at the amino positions of aromatic amino acids, phenylalanine, tyrosine and tryptophan. It does not attack ester bonds.

Chymotrypsin : It attacks on the carboxyl end of the above amino acids. It also hydrolyses esters.

Trypsin : It splits only those peptide bonds on the carboxyl sides of the basic amino acids, lysine and arginine.

Thrombin is even more specific. It cleaves only those bonds between arginine and glycine. It has no effect on the hydrolysis of other peptide linkages. It would not even attack the bonds between glycine and arginine.

In an oxidation-reduction reaction, we observe enzyme specificity with regard to both the reagents *i.e.*, both the substance oxidised and that reduced. Certain dehydrogenases will transfer hydrogen from substrates for which they are specific to coenzyme-I and not coenzyme-II. Some other dehydrogenases are specific to coenzyme-II and will not transfer hydrogen to coenzyme-I.

Enzyme specificity is very important in chemical reactions taking place in the cell. It ensures that proper reaction occurs in the proper place at the proper time. Enzyme specificity, therefore, plays a crucial role in the metabolism of the cell.

Enzymes permit reactions to take place in the living organisms which otherwise would not take place at an appreciable rate at ordinary temperature and may thereby exert a directive effect on the metabolism of the cell. For example, by accelerating certain pathways to a greater extent than other, they may virtually direct the mechanism along one of the possible routes.

D. Stereospecificity

Many enzymes show stereospecificity, *i.e.*, some enzymes act on only one of a pair of optical isomers. In addition some enzymes are involved in asymmetric syntheses producing only one of the two possible optical isomers. The following examples show stereospecificity among enzymes:

(*i*) Arginase catalyses the hydrolysis of L-arginine to ornithine and urea.

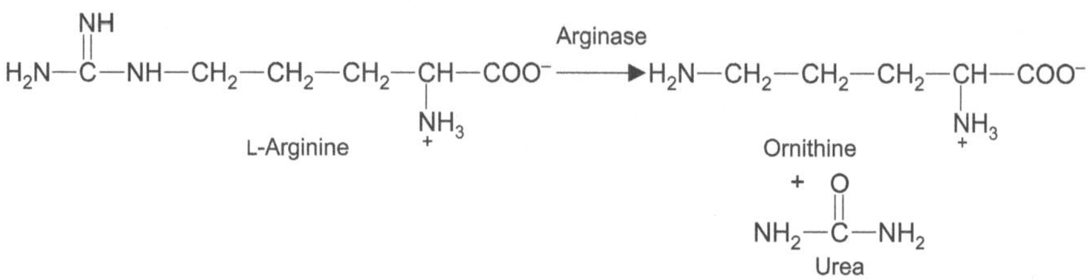

Arginase has, however, no effect on the rate of hydrolysis of D-arginine.

(*ii*) Pyruvic acid is converted to lactic acid by two different enzymes, each producing only one of the optical isomers.

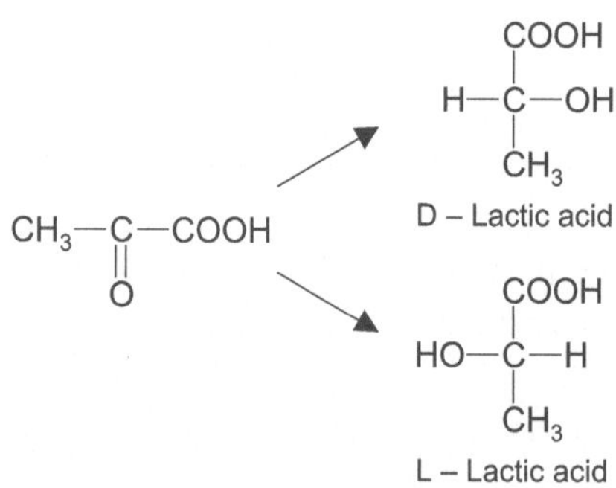

(*iii*) Fumarase catalyses addition of water to fumarate to give malate whereas maleate remains unaffected.

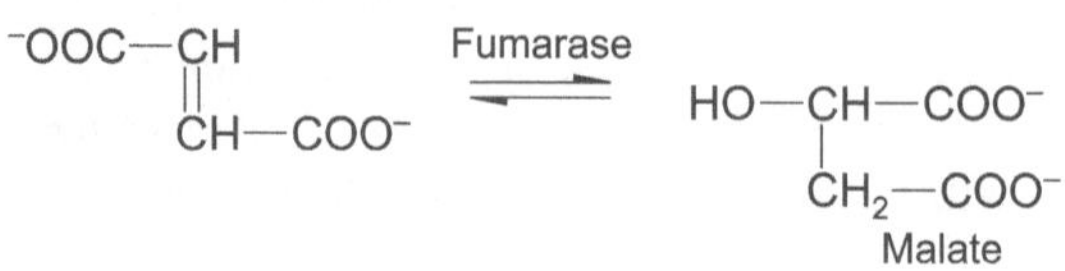

Maleate

(*iv*) Glycerol is converted to L-phosphoglycerol by the enzyme glycerolkinase. Phosphate group comes from the ATP molecule.

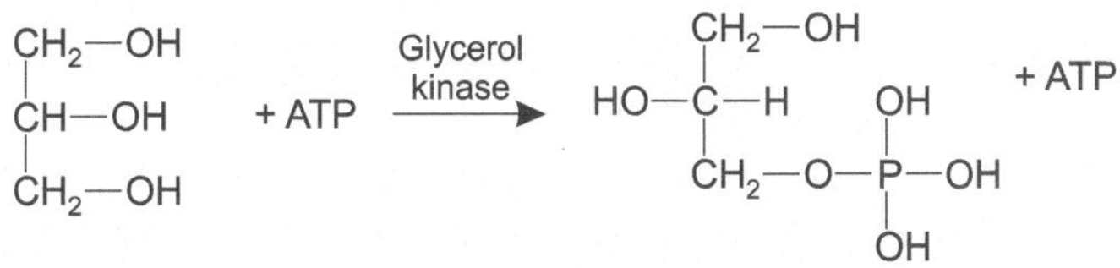

Glycerol L - Phosphoglycerol

(*v*) During citric acid metabolism, a series of enzymatic reactions occur in which we observe the selective action of an enzyme on only one of two chemically identical groups in a compound. For example, oxaloacetic acid is converted to citric acid by the condensing enzyme. If the carboxyl group adjacent to the —CH_2 group is labelled, the citric acid obtained will have a labelled position only in one of its terminal carboxyl groups.

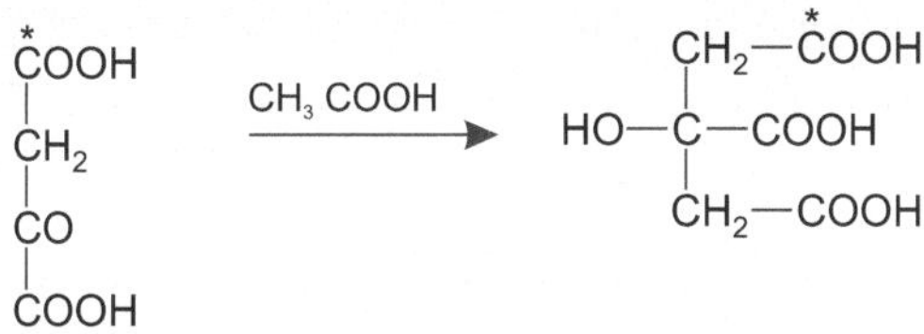

Oxaloacetic acid Citric acid

This citric acid can be converted to α-ketoglutaric acid (via isocitric acid) which contains C^{14} in the α-carboxyl group.

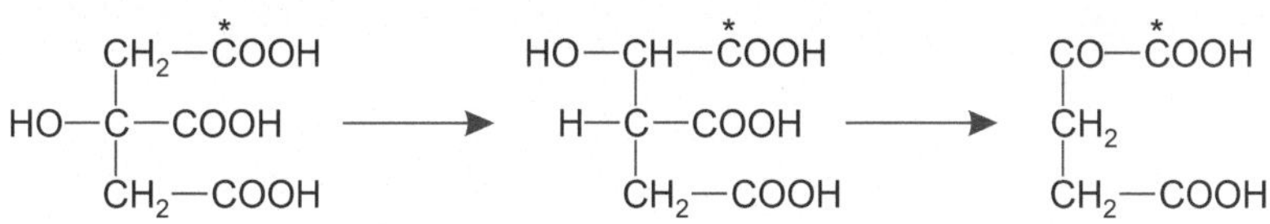

Citric acid Isocitric acid α-Ketoglutaric acid

In citric acid, we have two identical —CH_2—COOH groups, but the enzyme selectively attacks only one particular —CH_2—COOH group.

3.4 FACTORS INFLUENCING ENZYME ACTIVITY

A. Effect of Temperature on Enzyme Activity

The catalytic properties of enzymes are dependent on two features :

(*i*) that the enzyme is able to form an intermediate complex with the substrate and

(*ii*) that the protein part of the molecule is preserved in its native state.

Factors which prevent either of these conditions will destroy catalysis. Thus, all substances or conditions which cause denaturation of an enzyme will inevitably destroy its catalytic activity.

We know that heat supplies kinetic energy to reacting molecules causing them to move more rapidly. The chances of molecular collision taking place are thus increased at higher temperatures, so it is more likely that enzyme-substrate complexes will be formed. Therefore, increase in temperature will increase the rate of an enzyme-catalysed reaction. As we increase the temperature, heat energy also increases the vibrations of the atoms which make up the enzyme molecules. If the vibrations become too violent, chemical bonds in the enzyme break and the three-dimensional structure is lost. Therefore, above a certain temperature denaturation will take place and the enzyme loses its activity.

Enzyme-catalysed reactions often appear to have **optimum temperature** at which the reaction proceeds most rapidly. If the temperature is raised beyond this point, the reaction rate decreases due to denaturation of the enzyme (Fig. 3.1). This temperature is not necessarily that at which the enzyme is most stable. It is the resultant of the two contrary processes:

1. the usual increase in reaction rate with increase in temperature and
2. the increasing rate of thermal denaturation of enzyme above a critical temperature.

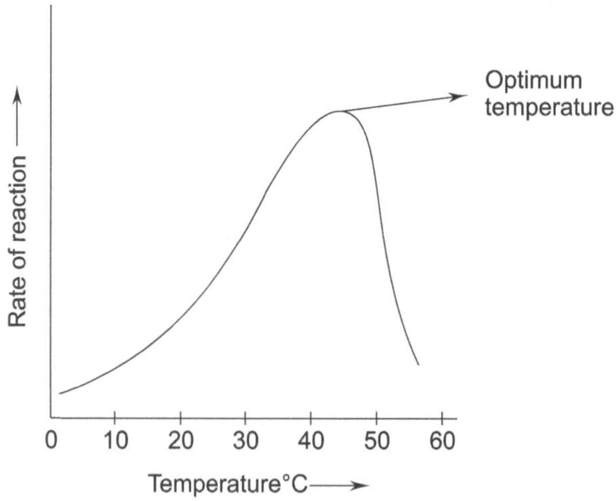

Fig. 3.1 Effect of temperature on enzyme-catalysed reaction.

The rate of most enzymatic reactions approximately doubles for every 10°C rise in temperature. The term temperature coefficient (Q_{10}) is used to express the effect of a 10°C rise in temperature on the rate of a chemical reaction.

$$Q_{10} = \frac{\text{rate of reaction at } (t+10)^\circ C}{\text{rate of reaction at } t^\circ C}$$

Optimum temperature for enzymes in the human body is 37°C, the Q_{10} for enzyme-catalysed reactions is 2.

Although most enzymes are inactivated at temperature above 55°C, some are quite stable and retain activity at much higher temperatures, *e.g.*, enzymes of various species of bacteria inhabiting hot springs are active at temperatures exceeding 85°C. Some enzymes, such as ribonuclease lose activity on heating but quickly regain it on cooling indicating that their unfolded polypeptide chain quickly reverts back into its natural conformation.

B. Effect of pH on Enzyme Activity

The symbol pH refers to concentration of hydrogen ions in solution. The activity of an enzyme varies with the pH of the medium. Most enzymes have a characteristic pH at which their activity is maximum. Above or below this pH the activity declines. When we plot activity versus pH most enzymes yield bell-shaped curve with a more or less sharply defined maximum. That means enzymes have maximum activity at **optimum pH.** Figure 3.2 illustrates the relationship between the pH and activity of the enzyme invertase.

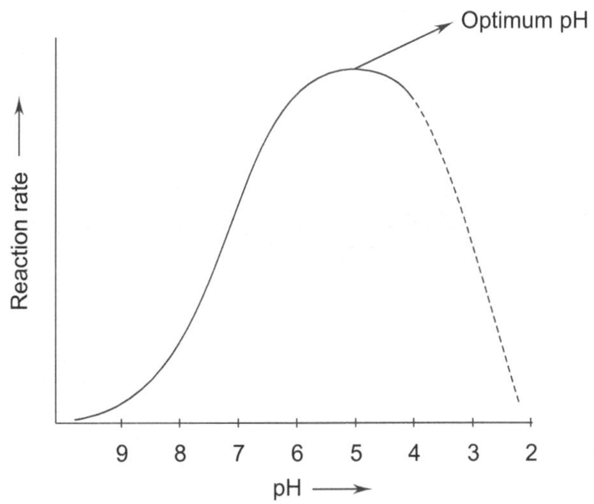

Fig. 3.2 Relationship between the pH and activity of enzyme.

We give below in Table 3.2, the optimum pH for some typical enzymes. The optimum pH of an enzyme is not ncessarily identical with pH of its normal intracellular surroundings.

Even small changes in pH can have a great effect on enzyme activity. Small changes in pH means relatively large changes in $[H^+]$. A change of 1 on the pH scale involves a ten-fold increase or decrease in $[H^+]$, while a change in pH of 2 represents a hundred-fold change in [H^+]. The concentration of [H^+] affects the stability of the electrovalent bonds which help to maintain the tertiary structure of protein molecules. Extremes of pH cause the bonds to break resulting in enzyme denaturation.

The affinity of an enzyme for its substrate may be altered by variations in pH. Changes in [H^+] can alter the ionisation of the amino acid side chains at the active centres of enzymes. Ionisation of the substrate molecules can also be affected. The formation of enzyme-substrate complexes depends on the active centres and substrate molecules having opposite electrostatic charges. If the charges are altered by changes in pH, some enzymes fail to function.

TABLE 3.2. Optimum pH for some enzymes

Enzyme	*Optimum pH*
Lipase (pancreas)	8.0
Lipase (stomach)	4 - 5
Lipase (castor oil)	4.7
Pepsin	1.5 - 1.6
Trypsin	7.8 - 8.7
Urease	7.0
Invertase	4.5
Maltase	6.1 - 6.8
Amylase (pancreas)	6.7 - 7.0
Amylase	4.6 - 5.2

It is clear from the Table 3.2, that for every enzyme there is an optimum pH at which the reaction it catalyses proceeds most rapidly. Most enzymes work within a pH range of 5 – 9 and catalyse reactions most efficiently at pH 7. There are, however, some exceptions. For example, pepsin and rennin secreted in the mammalian stomach work best at pH 1.5 – 2.5. Alkaline phosphatase in the kidneys has an optimum pH of 10 (Fig. 3.3).

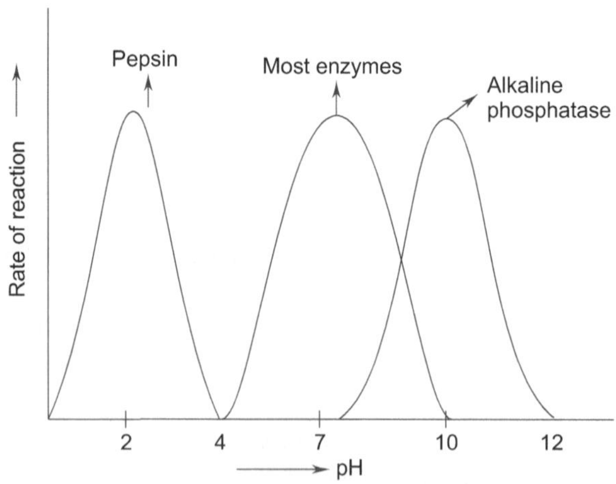

Fig. 3.3.

C. Effect of Substrate Concentration

If the concentration of the enzyme is kept constant and the substrate concentration is varied, we get the reaction profile as shown in Fig. 3.4. At low concentration of the substrate there is a linear relationship between the reaction rate and substrate concentration *i.e.,* reaction rate increases with increase in substrate concentration.

In these conditions, the ratio of the enzyme to substrate molecules is high. Some active sites are always free for the substrate molecules to bind with the enzyme. With increasing substrate concentration a point is reached when a further increase in substrate concentration does not cause the reaction to go any faster *i.e.,* the reaction velocity remains unaffected. The enzyme to substrate ratio is then lower and there are more substrate molecules present than there are free active centres with which to bind. Adding more substrate will not make the reaction go more quickly. The third phase comes when the substrate concentration becomes very high the enzyme activity is inhibited by high concentration of the substrate and the reaction rate declines.

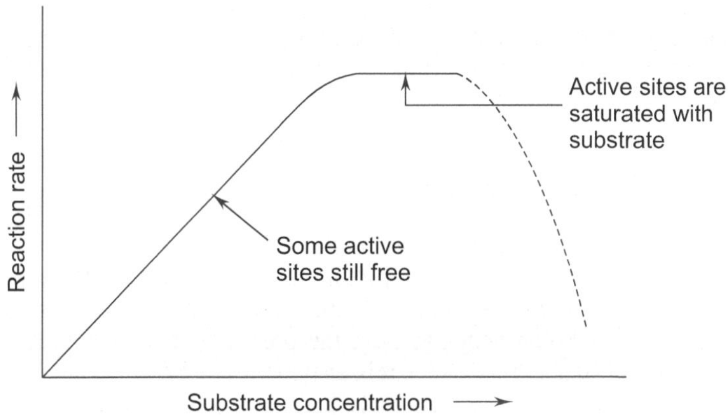

Fig. 3.4.

D. Effect of Enzyme Concentration

Enzymes catalyse reactions rapidly at very low enzyme concentrations. If the substrate concentration is maintained constant, the reaction rate will increase as concentration of the enzyme is increased. This is because enzyme molecules form complexes with substrates only very briefly. The products of the reaction are quickly released and the enzyme is then available for further activity (Fig. 3.5). This relationship holds good over a wide range of enzyme concentrations.

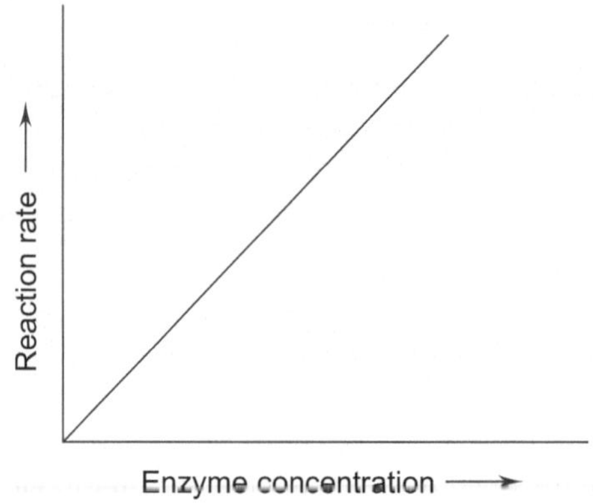

Fig. 3.5.

The rate at which enzymes use substrates is described as the turnover number. For some enzymes the turnover number is very high. A molecule of catalase for example can break down 40,000 molecules of hydrogen peroxide into water and oxygen every second.

The larger the number of enzyme molecules present the greater is the amount of substrate used in a given period of time provided that there is an excess of substrate available.

3.5 COFACTORS - COENZYMES

We have seen that most enzymes are simple globular proteins. Some others are conjugated proteins which have a non-protein fraction called the prosthetic group. A prosthetic group is an essential co-factor attached to the protein part of a conjugated enzyme. Some enzymes need the presence of co-factors. Co-factors are non-protein substances that take part in enzymatic reactions and are regenerated for further reaction. Metal ions frequently play such a role. They are one of two important classes of co-factors. The other important class is that of **coenzymes**. Coenzymes are organic non-protein substances which are not bonded to enzyme molecules like other prosthetic groups.

A prosthetic group is an essential co-factor attached to the protein part of a conjugated enzyme. If the prosthetic group is removed the enzyme fails to function. Mineral ions must be mixed with reactants before some enzymes will work. It is partly for these reasons that minerals are essential to living organisms.

Metal ions are Lewis acids *i.e.*, electron-pair acceptors. Therefore, they can act as Lewis acid-base catalysts. We know that they can form co-ordination compounds by behaving as Lewis acids. These coordination compounds are an important part of metal ions in biological systems. These coordination compounds formed by metal ions tend to have specific geometries, which help in positioning the groups involved in a reaction for optimum catalysis. Zinc, iron and magnesium act as enzyme activators in this way. *e.g.*, Zn (II) in carboxypeptidase and Fe (III) in haemoglobin.

Several important coenzymes are vitamins and their derivatives. Nicotinamide adenine dinucleotide (NAD) and its phosphate ester are derived from nicotinamide, a B-group vitamin. Riboflavin (Vitamin B_2) forms part of flavin-adenine dinucleotide, (FAD). Many of these coenzymes are involved in oxidation-reduction reactions which provide energy for the organism. Others serve as group transfer agents in metabolic processes. Some coenzymes with their reactions and vitamin precursors are given in Table 3.3.

TABLE 3.3. Some Coenzymes with their Reactions and Vitamins Precursors

Coenzyme	*Reaction Type*	*Vitamin Precursor*
Biotin	Carboxylation	Biotin
Coenzyme A	Acyl transfer	Pantothenic acid
Flavin coenzymes	Oxidation-reduction	Riboflavin (B_2)
Lipoic acid	Acyl transfer	
Nicotinamide-adenine coenzymes	Oxidation-reduction	Niacin
Pyridoxal phosphate	Transamination	Pyridomine (B_6)
Thiamine pyrophosphate	Aldehyde transfer	Thiamine (B_1)

Nicotaiamide adenine dinucleotide (NAD^+) is a coenzyme which takes part in many oxidation-reduction reactions Its structure (Fig. 3.6) has three parts - a nicotinamide ring, an adenine ring and two sugar phosphate groups linked together.

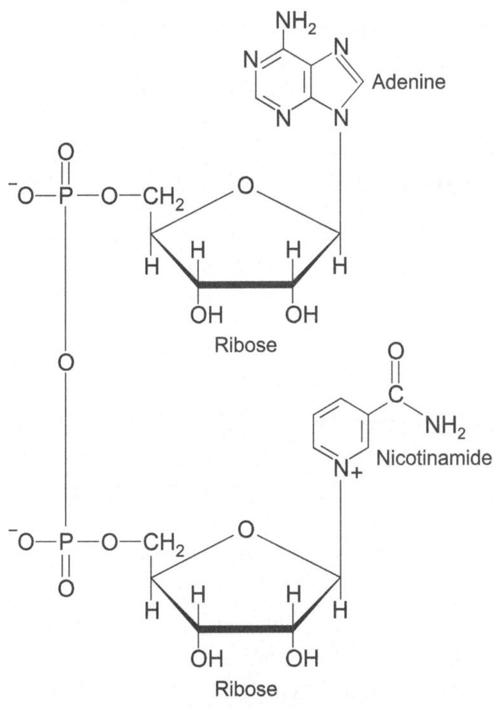

Fig. 3.6.

The phosphorylated forms of the B_6 vitamins, pyridoxal, pyridoxamine and pyridoxine are important coenzymes. They are involved in the transfer of amino groups from one molecule to another, an important step in the biosynthesis of amino acids.

Pyridoxal-5-phosphate is a versatile coenzyme which takes part in a variety of reactions required for the synthesis and catabolism of α-amino acids.

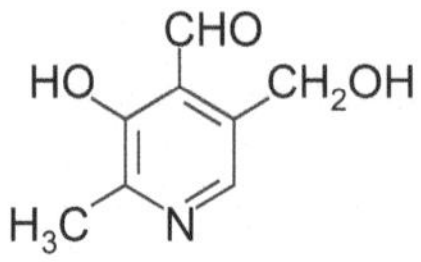

Pyridoxal

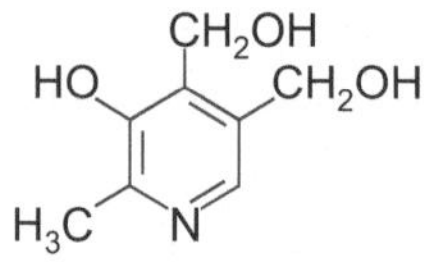

Pyridoxine

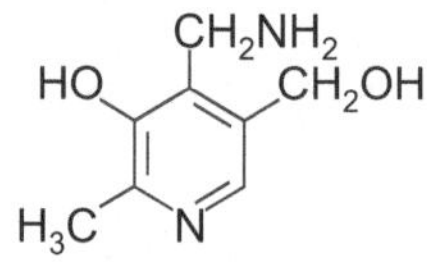

Pyridoxamine

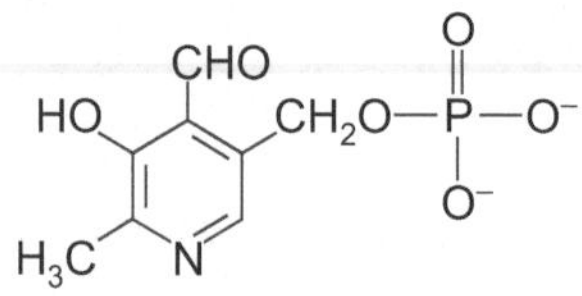

Pyridoxal-5-phosphate

3.6 ENZYME INHIBITORS

An inhibitor, as the name implies, is a substance that interferes with the action of an enzyme and slows the rate of a reaction. Many substances inhibit the activity of enzymes. Enzymes are sensitive to a variety of chemical reagents that react with groups at or near the surface of the protein, producing partial or complete inhibition of chemical activity. Inhibitors fall into two categories – reversible and non-reversible.

A. Reversible Inhibitors

Reversible inhibitors are substances which prevent enzymes from combining with substrates. Activity of the enzyme is restored when the inhibitor is removed. A reversible inhibitor can bind to the enzyme and subsequently be released, leaving the enzyme in its original condition.

There are two major classes of reversible inhibitors –

1. **Competitive Inhibitors:** These compete with the normal substrate molecule for combining with the active site on the enzyme. They can bind to the active site and block the substrate from binding to the enzyme. Thus, they affect enzyme action by becoming attached to the active centres stopping the substrate from binding to the enzyme. Such inhibitors are very similar in structure to the substrate.

   A well-known example of this behaviour is inhibition of the enzyme succinate dehydrogenase by malonic acid. Sucinate dehydrogenase catalyses the oxidation of succinate to fumarate in the Kreb's cycle. In the presence of malonate the reaction rate is slowed down Malonate has a molecular structure very similar to that of succinate. Inhibition occurs because the active centre of some of the enzyme molecules become occupied by malonate rather than by succinate. In fact, both malonate and succinate are competing for the active centre of the enzyme.

The degree of inhibition by a competitive inhibitor is less if ratio of substrate to inhibitor is high.

2. **Non-Competitive Inhibitors:** In this case, the inhibitor becomes attached to the enzyme at a site other than the active site. As a result of this binding a change is caused in the structure of the enzyme, especially around the active site.

The substrate is still able to bind to the active site, but the enzyme cannot catalyse the reaction when the inhibitor is bound to it. Thus, the enzyme, the substrate or possibly both become changed so that enzyme activity stops.

Disulphide bridges are important in maintaining the tertiary structure of enzyme molecules. If the disulphide bridges are broken the three-dimensional shape of the enzyme changes. The ions of heavy metals such as mercury, silver and copper affect enzymes in this way. Hg^{2+}, Ag^{+} and Cu^{2+} ions combine with thiol (—SH) groups in enzymes. They denature enzyme molecule and thus, inhibit enzyme activity. Cyanide is another non-competitive inihibitor. It blocks the action of some enzymes by combining with iron which may be present in a prosthetic group or which may be required as an enzyme activator. For this reason, salts of heavy metals and cyanide are potent poisons to living organisms. However, non-competitive inhibitors do not bind strongly to enzymes and can be removed by dialysis. Enzyme activity is then restored.

B. Non-reversible Inhibitors

Non-reversible or irreversible inhibitor reacts with the enzyme to produce a protein that is not enzymatically active and from which the original enzyme cannot be regenerated. The enzymes undergo irreversible inactivation when they are treated with agents capable of permanently modifying a functional group required for catalysis, making the enzyme molecule inactive.

Organosphosphorus insecticides such as malathion are good examples of non-reversible inhibitors. They become firmly bound to active centres so that substrate molecules cannot bind to enzymes and activity of enzyme is permanently stopped. Insecticides of this type inactivate the enzyme cholinesterase, which is essential for the functioning of the nervous system.

3.7 MECHANISM OF ENZYME CATALYSIS

Enzymes are soluble in water and work in aqueous solution in living cells. They are sometimes described as organic catalysts. A catalyst is a substance which accelerates the rate of a chemical reaction. Many of the biochemical reactions catalysed by enzymes are reversible reactions. An example of a reversible reaction is as follows:

$$\underset{\text{reactants}}{A + B} \rightleftharpoons \underset{\text{product}}{C}$$

At equilibrium the rate at which A and B are converted to C may or may not be equal to the rate at which C is converted back to A and B. The position of equilibrium depends on the energy difference between the reactants and product (Fig. 3.7).

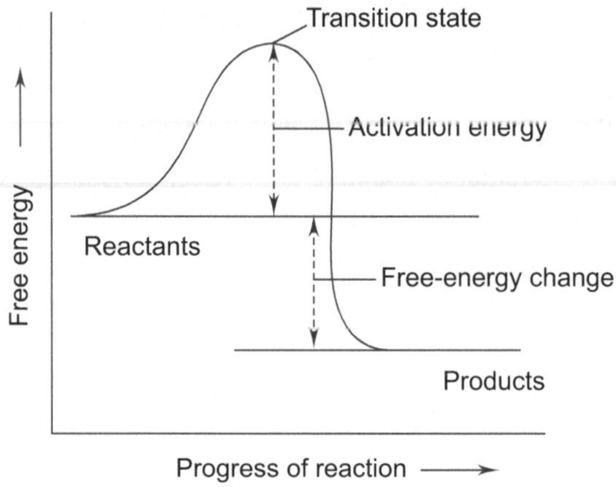

Fig. 3.7 Activation energy profile for a typical reaction.

Enzymes, like all catalysts, speed up reactions but they cannot alter the equilibrium constant or the free energy change. The reaction rate depends on the free energy of activation or activation energy *i.e.*, the energy required to initiate the reaction. The activation energy for an uncatalysed reaction is higher than that for a catalysed reaction. That means an uncatalysed reaction requires more energy to get started than that required for catalysed reaction (Fig. 3.8).

Enzymes do not alter the direction of a reaction ; they speed up the rate at which equilibrium is reached. In doing so, they can catalyse reversible reactions in either direction provided it is energetically feasible.

Biochemical reactions involve the formation or the destruction of chemical bonds. When two or more reactants are joined, chemical bonds are formed. When a complex molecule is split into simpler components, chemical bonds are destroyed. In both the cases, energy is required to bring about the changes. That energy is the activation energy. Heat can be used as a source of activation energy. Many chemical reactions do not proceed quickly unless the reactants are raised to relatively high temperatures. In living cells, however, reactions take place rapidly at relatively low temperatures. Enzymes lower the amount of activation energy needed, making it possible for reactions to occur at temperatures which are otherwise energetically unfavourable.

Enzymes catalyse reactions by lowering the energy of activation for a chemical reaction. Thus, they act in a similar way to an increase in temperature *i.e.*, by increasing the proportion of colliding molecules which react. Like inorganic catalysts, enzymes accelerate reactions but do not affect the position of equilibrium.

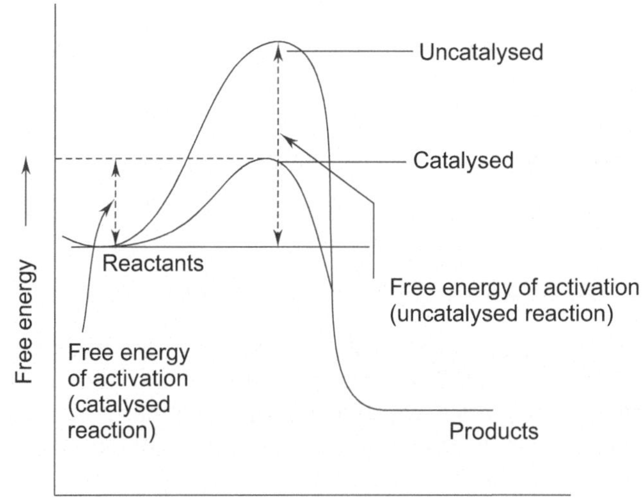

Fig. 3.8 A comparison of activation energy for catalysed and uncatalysed reactions.

It is not always possible to recover an enzyme from the reaction mixture unchanged at the end of the reaction, but the essential feature is that it should not affect the final equilibrium mixture *i.e.*, it does not affect the free energy of the reaction.

Michaelis and **Menten** first postulated that enzymes act like inorganic catalysts and follow the same principle as **Oswald's law of chemical catalysis**. They suggested that the energy 'hump', which must be overcome for reactions to take place, was lower for the enzyme-substrate complex than for the free substrate in the absence of enzyme. The enzyme-substrate complex then breaks down to give the products and free enzyme as shown below:

$$\text{Enzyme + substrate} \rightleftharpoons \text{Enzyme – substrate complex}$$
$$\updownarrow$$
$$\text{products + enzyme}$$

As shown above reactions are reversible. The enzyme catalyses both the forward and reverse reactions. The enzyme does not change the extent of reaction. It only changes the rate at which the reaction comes to an equilibrium. An equilibrium may be shifted towards left or right by the usual factors such as temperature, concentration of the reactants and products or removal of a reactant or product.

The enzyme-catalysed reaction may be represented in the following way:

$$E + S \underset{K_{-1}}{\overset{K_{+1}}{\rightleftharpoons}} ES \xrightarrow{K_2} E + \text{products}$$

This is the Michaelis-Menten rate equation for reactions catalysed by enzymes where K_{+1} is the rate constant for the formation of the enzyme-substrate complex, ES

K_{-1} is the rate constant for the reverse reaction.

K_2 is the rate constant for the formation of the product P.

3.8 HOW DO SUBSTRATES BIND TO ENZYMES ?

We have seen that in an enzyme catalysed reaction, the enzyme binds to the substrate to form a complex. The formation of the complex leads to the formation of the transition-state species, which then forms the product. A substrate binds, usually by non-covalent interactions, to a small portion of the enzyme called the **active-site**. This active site is frequently situated in a cleft or crevice in the protein which consists of certain amino acids that are essential for enzymatic activity.

The catalysed reaction takes place at the active site usually in several steps.

The **first** is the binding of the substrate to the enzyme at the active site.

After the substrate is bound and the transition state is subsequently formed, catalysis can occur. This means that bonds must be rearranged.

In the **third** step, bonds are broken and new bonds are formed, the substrate is converted into the product.

Then the product is released from the enzyme. The enzyme can now catalyse the reaction of more substrate to form more product.

Two important models have been developed to describe the binding process.

A. The Lock and Key Model

This model was proposed by **Emil Fischer** in 1894 to explain how enzymes and substrates interact when mixed. Fischer suggested that enzyme and substrate molecules combine to form an enzyme-substrate complex before the products of the reaction are released. In doing so the substrate must fit a portion of the enzyme *i.e.*, active site in a similar way just as a key must fit a lock in order to open it. As enzyme and substrate molecules are three-dimensional structures, the formation of an enzyme-substrate complex requires that shapes of the reactants and the active centres of enzymes are complementary. Otherwise, the enzyme and substrate cannot unite (Fig. 3.9). It is comparable to the way in which a lock can only be opened by a key of a specific shape.

The specificity of enzymes supports the lock and key model. As different locks have specific keys, similarly specific enzymes act on different substrates.

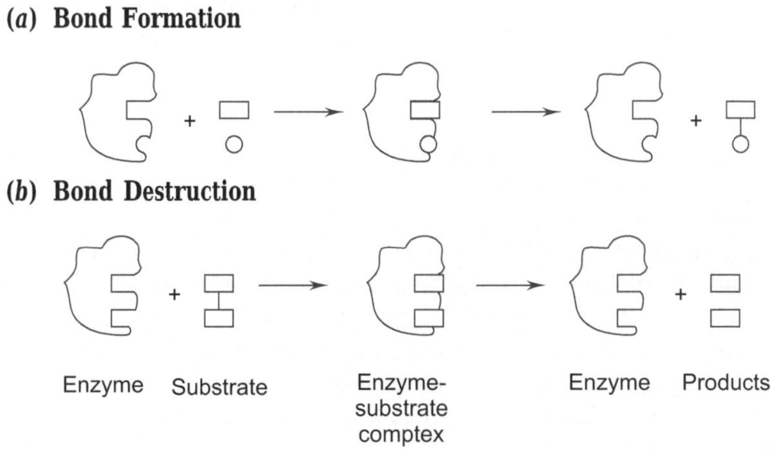

Fig. 3.9 The lock and key model of enzyme action.

B. Induced Fit Model

The above lock and key model is now largely of historical interest because it does not take into account an important property of the proteins *i.e.,* conformational flexibility. This mechanism was proposed by **Koshland**. The induced fit model takes into account the three dimensional flexibility of proteins. According to this model, the shape of the active centre is changed when a substrate molecule binds to such enzymes. The binding of the substrate induces a conformational change in the enzyme that results in a complementary fit after the substrate is bound (Fig. 3.10).

The binding site has a different three-dimensional shape before the substrate is bound. The shape changes after the enzyme-substrate complex is formed. This model gives more satisfactory explanation for the lowered activation energy that occurs with an enzyme catalysed reaction.

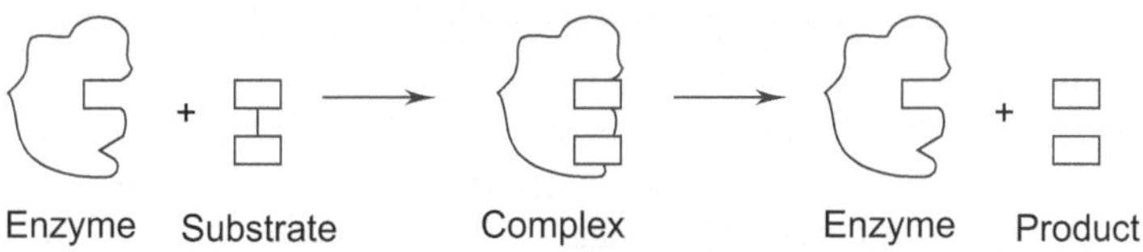

Fig. 3.10 The induced-fit Mechanism.

3.9 ALCHOLIC FERMENTATION

A. Historical Background

Although alcoholic fermentation was known to mankind since prehistoric times, but nothing was known until 1860. During this period, the French Chemist, **GayLussac** had described the production of alcohol from sugar by the following equation :

$$C_6H_{12}O_6 \longrightarrow 2CO_2 + 2CH_3{-}CH_2{-}OH$$

In the middle of the nineteenth century two schools of controversy developed – one group considered the process to be a chemical one, while the other group considered it to be a process due to the activities of living organisms. At that time **Louis Pasteur** was studying how lactic acid is produced in milk and how acetic acid is formed in certain wines. At that time there was no clear concept of enzymes, so he worked only with whole organisms, namely yeast cells. From his experiments he concluded that fermentation takes place anaerobically only in the presence of certain microorganisms. When these organisms were excluded, no fermentation occurred. Thus, fermentation is a physiological process that is closely bound up with the life of cells.

Earlier **Lavoisier** had concluded that all living things need oxygen for respiration. This was contradicted by Pasteur and that fermentation is the result of life activities being carried on in the absence of air. He was also able to show that aerobic metabolism was a much more efficient way to degrade sugars than anaerobic metabolism. We will see under aerobic metabolism that we get much more energy from glucose if oxygen is available.

In 1897, **Buchner** was interested in making protoplasm extracts from yeast to be injected into animals. He wanted to preserve this extract. Ordinary antiseptics could not be used, so he tried the usual kitchen chemistry of adding large amounts of sugar. Sucrose was rapidly fermented by the yeast juice. So, he observed fermentation in the presence of living cells for the first time. Yeast was later found to be capable of fermenting a large number of sugars, such as glucose, fructose, mannose, sucrose and maltose. Glucose was converted by the juice into ethyl alcohol and carbon dioxide according to Gay-Lussac equation.

In 1905, **Harden** and **Young** studied the activity of yeast juice. When they added fresh yeast juice to a solution of glucose at pH 5, fermentation began almost at once. The rate of carbon dioxide soon fell off, they restored it by adding inorganic phosphate. Again the rate of fermentation dropped as percentage of phosphate declined. So, they suspected the formation of an organic phosphate. Soon, they were able to isolate fructose diphosphate, which was fermented very actively by fresh yeast juice. It was concluded by them that this substance was probably an intermediate in alcoholic fermentation.

Robison was able to isolate glucose-6-phosphate and fructose-6-phosphate. Since both of these sugar phosphates could be fermented it became clear that all these sugar phosphates are intermediates in alcoholic fermentation. Later studies proved that fermentation of glucose by yeast juice is very similar to the anaerobic breakdown of glycogen by muscle extracts. Further, Harden and Young found that yeast juice loses its activity when it is dialysed *i.e.*, when the juice is placed in a selective permeable bag surrounded with water. The dialysed juice did not ferment but the activity could be restored by adding the material which had diffused into water. Material outside the bag was found to be quite stable, whereas the contents of the bag could be easily destroyed by heat. The large molecules inside the bag, which we now know, are enzymes and the dialysable factors, that move away from the catalysts inside the bag, are called coenzymes. Thus, yeast juice was thought to contain zymase, which is composed of non-dialysable thermolabile enzyme and a coenzyme which is dialysable and thermostable. Zymase is today known to be a complex mixture of many enzymes.

B. The Overall Pathway in Alcoholic Fermentation

If glucose and inorganic phosphate are added to dialysed juice from yeast no fermentation occurs and no sugar phosphate esters are formed showing that one of the coenzymes must play a role in the reaction. If ATP is added to the dialysed juice, phosphorylation of the sugar is begun and hexose diphosphate can be isolated. In alcoholic fermentation one molecule of glucose is converted to fructose-1, 6-diphosphate, which eventually gives rise to two molecules of pyruvate. Pyruvate loses carbon dioxide producing acetaldehyde which in turn is reduced to give ethanol.

All these reactions can be divided into three phases.

C. Conversion of Glucose to Glycoraldehyde-3-phosphate

It involves first five steps :

Step 1. Glucose is phosphorylated to give glucose-6-phosphate

The enzyme that catalyses this reaction is hexokinase. This reaction uses ATP and gets converted to ADP. The term kinase is applied to the class of ATP-dependent enzymes that transfer a phosphate group from ATP to a substrate. The substrate of hexokinase is not necessarily glucose ; rather it can be any one of a number of hexoses such as glucose, fructose and mannose. Mg^{2+} is the cofactor.

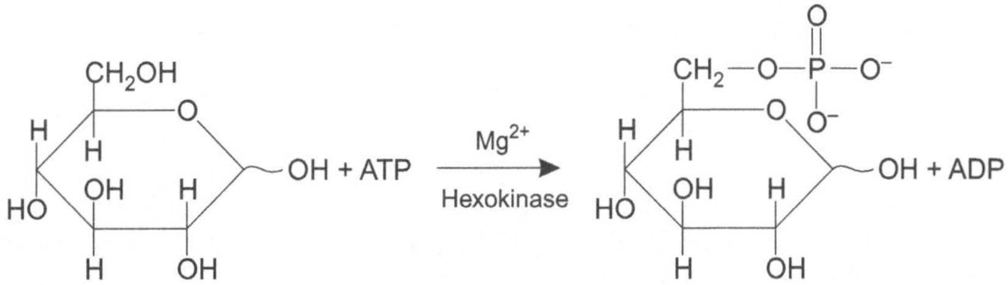

Step 2. Glucose-6-phosphate isomerises to give fructose-6-phosphate

Glucosephosphate isomerase is the enzyme that catalyses this reaction. The C-1 aldehyde group of glucose-6-phosphate is reduced to a hydroxyl group and C-2 hydroxyl group is oxidised to give the ketone group of the fructose-6-phosphate.

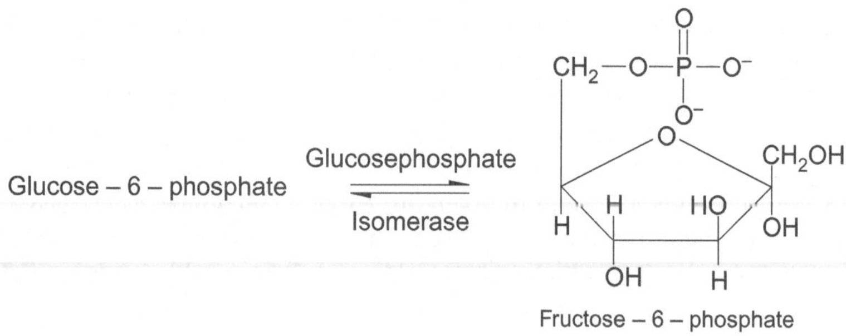

Step 3. Fructose-6-phosphate is further phosphorylated producing fructose-1, 6-bisphosphate

The reaction takes place in the presence of ATP. Phosphofructokinase is the enzyme and Mg^{2+} acts as cofactor. After fructose-1,6-bisphosphate is formed from the original sugar, no other pathways are available and the molecule must undergo the rest of the reactions of alcholic fermentation. This reaction is highly exothermic and irreversible.

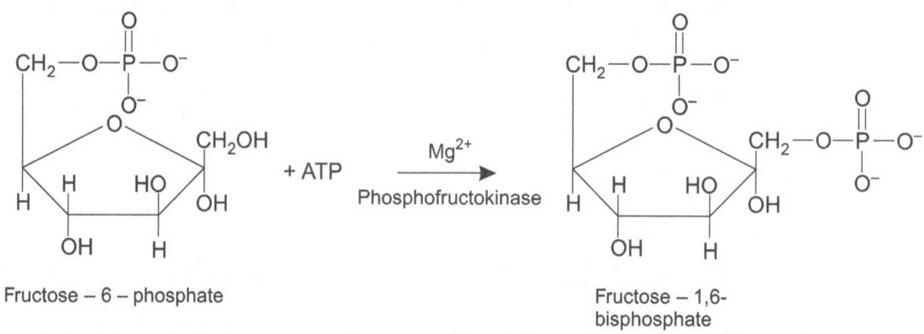

Step 4. Fructose-1,6-bisphosphate is split into two 3-carbon fragments

The cleavage reaction here is the reverse of an aldol condensation. The enzyme that catalyses it is called aldolase.

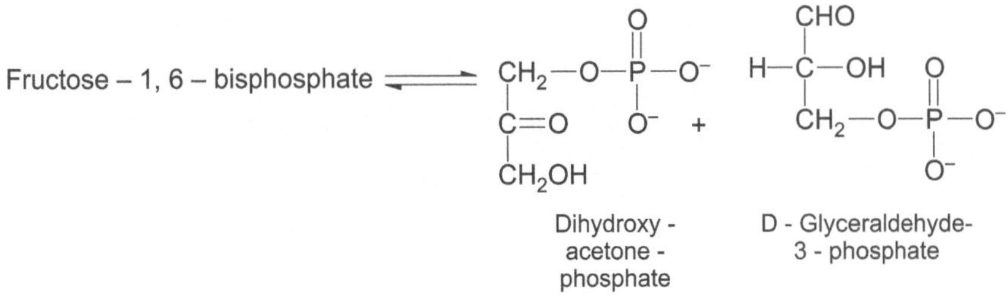

Step 5. The dihydroxyacetone phosphate is converted to glyceraldehyde-3-phosphate

The enzyme that catalyses the reaction is triosephosphate isomerase. One molecule of glyceraldehyde has already been produced by the aldolase reaction. We now have a second molecule of glyceraldehyde-3-phosphate produced in this reaction. The original molecule of glucose, which contains six carbon atoms, has now been converted to two molecules of glyceraldehyde-3-phosphate, which contains 3-carbon atoms.

$$\text{Dihydrooxycetone-phosphate} \xrightleftharpoons[\text{phosphateisomerase}]{\text{Triose}} \text{Glyceraldehyde-3-phosphate}$$

In the first five steps, one molecule of glucose is split into two 3-carbon compounds *i.e.*, D-glyceraldehyde-3-phosphate. Two molecules of ATP are required for these reactions (Fig. 3.11).

D. The Conversion of Glyceraldehyde-3-Phosphate to Pyruvate

Uptil now glucose has been converted to two molecules of glycraldehyde-3-phosphate. We have not seen any oxidation reaction yet, but now we shall encounter them. In this phase of the pathway, ATP is produced instead of being used.

Step 6. The oxidation of glyceraldehyde-3-phosphate to 1,3-bisphosphoglycerate

Here NAD^+ is the coenzyme. The reaction involves the addition of a phosphate group to glyceraldehyde-3-phosphate as well as an electron-transfer reaction from glyceraldehyde-3-phosphate to NAD^+ *i.e.*, NAD^+ is reduced to NADH

$$NAD^+ + 2H^+ + 2e \longrightarrow NADH + H^+$$

The enzyme that catalyses this conversion is glyceraldehyde-3-phosphate dehydrogenase.

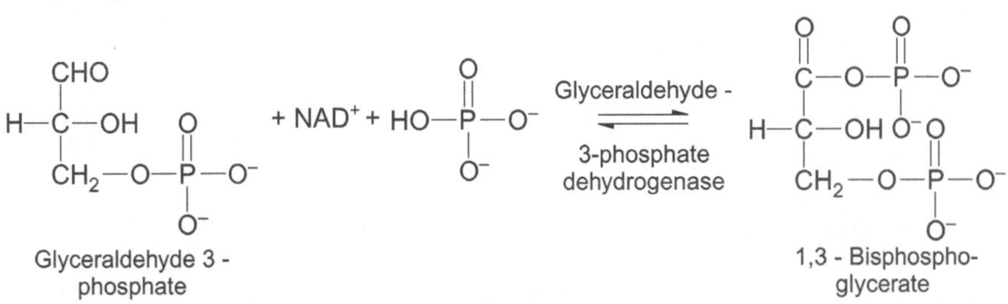

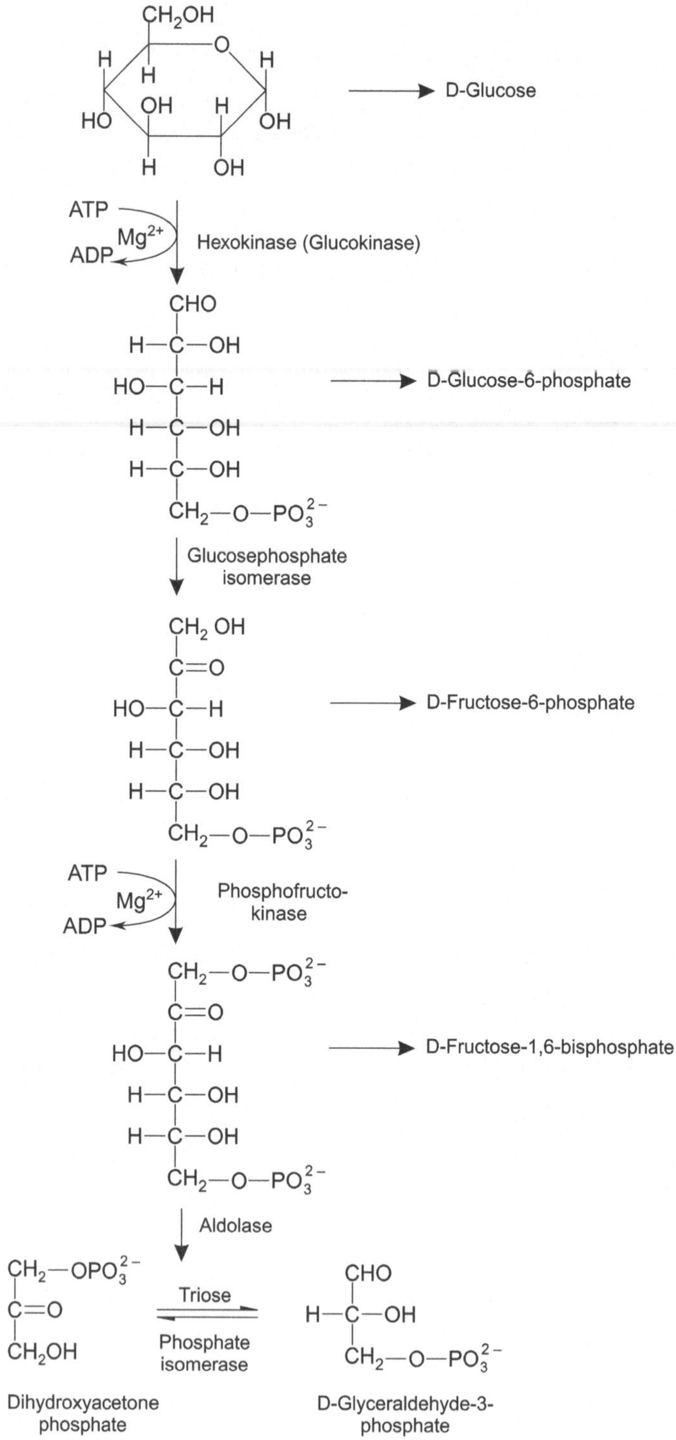
D-Glucose
ATP
Mg2+
ADP
Hexokinase (Glucokinase)
D-Glucose-6-phosphate
Glucosephosphate isomerase
D-Fructose-6-phosphate
ATP
Mg2+
ADP
Phosphofructo-kinase
D-Fructose-1,6-bisphosphate
Aldolase
Triose
Phosphate isomerase
Dihydroxyacetone phosphate
D-Glyceraldehyde-3-phosphate

Fig. 3.11.

Step 7. In this step, ATP is produced by phosphorylation of ADP and 3-phosphoglycerate is formed

The enzyme that catalyses this reaction is phosphoglycerate kinase.

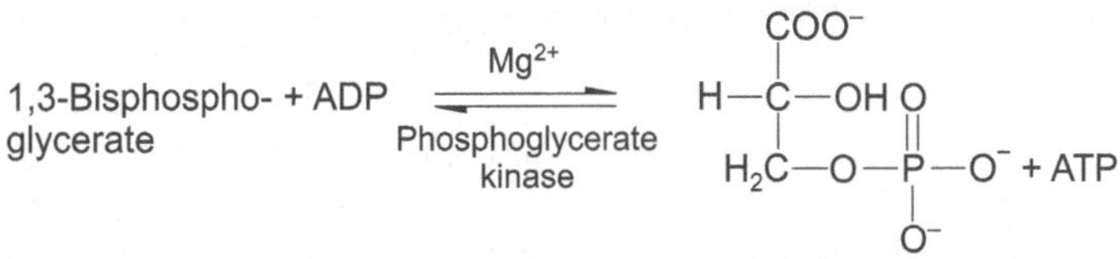

3-Phosphoglycerate

Step 8. The phosphate group is transferred from carbon 3 to carbon 2 of the glyceric acid backbone

The enzyme that catalyses this reaction is phosphoglyceromutase.

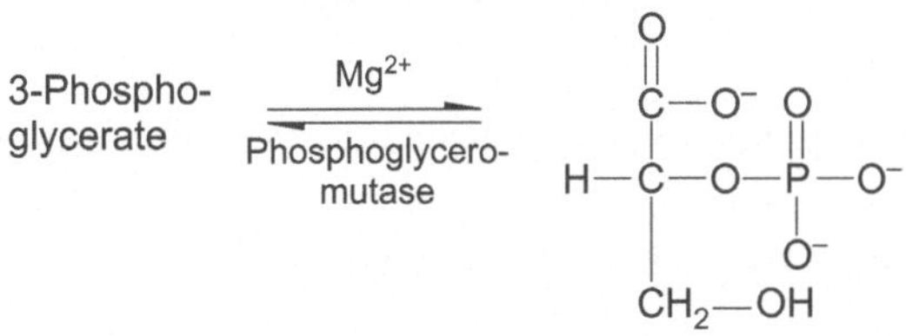

2-Phosphoglycerate

Step 9. The 2-phosphoglycerate loses one molecule of water producing phosphoenol pyruvate

This reaction does not involve electron transfer, it is a dehydration reaction. Enolase, the enzyme that catalyses the reaction, requires Mg^{2+} as a cofactor. The water molecule that is eliminated binds to Mg^{2+} in the course of reaction.

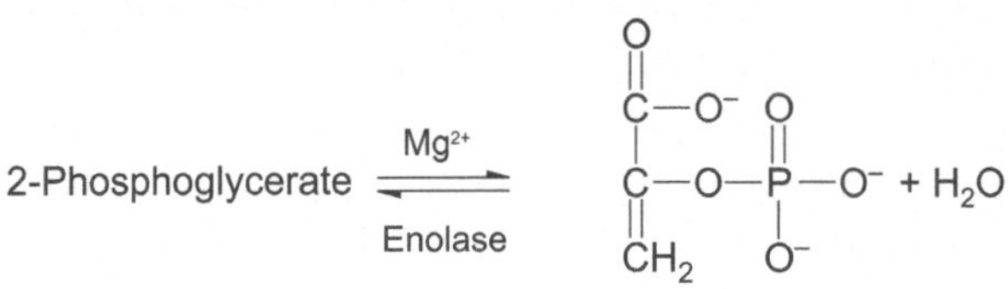

Phosphoenolpyruvate

Step 10. Phosphoenolpyruvate transfers its phosphate group to ADP, producing ATP and pyruvate

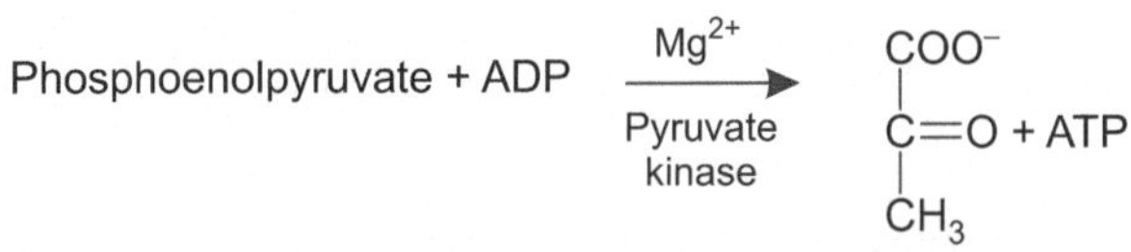

Pyruvate

Here double bond shifts to oxygen on carbon 2 and a hydrogen shifts to carbon 3.

Pyruvate kinase is the enzyme that catalyses this reaction.

Thus, in the second phase of alcoholic fermentation glyceraldehyde-3-phosphate is converted to pyruvate. The reactions are summarised in Fig. 3.12. These reactions yield four molecules of ATP, two for each molecule of pyruvate produced.

E. The Conversion of Pyruvate to Acetaldehyde and then to Ethanol

Pyruvate is decarboxylated to produce acetaldehyde. The enzyme that catalyses this reaction is pyruvate decarboxylase. This enzyme requires Mg^{2+} and a cofactor, thiamine pyrophosphate (TPP) (Thiamine is vitamin B_1). Carbon dioxide splits off, leaving a two carbon fragment covalently bonded to TPP. Thus, acetaldehyde is produced. This step is irreversible.

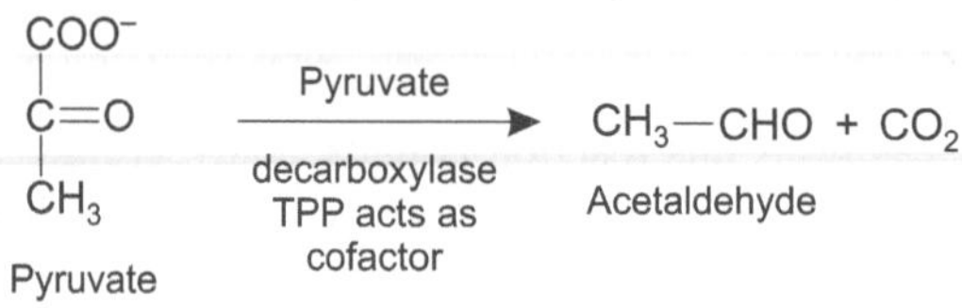

The carbon dioxide produced is responsible for the bubbles in beer and in sparkling wines.

Acetaldehyde is then reduced to produce ethanol and at the same time one molecule of NADH is oxidised to NAD^+, for each molecule of ethanol produced.

$$CH_3CHO + NADH \xrightarrow[\text{dehydrogenase}]{\text{alcohol}} CH_3CH_2OH + NAD^+$$

This step provides recycling of NAD^+ and thus allows further anaerobic reactions.

The net reaction for alcoholic fermentation is

$$\text{Glucose} + 2ADP + 2Pi + 2H^+ \longrightarrow 2\ \text{Ethanol} + 2\ ATP + 2CO_2 + 2H_2O.$$

NAD^+ and NADH do not appear in the net equation. It is essential that recycling of NADH to NAD^+ takes place at this step so that further anaerobic oxidation takes place. Alcohol dehydrogenase is the enzyme that catalyses the conversion of acetaledehyde to ethanol. It is a NADH-linked dehydrogenase.

3.10 GLYCOLYSIS

A. Introduction

We have seen earlier that carbohydrates are the major energy sources. In plants energy is stored in the form of starches, while glycogen is the storage polysaccharide present in animals. Glycogen is found in animal cells in granules similar to the starch granules in plant cells. Glycogen granules are observed in well-fed liver and muscle cells.

Glycogen is a branched-chain polymer of α-D-glucose and in this respect it is similar to the amylopectin fraction of starch. Like amylopectin, glycogen consists

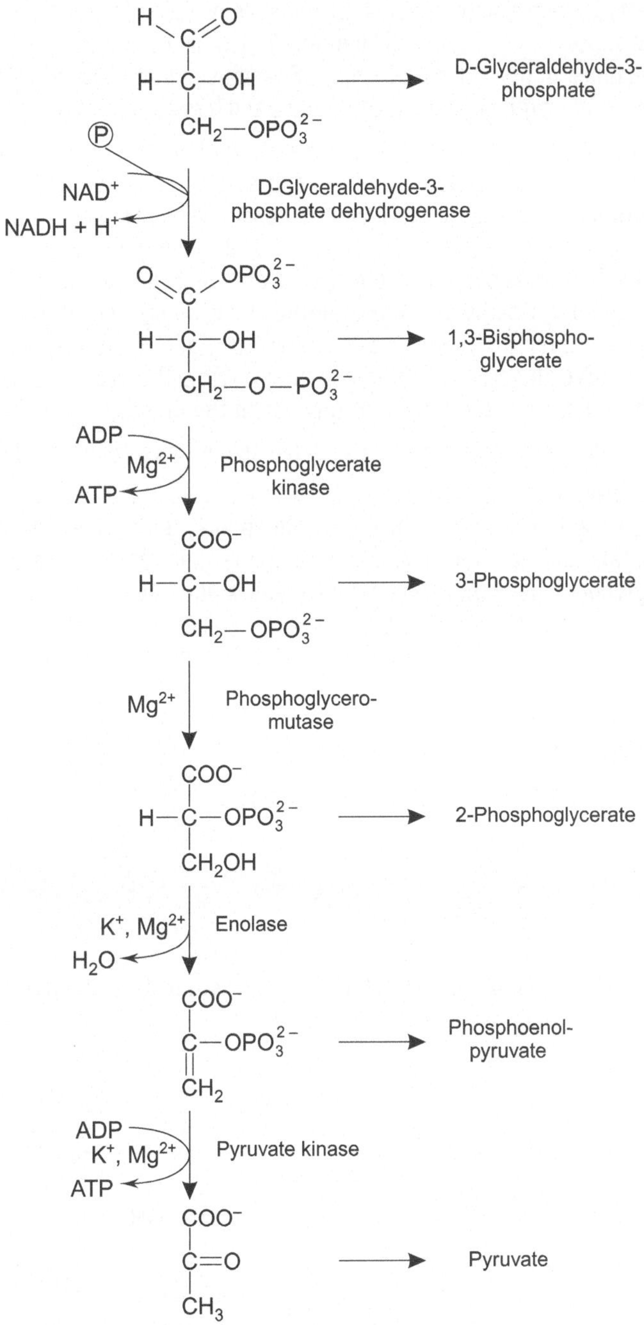
D-Glyceraldehyde-3-phosphate
NAD$^+$
NADH + H$^+$
D-Glyceraldehyde-3-phosphate dehydrogenase
1,3-Bisphosphoglycerate
ADP
Mg^{2+}
ATP
Phosphoglycerate kinase
3-Phosphoglycerate
Mg^{2+}
Phosphoglyceromutase
2-Phosphoglycerate
K^+, Mg^{2+}
H_2O
Enolase
Phosphoenolpyruvate
ADP
K^+, Mg^{2+}
ATP
Pyruvate kinase
Pyruvate

Fig. 3.12.

of α (1 $\rightarrow$ 4) linkages with α (1 $\rightarrow$ 6) linkages at the branch points. However, glycogen is more highly branched than amylopectin. Branch points occur about every 10 residues in glycogen and every 25 residues in amylopectin. In glycogen the average chain length is 13 glucose residues and there are 12 layers of branching.

When we digest a meal high in carbohydrates, we have a supply of glucose that exceeds our immediate needs. Therefore, excess glucose is stored in our body in the form of glycogen primarily in liver and muscle. When we need energy, various degradative enzymes break down glycogen to smaller polysaccharides or free glucose units. In muscle glucose-6-phosphate, obtained from glycogen breakdown, enters the glycolytic pathway directly rather than being hydrolysed to glucose first. However, liver glycogen breaks down to glucose-6-phosphate which is hydrolysed to give glucose. The release of glucose from the liver by this breakdown of glycogen replenishes the supply of glucose in the blood.

The conversion of glycogen to glucose-6 phosphate takes place in three steps.

In the **first reaction** each glucose residue cleaved from glycogen reacts with inorganic phosphate to give glucose-6-phosphate. This cleavage reaction is phosphorolysis and not hydrolysis. The enzyme that catalyses this reaction is known as glycogen phosphorylase, which cleaves one glucose unit at a time from the non-reducing end of a branch of glycogen to produce glucose-1-phosphate and the remainder of the glycogen molecule. This then enters the metabolic pathways of carbohydrate breakdown

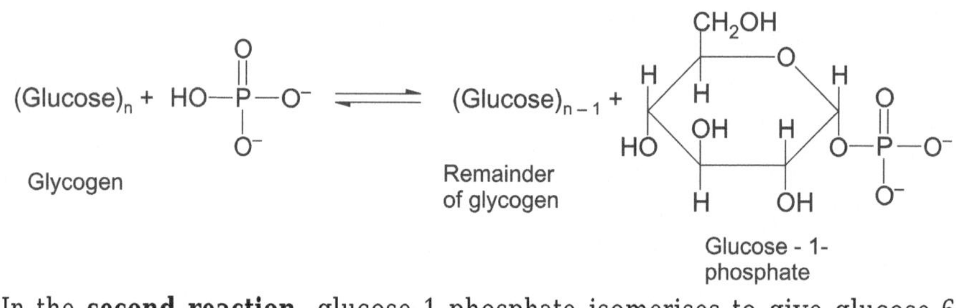

In the **second reaction**, glucose-1-phosphate isomerises to give glucose-6-phosphate by the enzyme phosphoglucomutase.

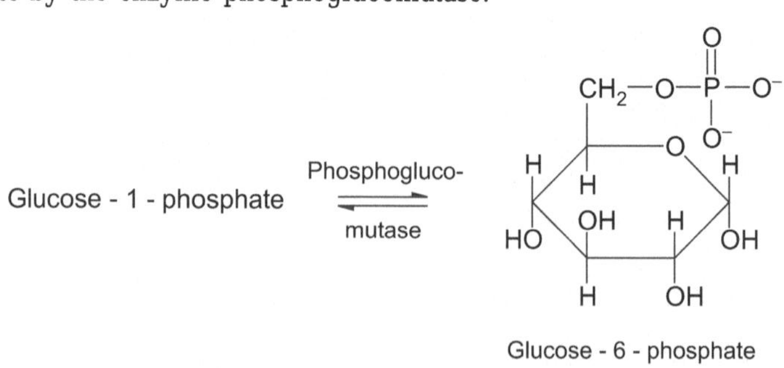

Complete breakdown of glycogen also requires debranching enzymes that degrade the α(1 $\rightarrow$ 6) linkages and thereby expose additinoal 1 $\rightarrow$ 4 glycosidic linkages to be attacked by phosphorylase.

When an organism needs energy quickly, glycogen breakdown is important. Muscle tissue can mobilise glycogen more easily than fat and can do so anaerobically. When we do low-density exercise such as jogging, fat is the preferred food, but as the intensity increases, muscle and liver glycogen becomes more important.

Thus, glycolysis is the first stage of glucose metabolism. It is an anaerobic process and gives only small amount of energy in the form of only two molecules of ATP. We will see that the complete aerobic oxidation of glucose to carbon dioxide and water *via* the citric acid cycle yields high energy equivalent of 30 – 32 molecules of ATP.

B. Overall Pathway in Glycolysis

We have seen the various steps in alcoholic fermentation. After the formation of glucose-6-phosphate, glycolysis and alcoholic fermentation follow a common pathway under anaerobic conditions until pyruvate is formed. In glycolysis, pyruvate is converted to lactic acid, which will eventually be exported from the muscle to the liver.

When pyruvate is formed, it can have one of several fates :-

Under aerobic conditions, pyruvate loses carbon dioxide and the remaining two carbon atoms become linked to coenzyme A as an acetyl group to form acetyl-CoA which then enters the citric acid cycle. There are two fates for pyruvate in anaerobic metabolism.

In organisms capable of alcoholic fermentation, pyruvate loses carbon dioxide producing acetaldehyde, which in turn is reduced to give ethanol.

In the second anaerobic metabolism, lactate is produced, especially in muscle. This is anaerobic glycolysis. The conversion of glucose to pyruvate is simply called glycolysis.

The various steps involving conversion of glucose to pyruvate have already been described under alcoholic fermentation. The same reactions take place in anaerobic glycolysis. We give below the conversion of pyruvate to lactate in muscle.

C. The Conversion of Pyruvate to Lactate in Muscle

This is the final reaction of anaerobic glycolysis *i.e.*, pyruvate is reduced to lactate in the presence of NADH. Lactate dehydrogenase is the enzyme that catalyses this reaction.

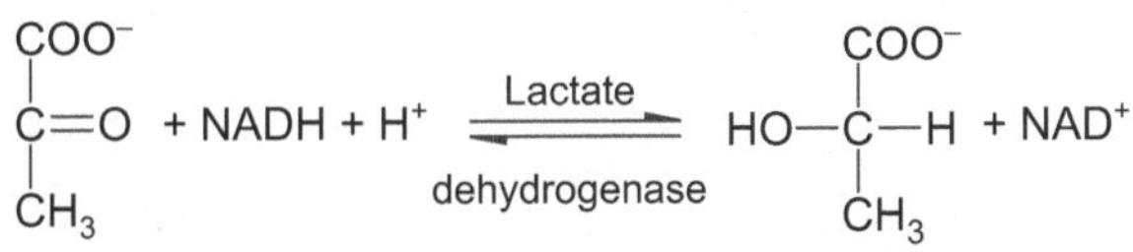

Pyruvate Lactate

The NADH produced from NAD^+ by the earlier oxidation of glyceraldehyde-3-phosphate is used up with no net change in the relative amounts of NADH and

NAD^+ in the cell. This regeneration is required under anaerobic conditions in the cell so that NAD^+ will be present for further glycolysis to take place.

D. Control Points in Glycolysis

Pathways can be shut down if an organism has no immediate need for their products. This saves energy for the organism.

In glycolysis, there are three control points. The first is the reaction of glucose to glucose-6-phosphate which is catalysed by hexokinase. This conversion is inhibited by glucose-6-phosphate itself.

The second control point is the production of fructose-1,6-bisphosphate. This step is catalysed by phosphofructokinase and is inhibited by ATP.

The last point is the conversion of phosphoenol pyruvate to pyruvate, catalysed by pyruvate kinase. This step is also inhibited by ATP (Fig. 3.13).

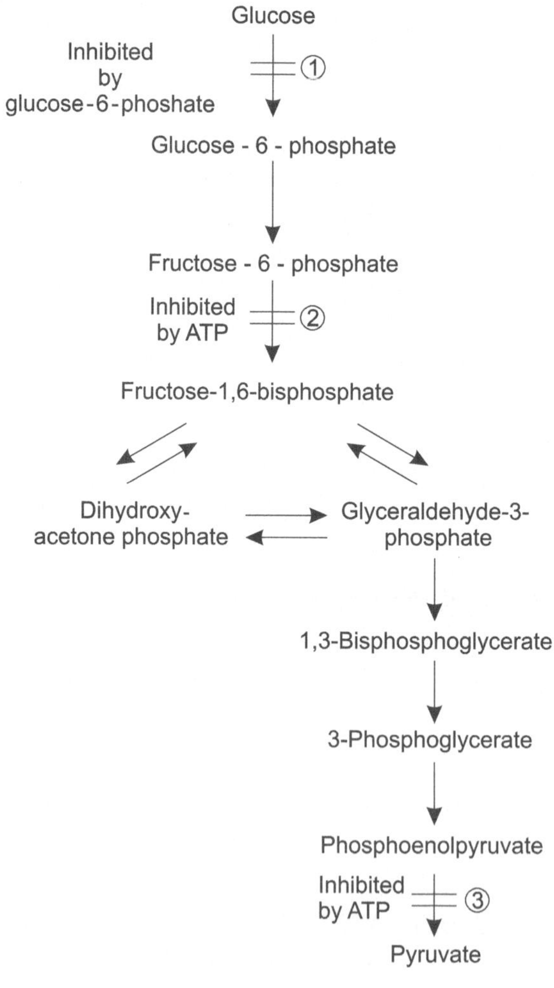

Fig. 3.13 Control Point in Glycolysis.

3.11 CITRIC ACID CYCLE

A. Introduction

We have seen that glycolysis releases only a very small amount of chemical energy which is available in the glucose molecule. Much more energy is released when glucose molecule is oxidised completely to CO_2 and water. This is **aerobic metabolism**. The evolution of aerobic metabolism by which nutrients are oxidised to carbon dioxide and water was an important step in the history of life on earth. Organisms can obtain far more energy from nutrients by aerobic oxidation than by anaerobic oxidation. We have seen earlier glycolysis produces only two molecules of ATP for each molecule of glucose metabolised. However, in complete aerobic oxidation each molecule of glucose produces about 30 to 32 molecules of ATP.

Three processes play roles in aerobic metabolism: the citric acid cycle, electron transport and oxidative phosphorylation. Here we shall discuss only citric acid cycle.

Metabolism consists of catabolism, which is the oxidative breakdown of nutrients and anabolism, which is reductive synthesis of biomolecules. The citric acid cycle is amphibolic *i.e.*, it plays a role in both catabolism and anabolism. The citric acid cycle, as the hub of metabolic pathways, serves to connect the breakdown and synthesis of proteins, carbohydrates and lipids. Most of the metabolites of major nutrients can be fed into the citric acid cycle as acetyl-CoA and then oxidised to produce energy.

The citric acid cycle operates inside the mitochondrion which acts as its engine room. Here, metabolic fuels – glucose derived from carbohydrates, amino acids derived from proteins and fatty acids from lipids – are fed into the cycle and will ultimately be oxidised to carbon dioxide and water. The energy is transferred to electron carriers and finally to the terminal electron acceptor-oxygen.

There are two other common names for the citric acid cycle. One is **Krebs cycle**, named after **Sir Hans Krebs**, who first investigated the pathway and was awarded Nobel Prize in 1953. The other name is **tricarboxylic acid cycle** (or TCA cycle) because of acids with three carboxyl groups.

B. Overall Pathway of the Citric Acid Cycle

The citric acid cycle takes place in mitochondria. Most of the enzymes of the citric acid cycle are present in the mitochondrial matrix.

Pyruvate produced by glycolysis is oxidised to one carbon dioxide molecule and one acetyl group which becomes linked to coenzyme A to give acetyl-CoA. It then enters the citric acid cycle. Two more molecules of carbon dioxide are produced for each molecule of acetyl-CoA that enters the cycle and electrons are transferred in the process. The electron acceptor in all cases but one is NAD^+, which is reduced to NADH. In one case the electron acceptor is FAD (flavin adenine dinucleotide). This takes up two electrons and two hydrogen ions to produce $FADH_2$. The electrons are passed from NADH and $FADH_2$ through several stages and the final acceptor is oxygen with water as the product.

There are eight steps in the citric acid cycle (Fig. 3.14), each catalysed by a different enzyme. Four of the eight steps – steps 3, 4, 6 and 8 – are oxidation steps.

The oxidising agent is NAD^+ in all except step 6, in which it is FAD. In step 5, a molecule of GDP (guanosine diphosphate) is phosphorylated to produce GTP.

In the first reaction of the cycle, the two carbon acetyl group condenses with four-carbon oxaloacetate to produce six-carbon citrate ion. In the next few steps citrate isomerises and then oxidative decarboxylation takes place to give α-keto-glutarate, which again is oxidatively decarboxylated to produce the four-carbon compound succinate. The cycle is completed by regeneration of oxaloacetate from succinate *via* the fumarate in several steps.

C. Conversion of Pyruvate to Acetyl-Co A

We have seen earlier that pyruvate is generated from glycolysis under anaerobic conditions. It moves from the cytosol into the mitochondrion via a specific transporter. Here pyruvate is converted to carbon dioxide and the acetyl portion of acetyl-Co A. An oxidation reaction precedes the transfer of the acetyl group to the Co A.

The whole process involves several enzymes, all of which are part of the pyruvate dehydrogenase complex. The overall reaction is

$$\text{Pyruvate} + \text{Co A—SH} + NAD^+ \longrightarrow \text{Acetyl-CoA} + CO_2 + H^+ + \text{NaDH.}$$

There is a —SH group at one end of the CoA, it is frequently shown in equations as CoA—SH. The reaction is catalysed by four cofactors, TPP, FAD, Mg^{2+} and lipoic acid.

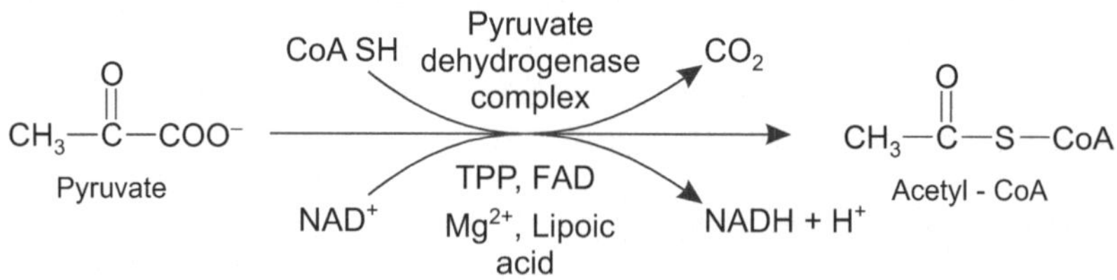

D. The Individual Reactions of Citric Acid Cycle

There are eight steps in the citric acid cycle. These reactions and the enzymes that catalyse them are listed in Table 3.4. Let us study these reactions one by one.

Step 1. Formation of Citrate: This is the first step in the citric acid cycle. It involves the reaction of acetyl-CoA and oxaloacetate to form citrate and CoA-SH.

It is a condensation reaction because a new carbon-carbon bond is formed. The reaction is catalysed by the enzyme citrate synthase

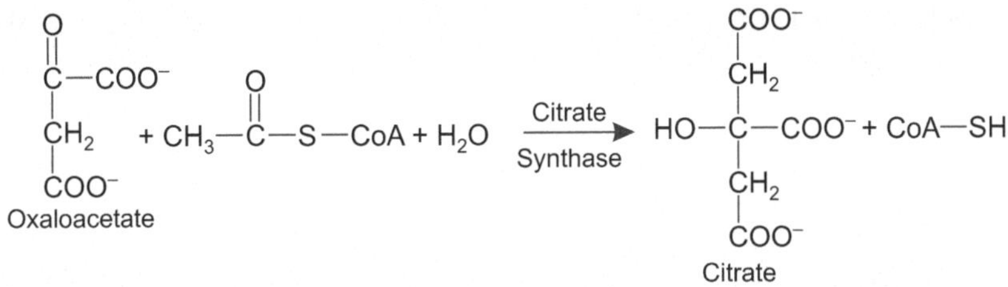

Step 2. Isomerisation of Citrate to Isocitrate: This reaction is catalysed by aconitase. The enzyme requires Fe^{2+} as cofactor. The enzyme is able to select one end of the citrate molecule in preference to the other.

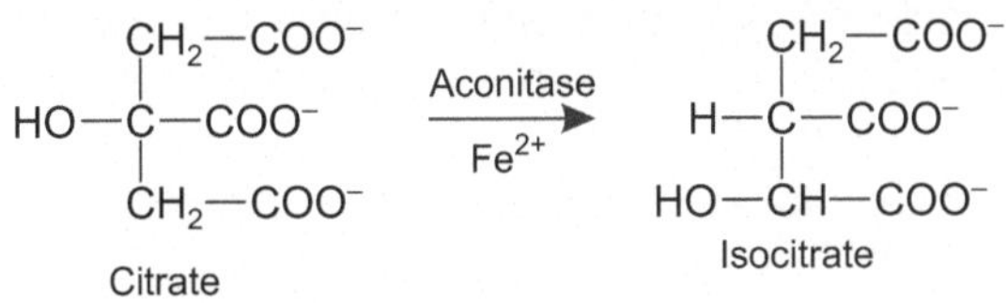

Table 3.4 The Reactions of Citric Acid Cycle

Step	*Reaction*	*Enzyme*
1.	Acetyl-CoA + Oxaloacetate + H_2O → Citrate + CoA-SH	Citrate synthase
2.	Citrate → Isocitrate	Aconitase
3.	Isocitrate + NAD^+ → α-ketoglutarate + NADH + CO2 + H^+	Isocitrate dehydrogenase
4.	α-Ketoglutarate + NAD^+ + CoA-SH → Succinyl-CoA + NADH + CO_2 + H^+	α-Keto-glutarate dehydrogenase
5.	Succinyl-CoA + GDP + P → succinate + GTP + CoA-SH	Succinyl-CoA synthetase
6.	Succinate + FAD → Fumarate + $FADH_2$	Succinate dehydrogenase
7.	Fumarate + H_2O → L-Malate	Fumarase
8.	L-Malate + NAD^+ → Oxaloacetate + NADH + H^+	Malale-dehydrogenase

Step 3. Formation of *α*-ketoglutarate and CO_2 (First Oxidation Step)

This is the oxidative decarboxylation of isocitrate to *α*-ketoglutarate and carbon dioxide. The enzyme that catalyses it is isocitrate dehydrogenase.

This is the first reaction in which NADH is produced.

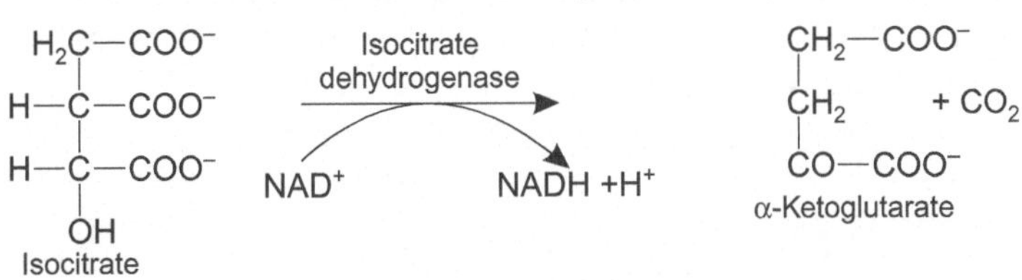

Step 4. Formation of Succinyl-CoA and CO_2 (Second Oxidation Step)

This reaction occurs in several stages and is catalysed by enzyme system called *α*-ketoglutarate dehydrogenase complex. The cofactors used are TPP, FAD, Mg^{2+} and lipoic acid.

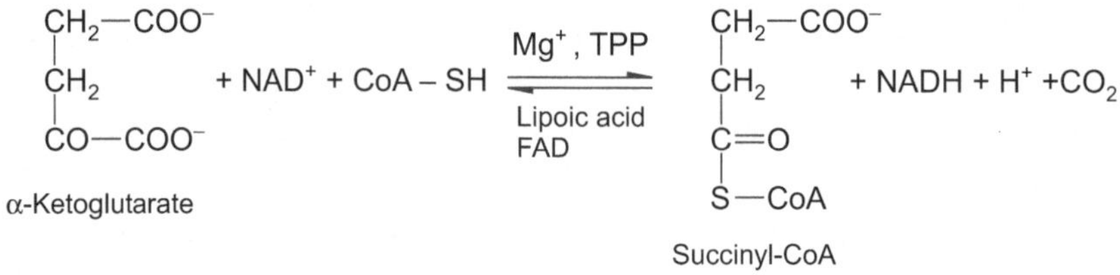

Step 5. Formation of Succinate

In this step, succinyl-CoA is hydrolysed to produce succinate and CoA-SH. The reaction is catalysed by the enzyme succinyl-CoA synthetase. The accompanying reaction is the phosphorylation of GDP to GTP. The energy required for the phosphorylation of GDP to GTP is provided by the hydrolysis of succinyl-CoA to produce succinate and CoA.

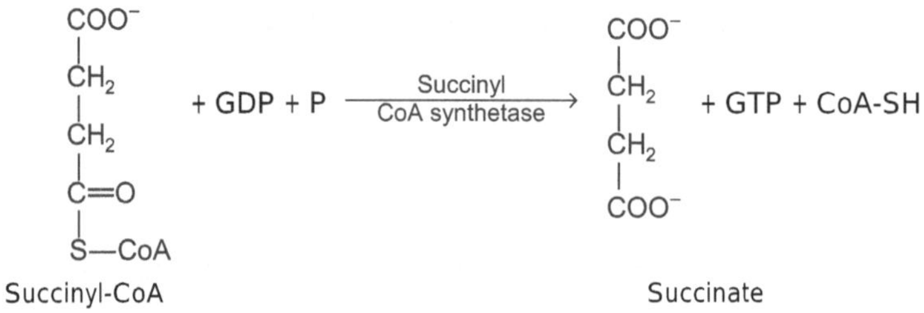

Step 6. Formation of Fumarate (Oxidation-Step)

Succinate is oxidised to fumarate, a reaction that is catalysed by the enzyme succinate dehydrogenase. In this reaction, FAD is the coenzyme which is reduced to $FADH_2$.

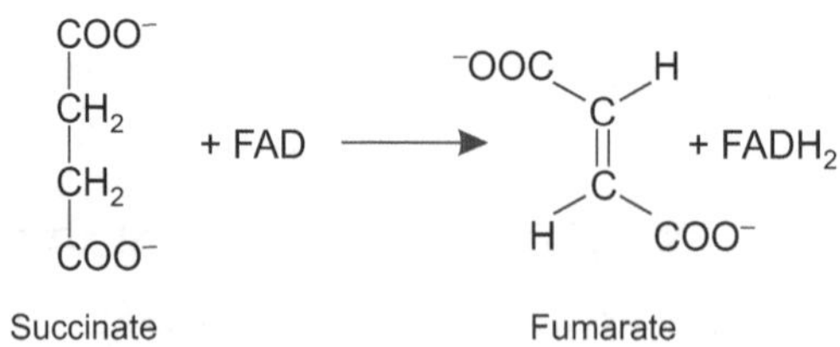

Step 7. Formation of L-Malate

This reaction is catalysed by the enzyme fumarase. Water is added across the double bond of fumarate in a hydration reaction to give malate. Here the reaction is stereospecific. It gives only L-malate.

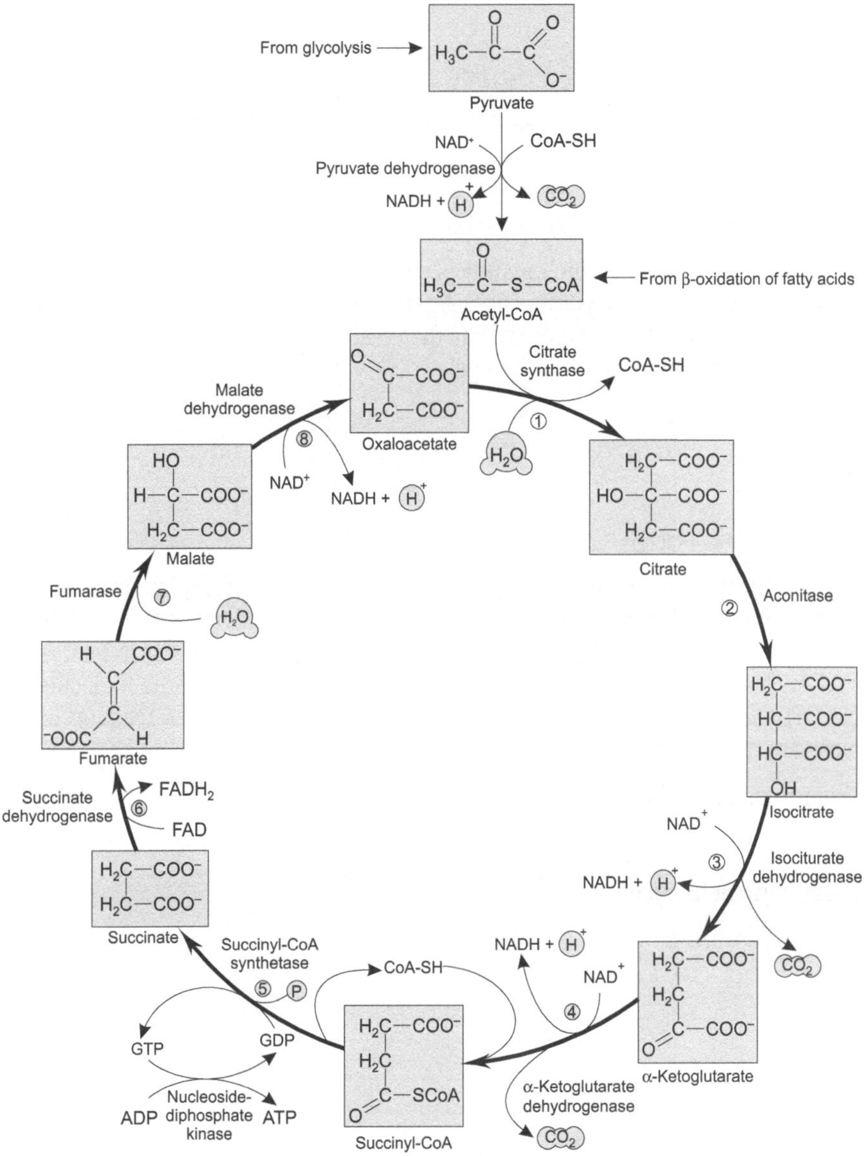

Fig. 3.14 Tricarboxylic acid cycle (Citric acid cycle) Krebs cycle.

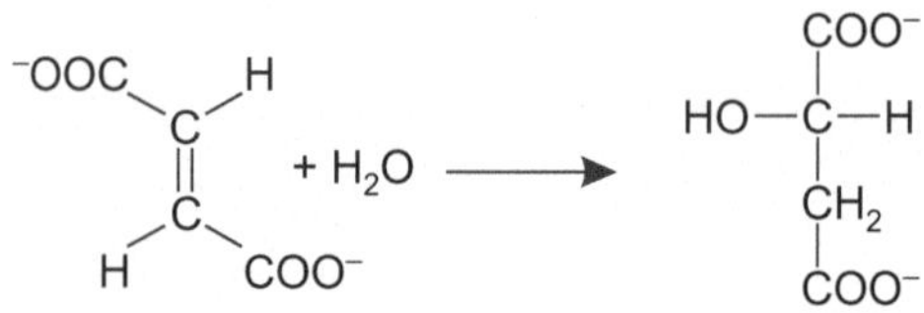

Fumarate L - Malate

Step 8. Regeneration of Oxaloacetate (Final Oxidation Step)

This is the final step in which malate is oxidised to oxaloacetate. The reaction is catalysed by malate dehydrogenase. Another molecule of NAD^+ is reduced to NADH. The oxaloacetate produced can then react with another molecule of acetyl-CoA to start another round of the cycle.

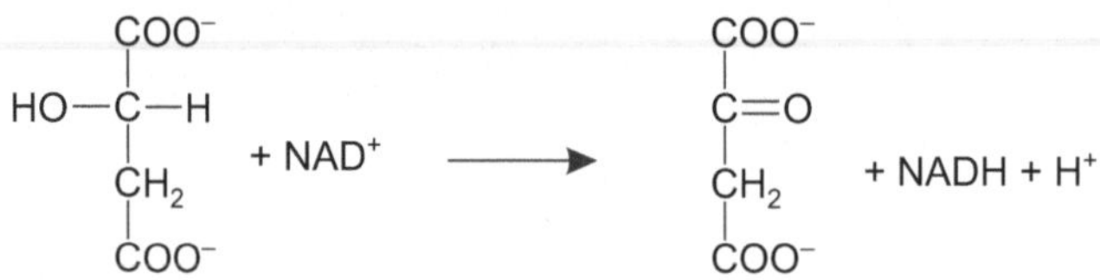

The oxidation of pyruvate by the pyruvate dehydrogenase complex and the citric acid results in the production of three molecules of CO_2. During this period one molecule of GDP is phosphorylated to GTP, one molecule of FAD is reduced to $FADH_2$ and four molecules of NAD^+ are reduced to NADH. Three molecules of NADH come from the citric acid cycle and one comes from the reaction of the pyruvate dehydrogenase complex.

From pyruvate dehydrogenase complex

$$\text{Pyruvate} + \text{CoA-SH} + NAD^+ \longrightarrow \text{Acetyl-CoA} + \text{NaDH} + CO_2 + H^+$$

From citric acid cycle

$$\text{Acetyl-CoA} + 3NAD^+ + FAD + GDP + P + 2H_2O \longrightarrow 2CO_2 + \text{CoA-SH} + 3\text{NaDH} + 3H^+ + FADH_2 + GTP$$

❑❑❑

CHAPTER 4

PYRIMIDINES, PURINES AND NUCLEIC ACIDS

Learning Objectives

In this chapter, we will study:

- Pyrimidines - their aromatic character, chemical properties and synthetic methods.
- Synthesis of uracil, thymine and cytosine.
- Purines - their structure and chemical reactions.
- Synthesis of purines, synthesis of adenine and guanine.
- Ureides
- Uric acid - its structure and synthesis.
- Caffeine - its isolation, structure and synthesis.
- Nucleosides and Nucleotides.
- Nucleic Acids - deoxyribonucleic acid and ribonucleic acid
- Structure of DNA - Conformational variations in DNA, denaturation and renaturation of DNA, DNA replication.
- Principal kinds of RNA and their structures.

4.1 INTRODUCTION

We know nucleic acids play an important role in all biological systems. They help in storing the genetic information and in protein biosynthesis.

There are two types of nucleic acids: ribonucleic acid (RNA) and deoxyribonucleic acid (DNA). Both are biopolymers in which the repeating monomer units are called nucleotides. A nucleotide is made up of one unit each of phosphate, a sugar and a heterocyclic base. The heterocyclic base is either a pyrimidine or a purine.

4.2 PYRIMIDINES

Pyrimidines are the compounds having the same parent heterocyclic nucleus – the pyrimidine (2). Pyrimidine is a purely synthetic compound. It does not occur in nature. Pyrimidine is a diazine. The diazines are a group of compounds formally derived from benzene and napthalene by the replacement of two carbon atoms of a six-membered ring by nitrogen.

The six-membered diazines are known as pyridazine (1), pyrimidine (2) and pyrazine (3). Out of these the pyrimidines are the most thoroughly studied class of compounds.

The corresponding diazines derived from naphthalene are known as cinnoline (4), phthalazine (5), quinazoline (6) and quinoxaline (7).

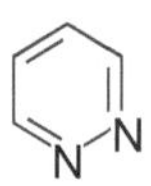

(1)

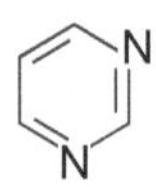

(2)

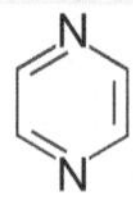

(3)

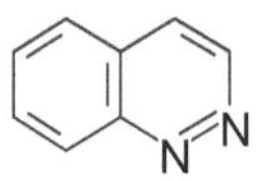

(4)

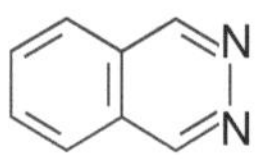

(5)

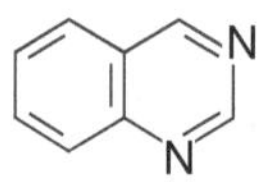

(6)

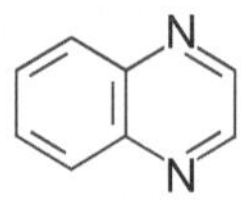

(7)

Pyridazines do not occur naturally. Pyrimidine natural products are quite important. The nucleic acids are essential constituents of all cells and thus of all living matter. The nucleic acids contain pyrimidine and purine bases.

The pyrimidine bases present in nucleic acids are uracil (8), thymine (9), cytosine (10). Synthetic derivatives of barbituric acid (11) are used as hypnotics. An example is that of veronal (12). Vitamin B_1 (13) also contains a pyrimidine ring.

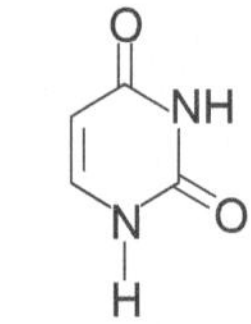

(8)
Uracil

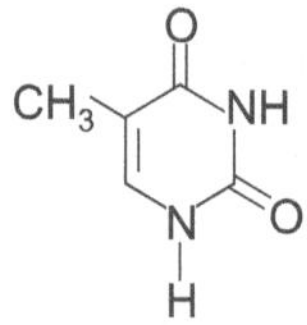

(9)
Thymine

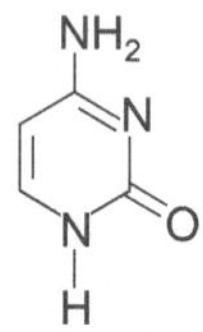

(10)
Cytosine

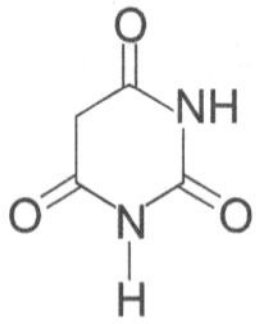

(11)
Barbituric acid

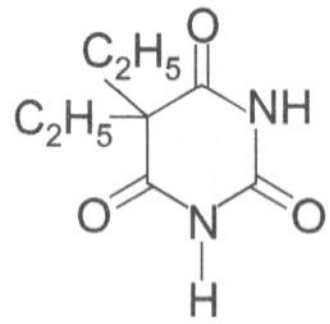

(12)
Veronal

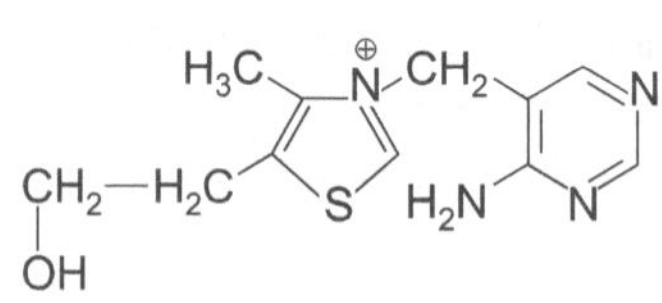

(13)
Vitamin B_1

Luminal is a hypnotic and is used in medicine

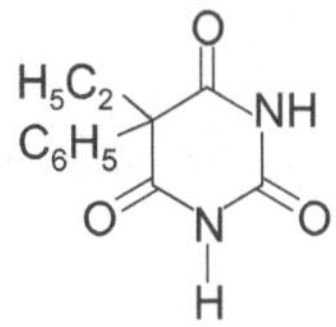

Luminal

The function and structure of a number of enzymes possessing the pyrimidine ring have been established. Purines can also be strictly considered as pyrimidine derivatives.

Deoxyribonucleic acid contains adenine (14) and guanine (15).

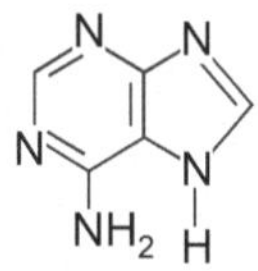

(14)
Adenine

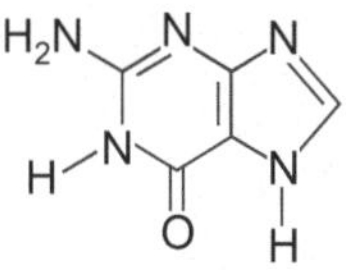

(15)
Guanine

4.3 AROMATIC CHARACTER OF PYRIMIDINE

All the parent diazine ring systems may be regarded as aromatic. This conclusion has been derived from the experimental observations of the C–C and C–N bond lengths which have been found to be intermediate in length between the values expected for the characteristic single and double bond in question.

In case of simple pyrimidine, the dimensions have not be determined yet, although a number of derivatives have been studied in this connection. An example is 4-amino- 2,6-dichloropyrimidine. The bond distances are given in structure (16). Out of the two C–C bond distances (1.40 Å and 1.35 Å) one is very close to C—C distance of benzene (1.39 Å).

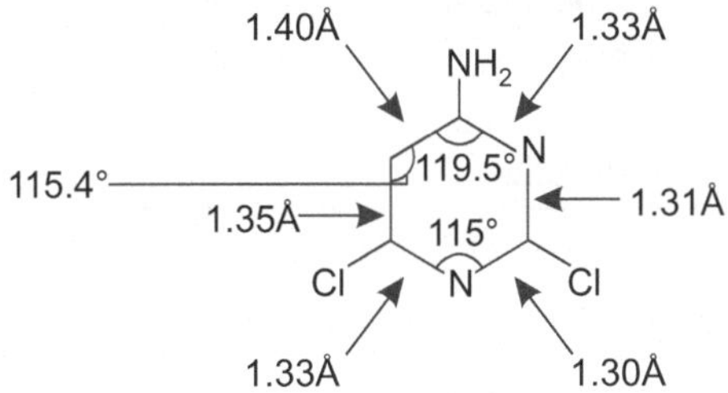

(16)

The general similarity of the size and shape of the pyrimidine ring to that of benzene and pyridine is consistent with its highly aromatic chemical characteristics. Thus, pyrimidine also undergoes the usual electrophilic substitution reactions (p.185). The differences between the bond angles and bond distances between the ring systems suggest that there is maximum distortion in the case of pyrimidine.

This is in agreement with the resonance energies calculated by molecular orbital methods. The values are benzene 36 K cal, pyridine = 31 K cal and pyrimidine = 26 K cal/mole.

Pyrimidine is best considered as a resonance hybrid of eight structures, out of which only two are uncharged structures.The π-electron densities at the 2, 4 and 5-positions have been calculated as 0.776, 0.825 and 1.103 respectively.

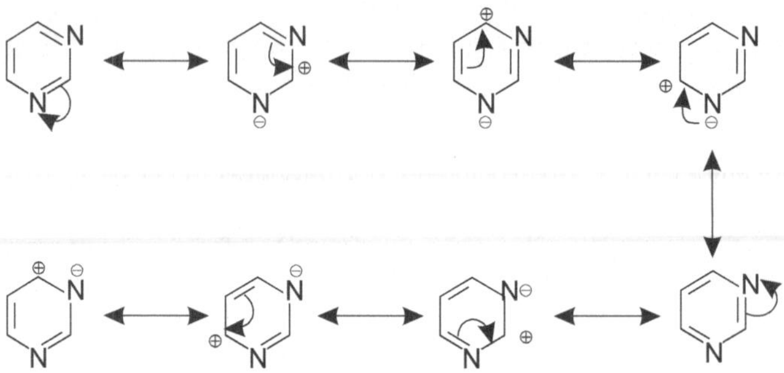

4.4 CHEMICAL PROPERTIES

An enormous amount of work has been done on the derivatives of pyrimidines but the chemical reactions of pyrimidine itself have been investigated only recently. This is partly because pyrimidine was not readily available. However, it can now be obtained quite easily.

1. Basic Character

The diazines, in general are weakly basic because of the added electron deficiency of the second nitrogen which conveys to the heterocyclic system.

The basic strengths of the diazines are given below. It is to be noted that pyrimidine is particularly weaker than pyridazine, whereas pyrazine is the weakest of all these.

Compound	pK_a
Pyridazine	2.3
Pyrimidine	1.3
Pyrazine	0.6

Comparing the basic strenghts of pyrimidine with other heterocyclic ring systems, we find that pyrimidine is much weaker base than pyridine (pK_a = 5.23) or imidazole (pK_a = 7.2)

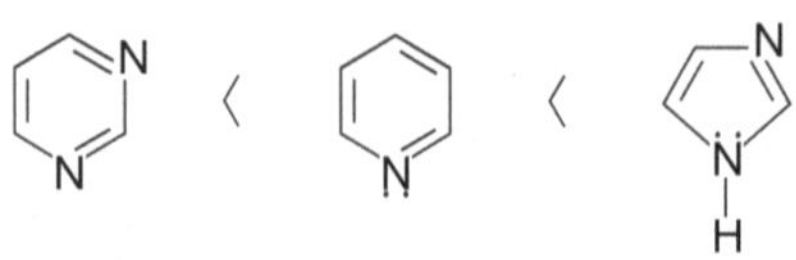

This is due to the fact that addition of a proton does not increase the possibilities of resonance.

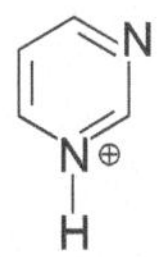

2. Methylation-Formation of Quaternary Salts

The quaternization of these substances has been studied to a limited extent. Only one of the nitrogen atoms of pyrimidine is alkylated by alkylating agents such as dimethyl sulphate, *e.g.*,

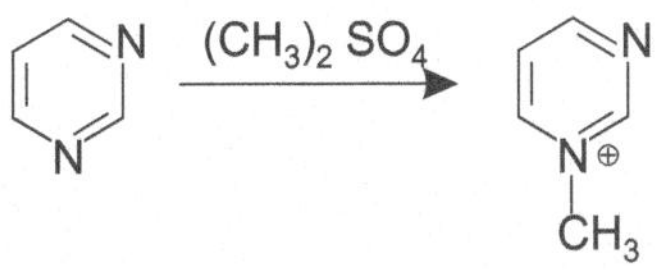

In case of substituted pyrimidines, the reaction with alkylating agents is subject to direction by the attached groups *e.g.*, 4-amino-2-methylpyrimidine is methylated slowly at N – 1.

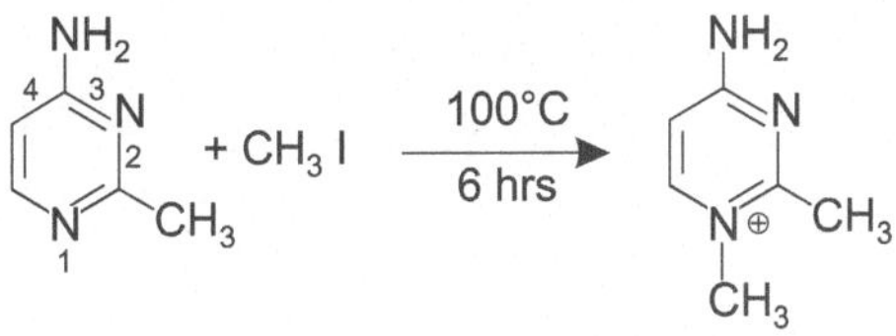

3. Electrophilic Substitution

We know that pyridine is somewhat resistant to electrophilic substitution. The substitution, however, takes place only under drastic conditions. The presence of a second nitrogen atom in the monocyclic diazines would be expected to increase further the electron deficiency of these heterocycles and thus make these compounds pyridazine, pyrimidine and pyrazine extremely resistant to attack by positively charged entities. This unreactivity is quite pronounced and ring systems are often destroyed when vigorous drastic conditions are used.

Pyrimidines are invariably attacked at the 5-position which should correspond to position 3 of pyridine. Thus, pyrimidine hydrochloride is brominated at position 5, but no other electrophilic substitution of pyrimidine itself has been reported. If activating groups such as hydroxyl and amino substituents or even the *N*-oxide are present at other positions (2, 4 or 6) in the molecule, then electrophilic substitution (nitration, nitrosation, diazocoupling etc.) usually occurs but only at position 5. Halogenation can also be effected when one or two activating groups are present.

We can explain this on the basis of stabilities of the intermediate complex when attack is at 4, 5 or 6-positions. The intermediate carbocation, obtained when attack is at 5-position, is more stable than those obtained from attack at the 4- or the 6-position (Fig. 4.1).

Attack at 5-position

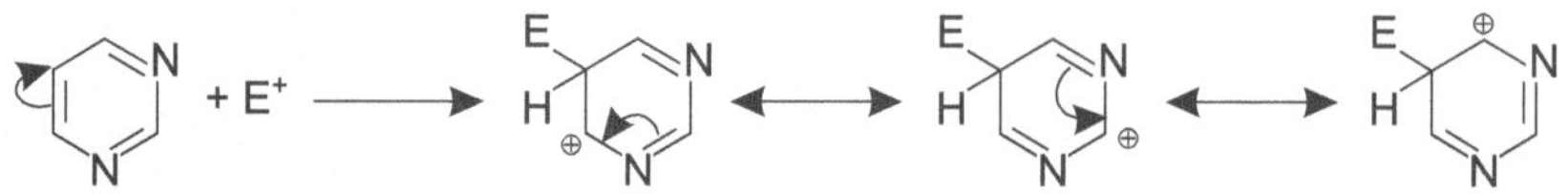

Attack at 4-position

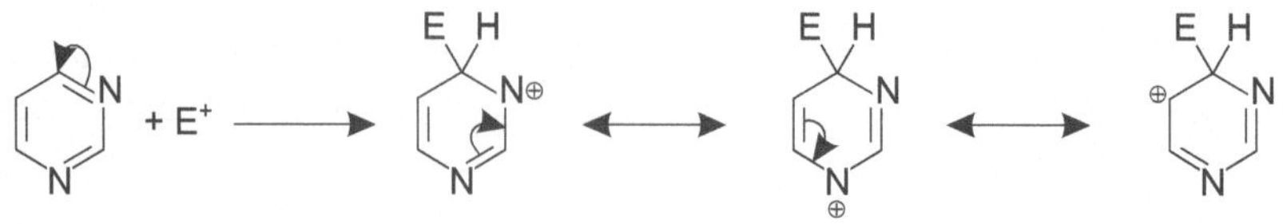

Attack at 6-position

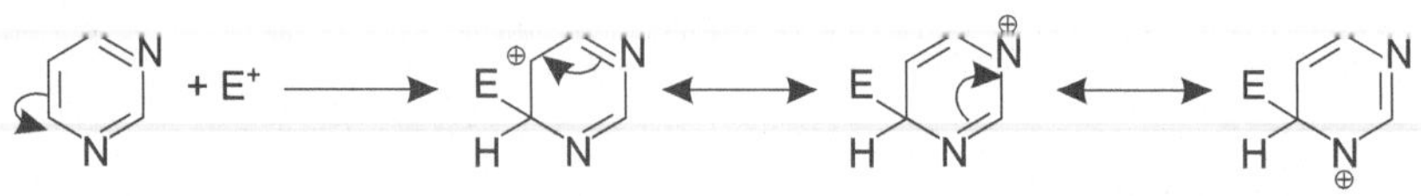

Fig. 4.1 Electrophilic attack at 4, 5 and 6-positions of pyrimidine.

Some examples are:

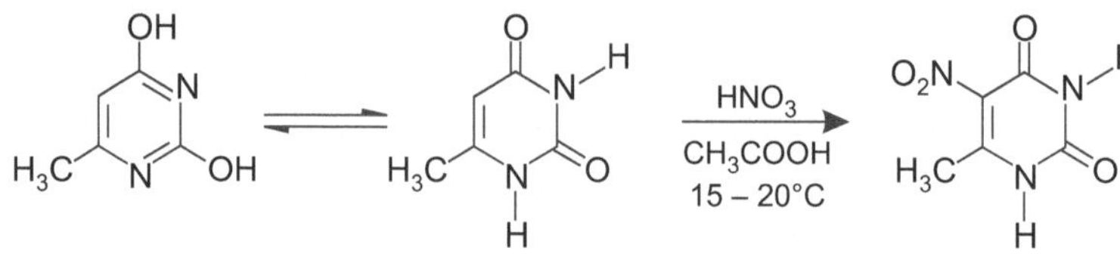

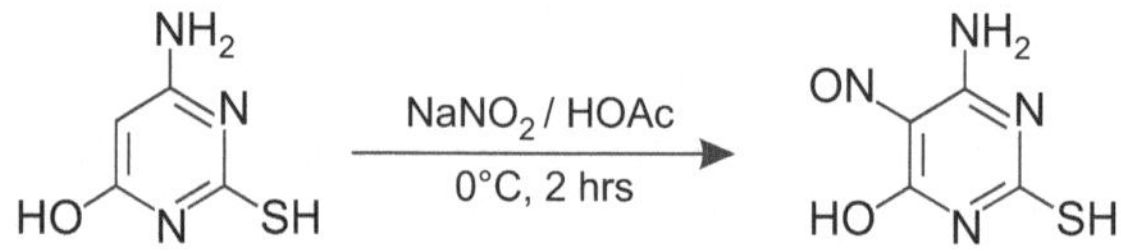

4. Nucleophilic Substitution

We know that in pyridine position 2 is attacked by nucleophilic reagents. Similarly, in diazine series positions α and γ to the nitrogen atoms are reactive towards nucleophilic reagents. This behaviour is similar to the behaviour of the analogous positions in pyridine, quinoline and isoquinoline.

In the few cases investigated, we find that pyrimidines are attacked by nucleophilic reagents such as sodamide and phenyl magnesium bromide. For example,

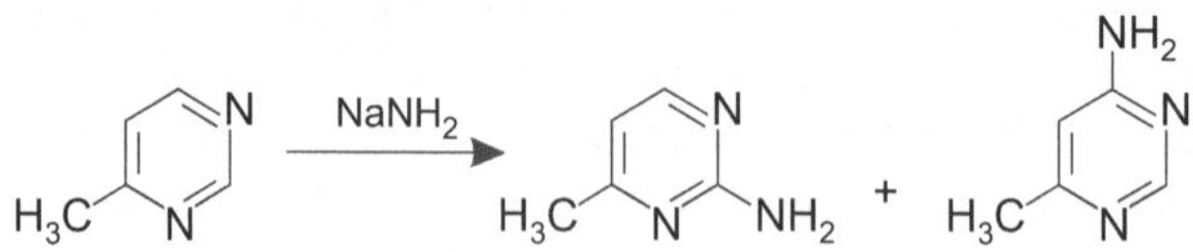

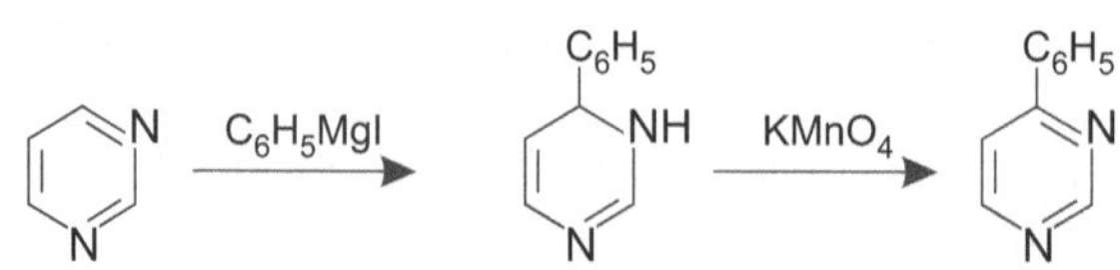

The presence of an activating group facilitates the nucleophilic substitution. Some reactions of uracil (2, 4-dihydroxy pyrimidine) are outlined below :

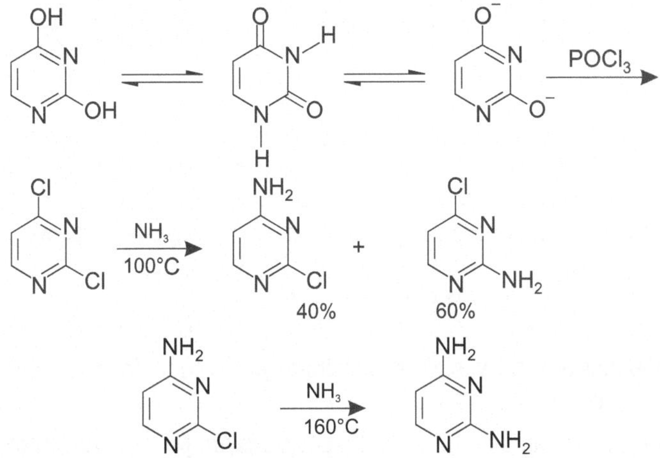

Halogen atoms at 2, 4 or 6-positions can also be replaced by —OH groups on hydrolysis with dilute mineral acid. Replacement by —OCH_3 group gives 50 : 50 mixture.

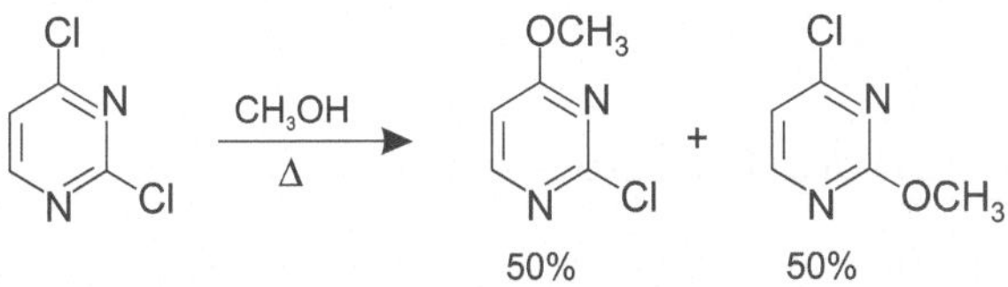

5. *N*-Oxide Formation

Mono-*N*-oxides are known for all the diazines and several of them are known to be capable of di-*N*-oxide formation. Pyrimidine forms only a mono-*N*-oxide which, unlike pyridine-*N*-oxide, has not been nitrated successfully.

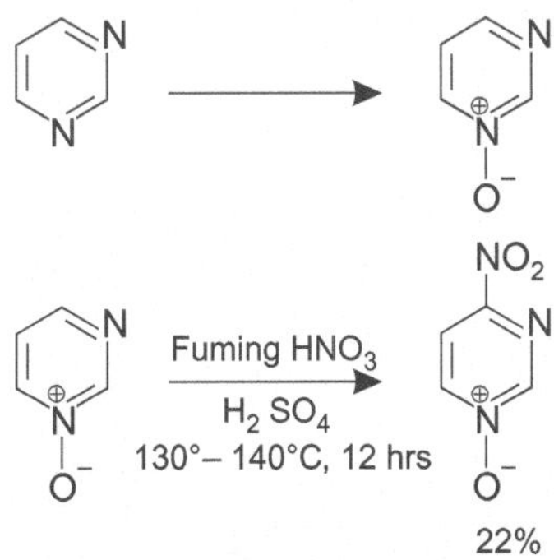

6. Hydrogenation

Hydrogenation of pyrimidine and its derivatives under acid conditions gives 1, 4, 5, 6-tetrahydropyrimidine which has the properties of an aliphatic amidine.

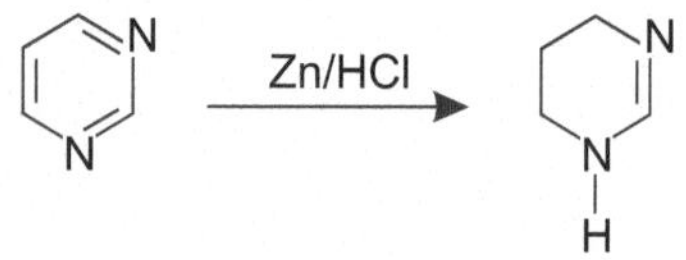

7. Side Chain Reactivity

In general, chemical properties of substituents at the so called activated α- and γ-positions of the diazine molecule display reactions similar to the reactions of pyridine derivatives. For alkyl groups at these sites possess acidic hydrogens and undergo condensation reactions in the presence of appropriate bases or Lewis acids.

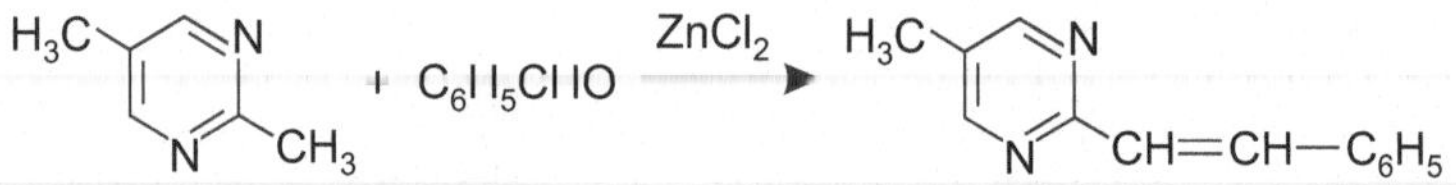

Hydroxy diazines generally exist in the keto forms. Aminodiazines, however, exist predominantly in the amino forms.

Carboxyl groups situated at the α- and γ-positions are prone to decarboxylation at elevated temperature

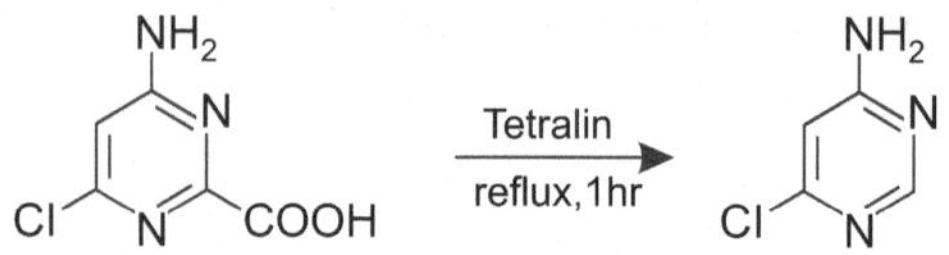

4.5 SYNTHETIC METHODS

1. The most common synthesis involves the condensation of a three carbon unit to a species having a N-C-N linkage. This general method is extremely versatile because of the large variety of molecules which undergo the cyclisation. The 3-carbon unit may be a β-dialdehyde, β-ketoaldehyde, β-ketoester, malonic ester, or β-ketonitrile. The N-containing unit may be a thiourea, amidine, urea or guanidine. For example,

A.

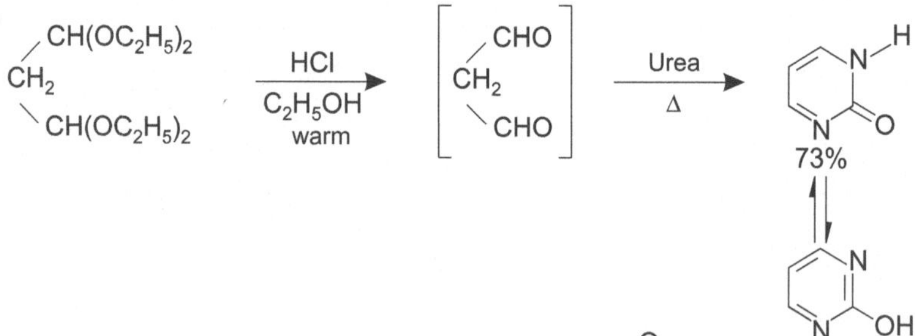

2-Hydroxy-pyrimidine

B.

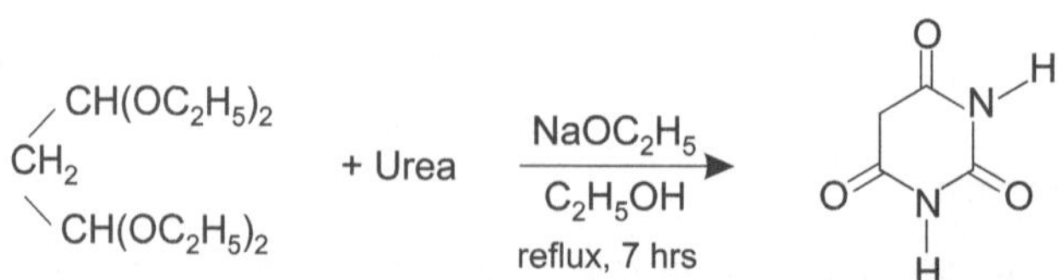

Barbituric acid

c.

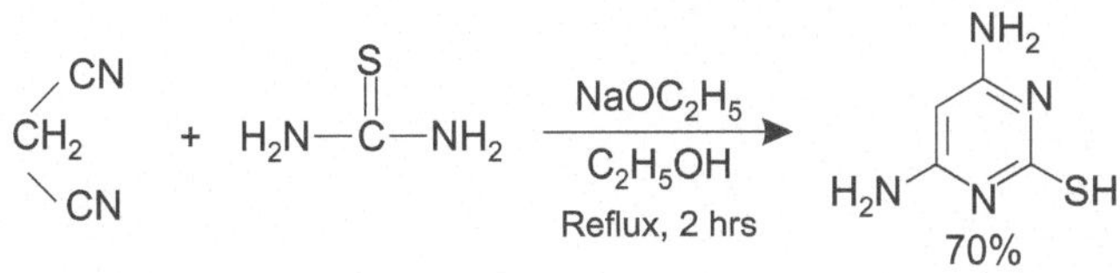

The malonic ester can be replaced by any β-keto ester.

A variation of the general method is to replace the ethyl malonate by ethyl cyanoacetate when the reaction takes place in two stages.

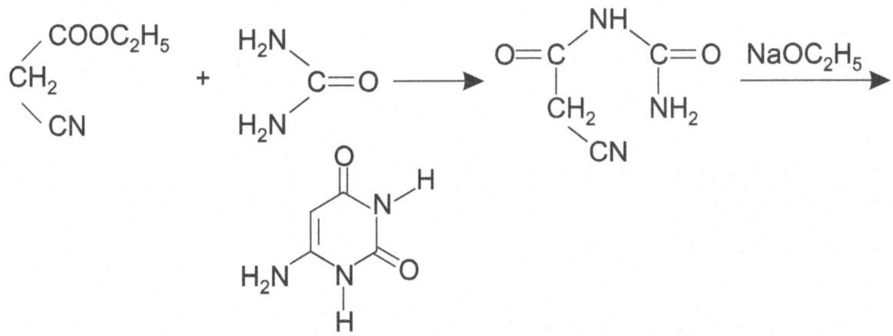

2. A second pyrimidine synthesis, which is of considerable use, involves the condensation of formamide with β-dicarbonyl compounds or their precursors at elevated temperatures. This affords pyrimidines unsubstituted at the 2-position

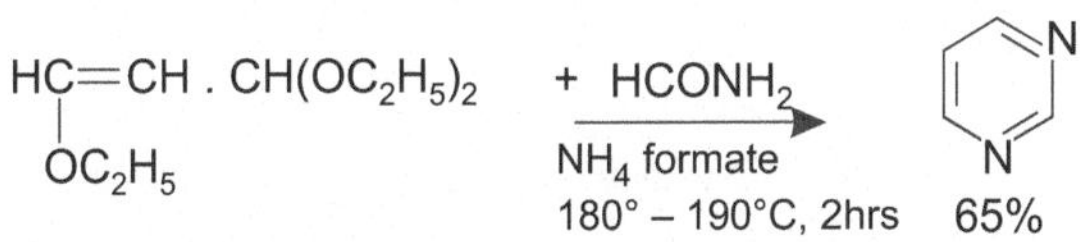

3. The third important method of pyrimidine synthesis consists in the addition of a C–N fragment to a molecule containing the sequence C–C–CN.

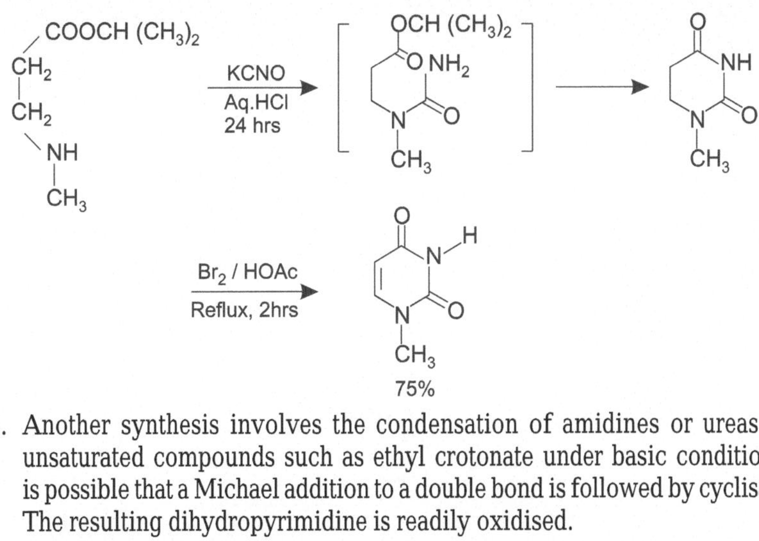

4. Another synthesis involves the condensation of amidines or ureas with unsaturated compounds such as ethyl crotonate under basic conditions. It is possible that a Michael addition to a double bond is followed by cyclisation. The resulting dihydropyrimidine is readily oxidised.

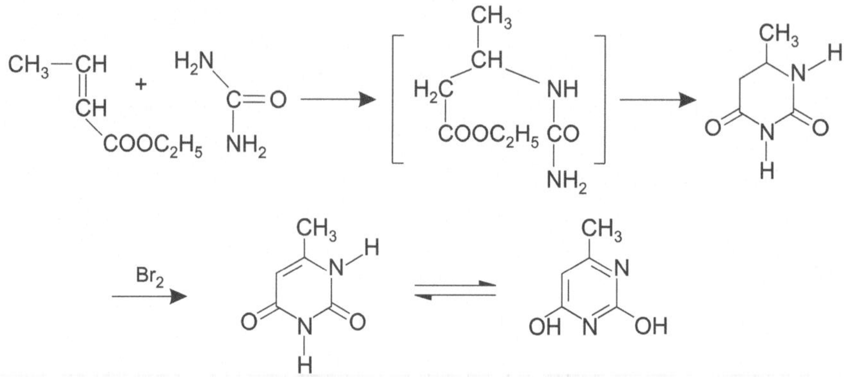

4.6 SYNTHESIS OF URACIL (2, 6-DIHYDROXYPYRIMIDINE)

1. Fischer and Roder Synthesis

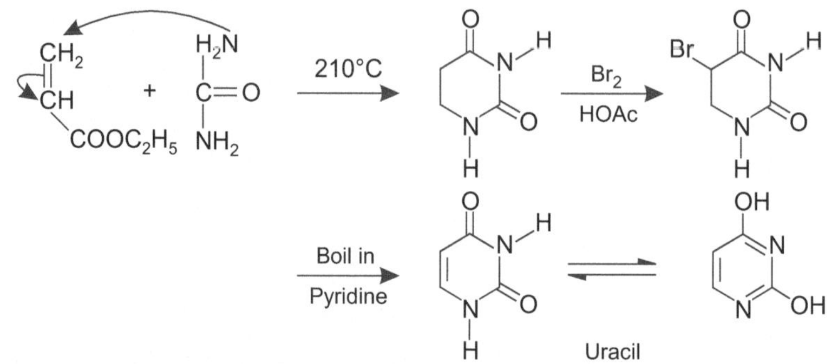

2. Wheeler and Liddle

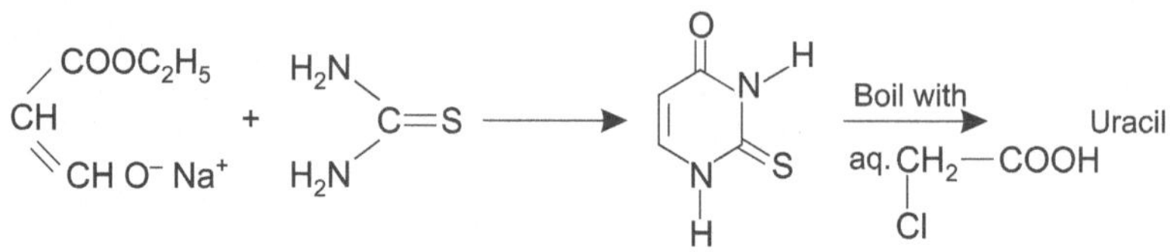

3. Davidson *et al*

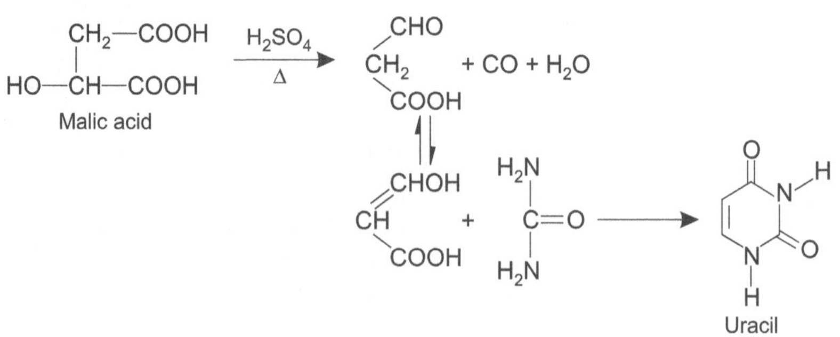

4.7 SYNTHESIS OF THYMINE (5-METHYL URACIL)

1. Fischer and Roder Synthesis

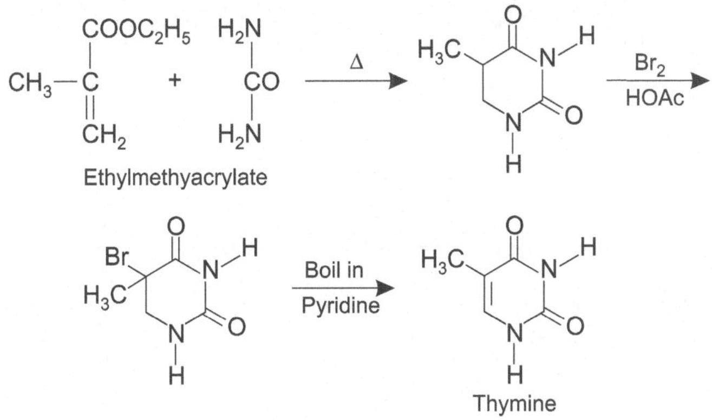

2. Wheeler and Liddle

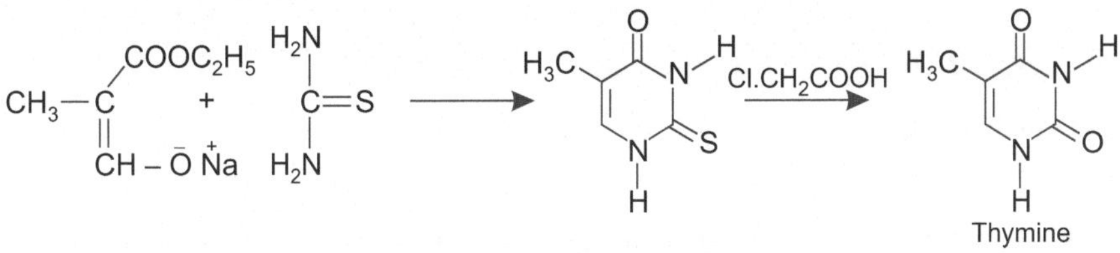

3. Bergmann *et al.*

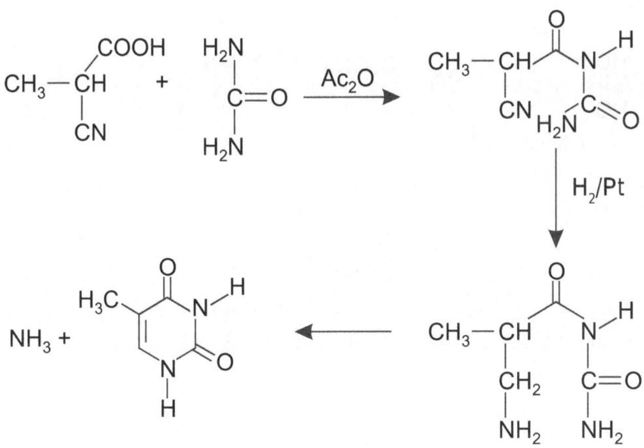

4.8 SYNTHESIS OF CYTOSINE (4-AMINOURACIL)

1. Wheeler and Johnson

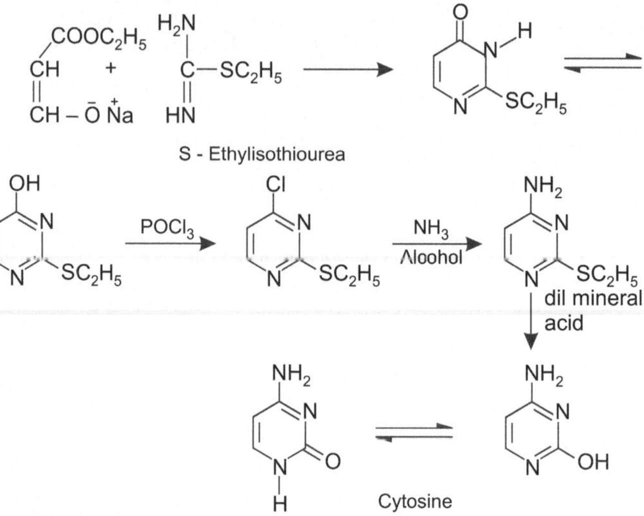

2. Tariso *et al.*

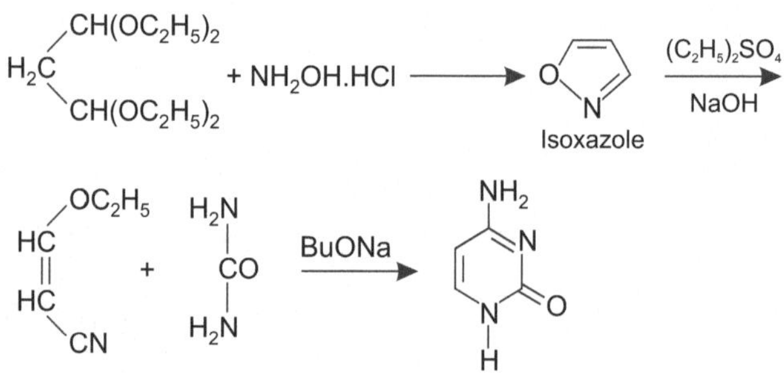

4.9 PURINES

The purine ring system is one of the most important unit present in living systems, but as in the case of pyrimidines the parent compound has not been found in nature. Theoretically, the purine results by fusion of pyrimidine and imidazole nuclei.

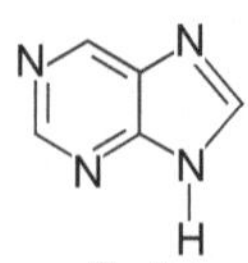

Purine

The purine unit is present in many natural products. Thus adenine (14) and guanine (15) are building blocks of RNA and DNA.

Attachment of either of the two sugar residues D-ribose (17) or 2-deoxy-D-ribose (18) at N-9 of adenine and guanine results in the formation of glycosides known as nucleosides. The structures of adenosine (19) and guanosine (20) are given below:

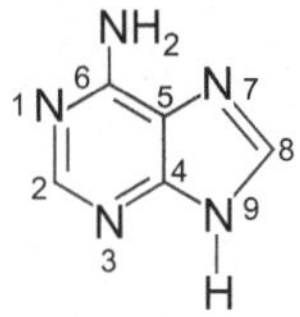

(14)
Adenine
(6-Amino purine)

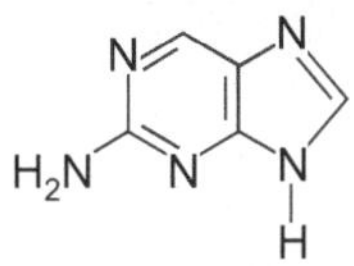

(15)
Guanine

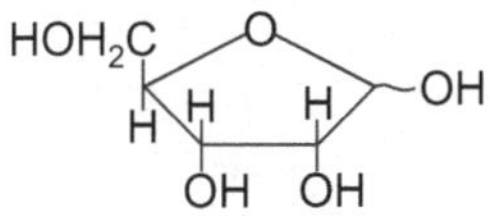

(17)

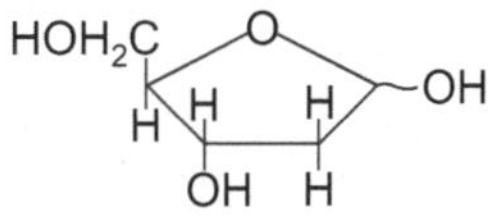

(18)

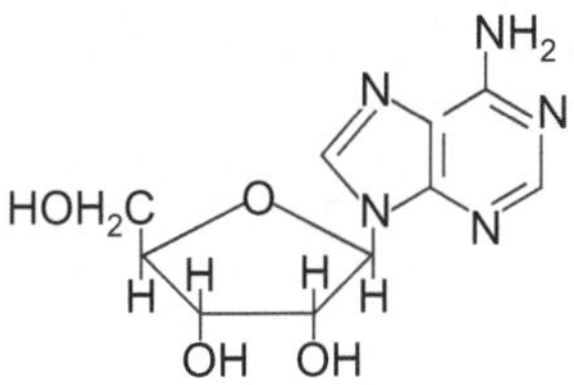

(19)
Adenosine

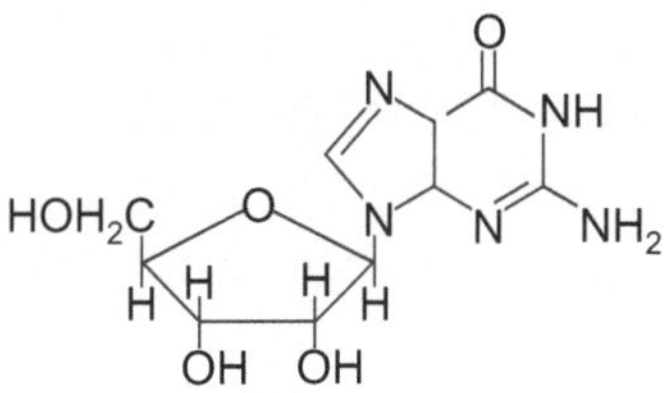

(20)
Guanosine

The purine ring system is also present in a number of coenzymes. The coenzymes adenosine mono-, di- and triphosphate (AMP, ADP, ATP) are biologically important nucleotides.

Several other purine based natural products constitute important components of several plants. Hypoxanthine and xanthine occur in tea extract and animal tissue, whereas caffeine and theophylline are constituents of tea leaves. Theobromine is found in cocoa beans.

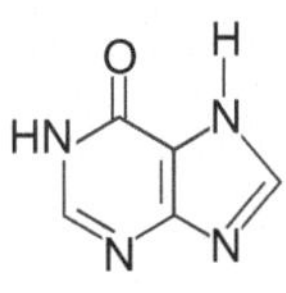

Hypoxanthine

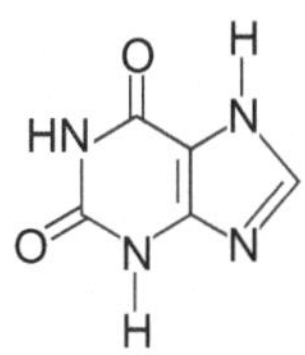

Xanthine

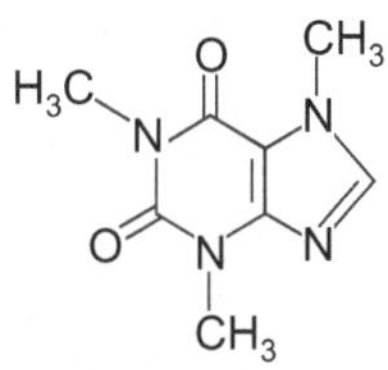

Caffeine

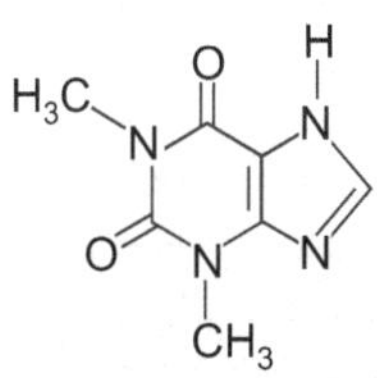

Theophylline

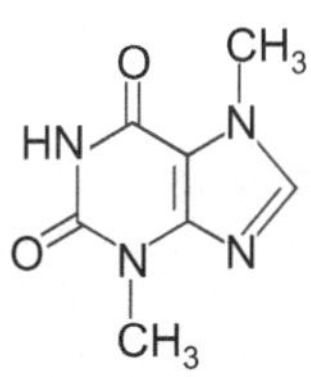

Theobromine

4.10 STRUCTURE AND CHEMICAL REACTIONS OF PURINE AND ITS SIMPLE DERIVATIVES

Purine itself is a colourless compound, melting point 212°C, which is very soluble in water but sparingly soluble in organic solvents. The structure of purine has not been examined by physical methods but dimensions of 6-aminopurine hydrochloride (*i.e.*, adenine) have been calculated by X-ray diffraction techniques. The C–C and C–N distances indicate that purine is aromatic in nature.

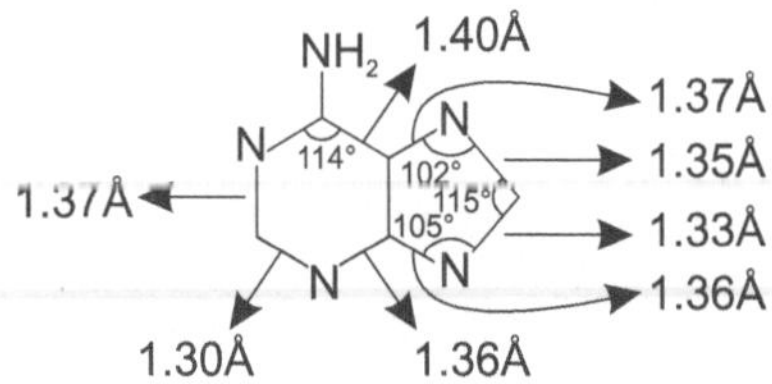

By analogy positions 2 and 6 of purine should be similar chemically to positions 2 and 4 of pyrimidine and position 8 to position 2 of imidazole.

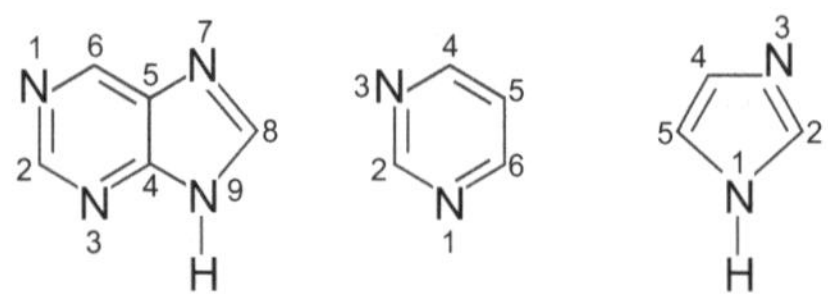

The results of molecular orbital calculations of π-electron densities at various positions of purine indicate that position 8 has the lowest electron density and therefore, would be expected to be least susceptible to electrophilic attack. However, in actual practice this is contrary to this fact.

Position	*Electron density*
2	0.902
6	0.907
8	0.895

Purine does not react with diazonium salts whereas 2, 6-dihydroxy purine couples at position 8 and 2, 8- and 6, 8-dihydroxypurines do not react at all.

Also some amino and hydroxy purines brominate at position 8. π-Electron densities alone can, therefore, be a misleading guide to the positions of attack of a reagent on a molecule.

This is best shown by considering the stabilities of the intermediates obtained after the electrophilic attack at various positions. We find that the intermediate carbocation formed is most stable when the attack is there at position 8 (Fig. 4.2.).

Attack at Position 8

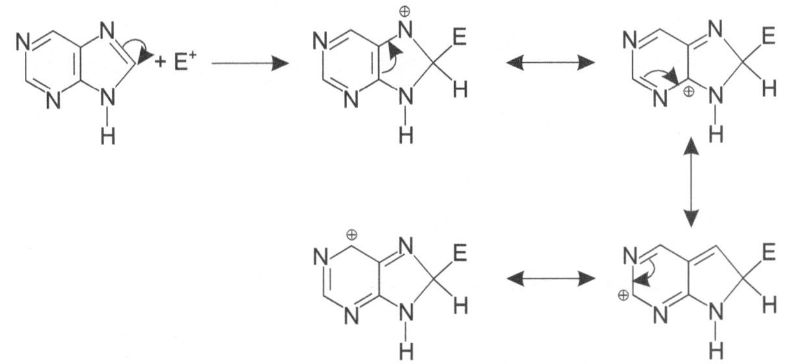

Attack at Position 2

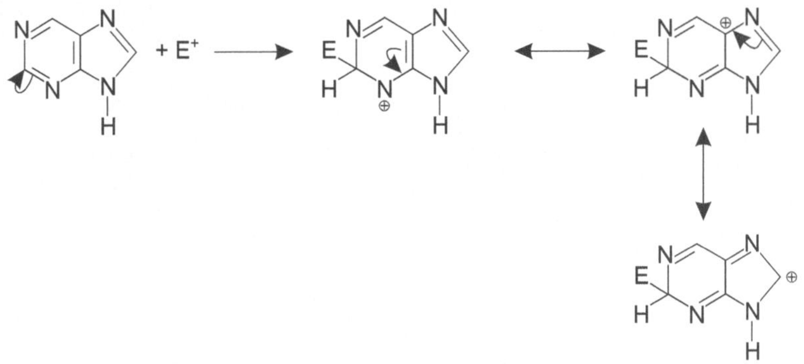

Attack at Position 6

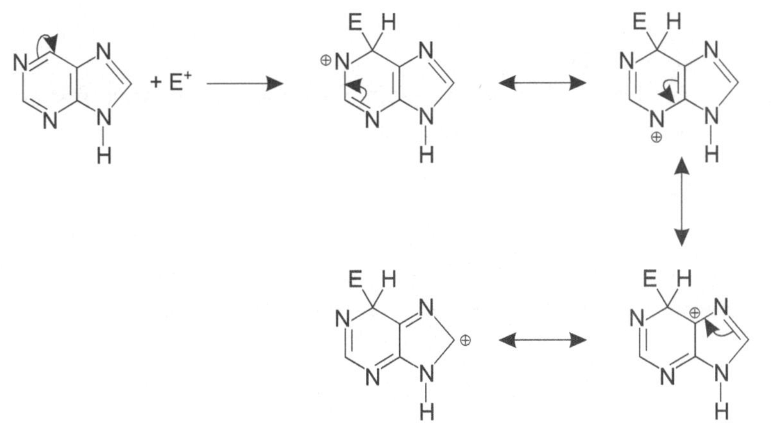

Fig. 4.2 Electrophilic attack at 8, 2 and 6-positions of purine

Basic Character

Purine (pK_a 2.4) is a stronger base than pyrimidine (pK_a 1.3) but weaker than imidazole (pK_a 7.2).

Diazomethane or dimethyl sulphate in the presence of alkali methylates purine in the 9-position.

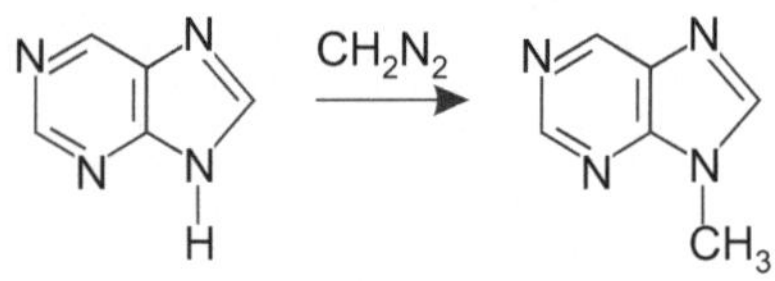

4.11 SYNTHESIS OF PURINE

1. Purine may be synthesised from uric acid as follows:

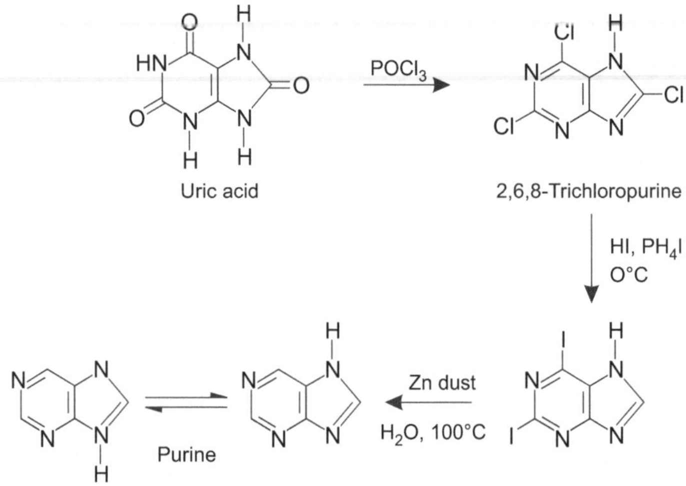

   Catalytic reduction of trichloropurine (H_2-Pd in aqueous sodium acetate) also gives purine.

2. In general synthetic methods, pyrimidine ring is synthesised first and then the imidazole ring is built up on this.

A. Traube's Method

In this 4,5-diaminopyrimidine is synthesised first which is then condensed with formic acid to produce the imidazole ring. When the formyl derivative or its sodium salt is heated, ring closure takes place to give purine.

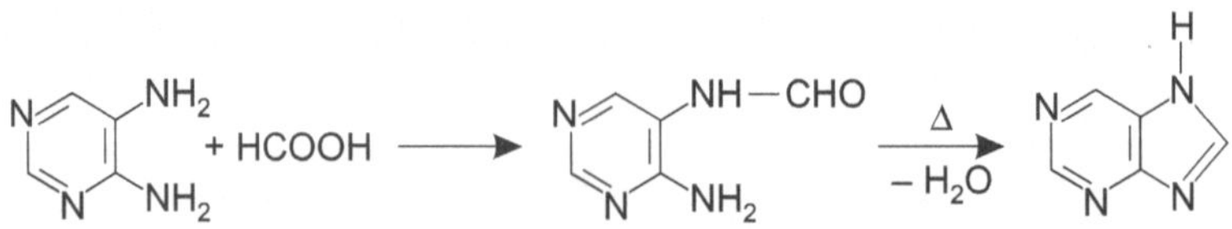

This synthesis leads to the preparation of purines that are unsubstituted in position 8.

8-Hydroxy purines may be prepared by using ethyl chloroformate instead of formic acid. Also, diamino pyrimidines may be fused with urea to produce 8-hydroxy purines.

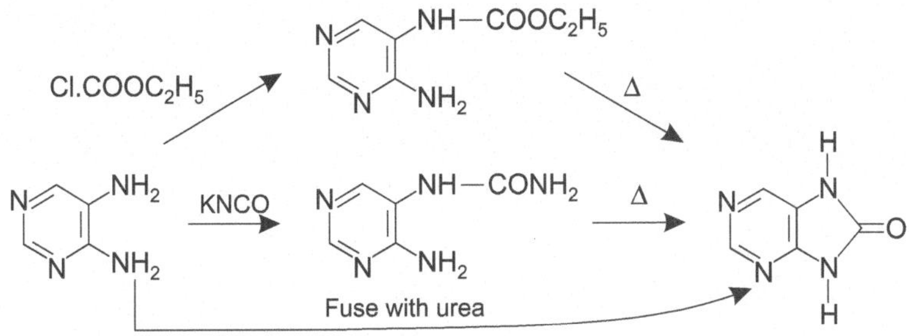

4.12 ADENINE (6-AMINOPURINE)

In addition to nucleic acids, adenine occurs in the pancreas of cattle and in tea extracts. General reactions of adenine are similar to that of purine. On treatment with nitrous acid, it can be converted into hypoxanthine (an oxypurine). Its structure was established by synthesis.

Synthesis of Adenine

1. Fischer's Method

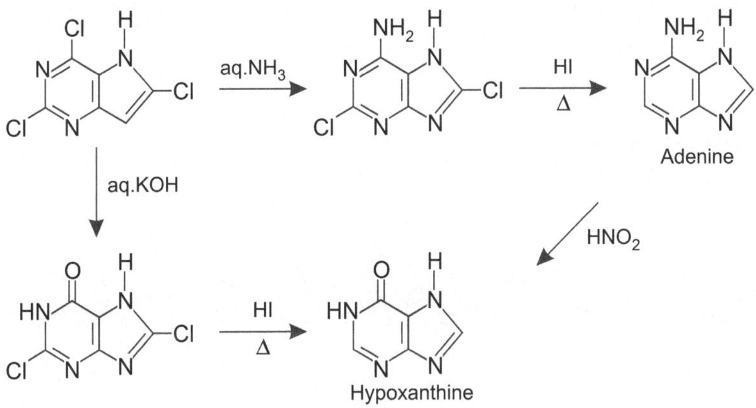

2. Traube Method

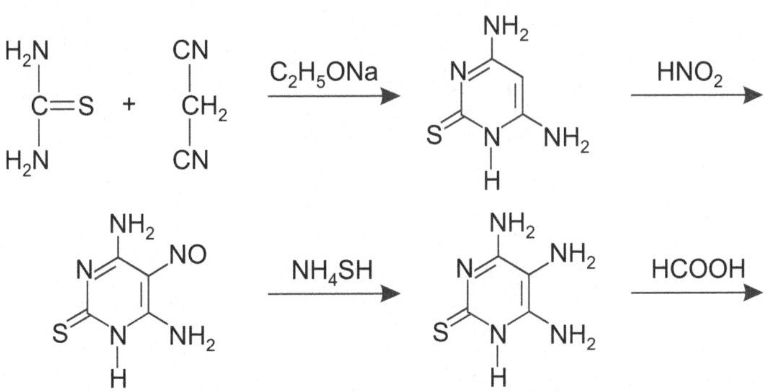

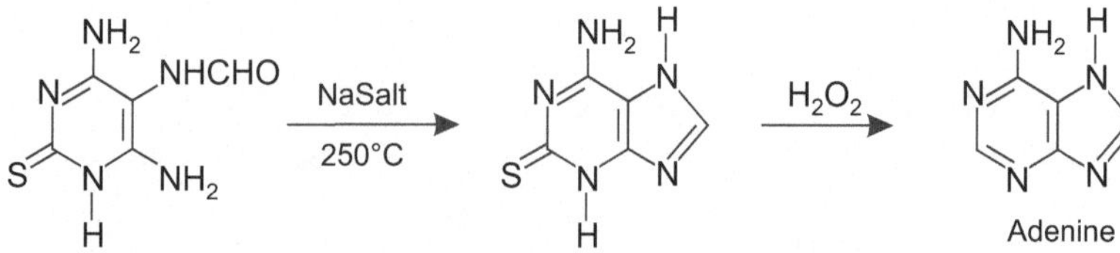

3. **Todd *et al.***

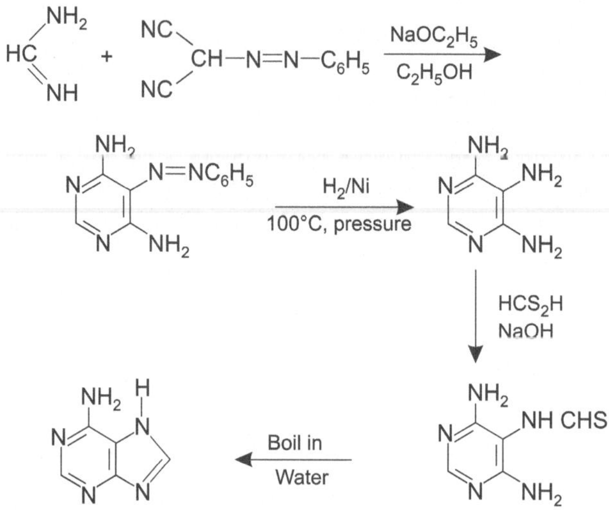

4. **Bredereck *et al.***
   They modified the synthesis given by Traube

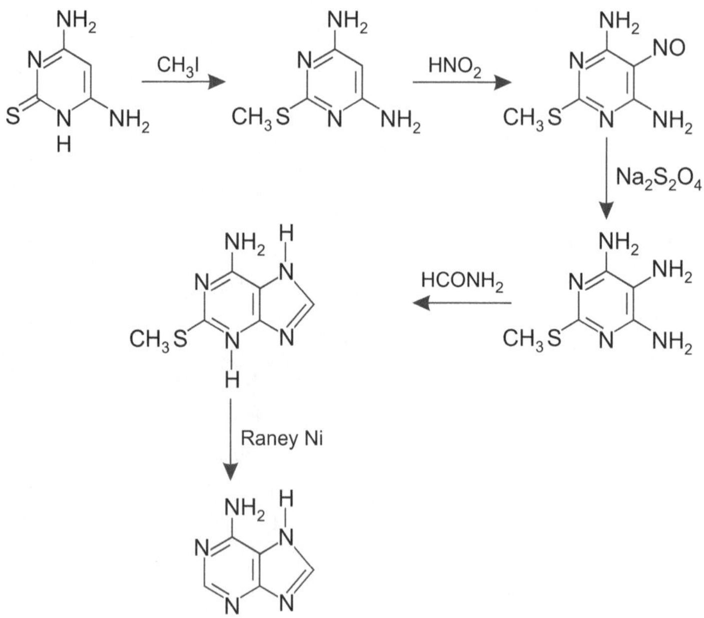

Synthesis of 4,5-Diaminopyrimidine

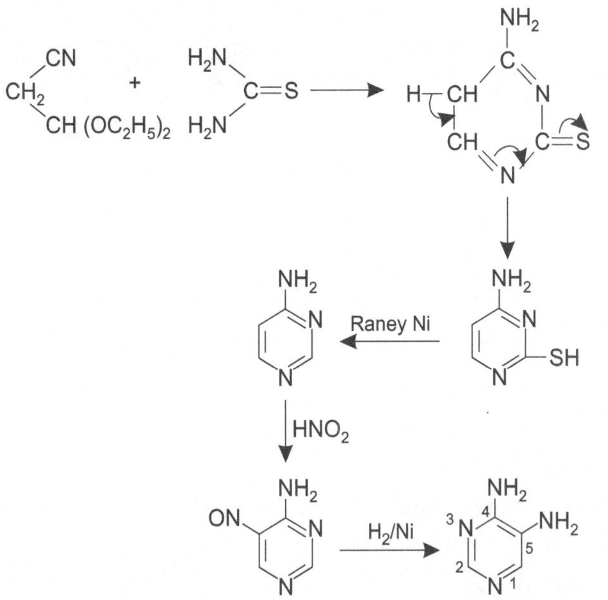

4,5-Diaminopyrimidine

4.13 GUANINE (2-AMINO - 6-HYDROXY PURINE)

In addition to nucleic acids, guanine occurs in the pancreas of cattle and in certain fish scales. It gives xanthine on treatment with nitrous acid. This conversion is also effected by boiling guanine with 25% hydrochloric acid.

Synthesis of guanine

1. **Fischer**

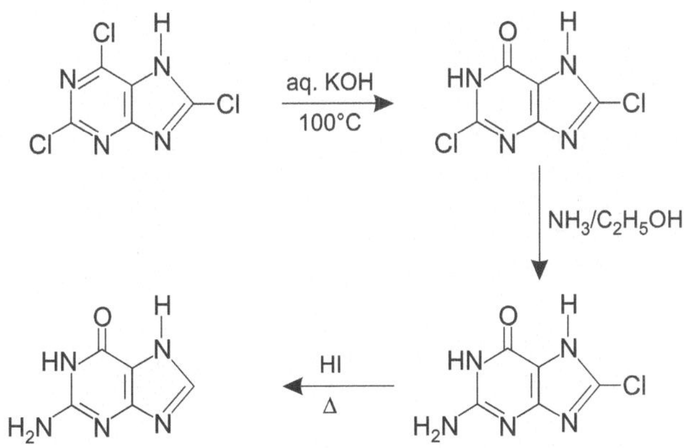

Guanine

2. **Traube**

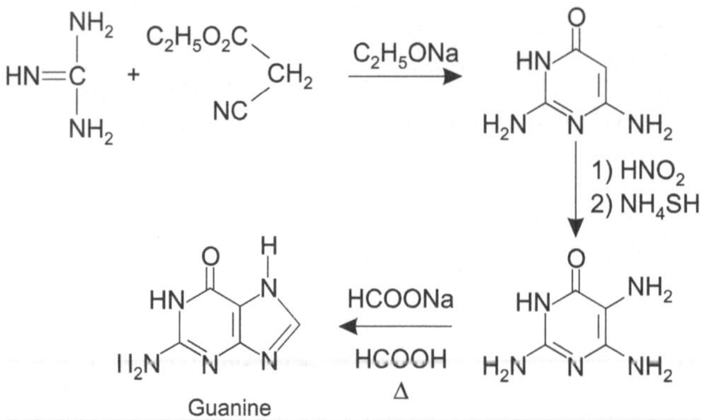

Guanine

4.14 UREIDES

Acyl derivatives of urea are called ureides, *e.g.*, acetyl urea in which one hydrogen of urea is replaced by an acetyl group.

They may be classified as:

A. Simple ureides
B. Cyclic ureides

A. Simple Ureides

They are generally prepared by the action of acyl chlorides or acid anhydrides of mono-carboxylic acids on urea. For example,

$$CH_3\,COCl + H_2N-\overset{O}{\overset{\|}{C}}-NH_2 \longrightarrow \underset{\text{Acetyl urea}}{CH_3\,CO-NH-\overset{O}{\overset{\|}{C}}-NH_2} + HCl$$

$$CH_3-CONH-\overset{O}{\overset{\|}{C}}-NH_2 + CH_3COCl \longrightarrow \underset{\text{Diacetylurea}}{CH_3-CONH-\overset{O}{\overset{\|}{C}}-NH-CO\,CH_3} + HCl$$

Simple ureides resemble amides *i.e.*, they are almost neutral in character and are hydrolysed by alkali into urea and the acid.

The halogen derivatives of some simple ureides have been used as hypnotics. An example is that of **bromural**, which is α-bromoisovaleryl urea.

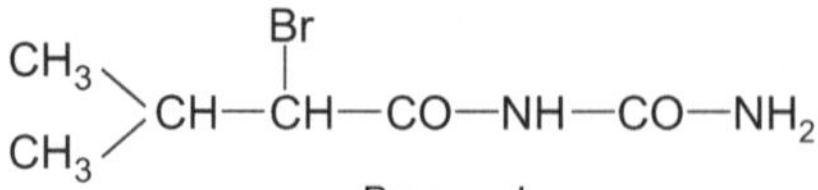

Bromural

B. Cyclic Ureides

They are prepared by the action of dicarboxylic acid derivatives with urea. For example, we get parabanic acid, which is oxalylurea obtained by the action of urea on oxalyl chloride

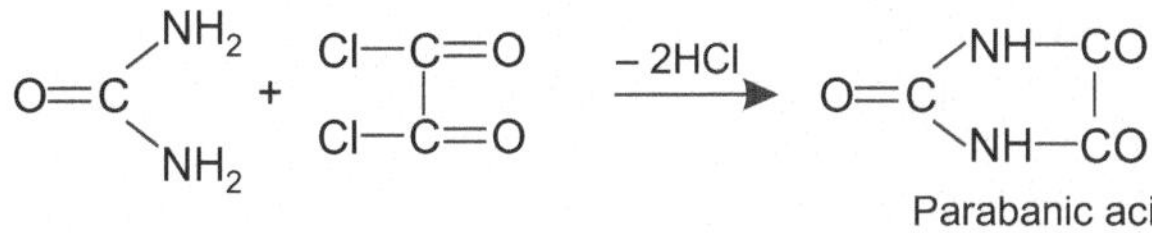

Parabanic acid

Barbituric Acid

It is an important pyrimidine derivative.

We get barbituric acid (malonyl urea) by the action of malonic ester with urea in ethanolic solution containing sodium ethoxide.

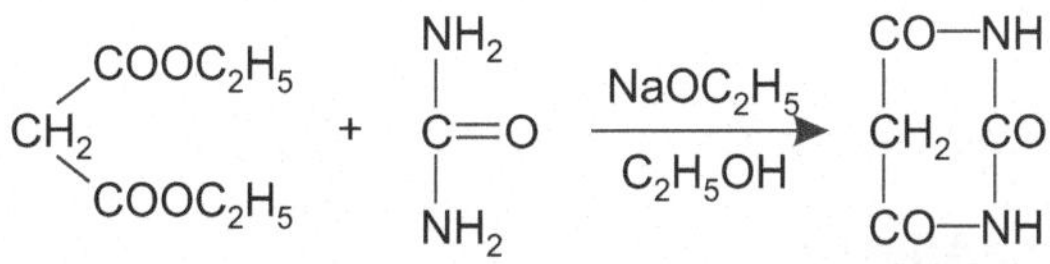

Barbituric acid

Barbituric acid is a solid, m.p. 250°C. It is not very soluble in water.

Cyclic ureides, unlike simple ureides, are acidic in character and form salts with bases. They are hydrolysed by bases to give urea and the corresponding acid.

Barbituric acid and its derivatives, particularly 5, 5-disubstituted derivatives, are used as hypnotics and sedatives. Two important examples are of barbitone (5, 5-diethyl barbituric acid) and phenobarbitone (5-phenyl, 5-ethyl barbituric acid).

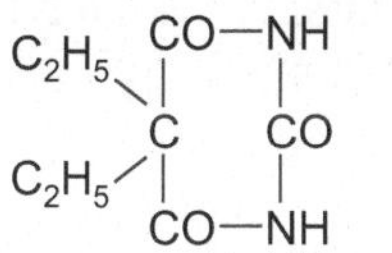

Barbitione

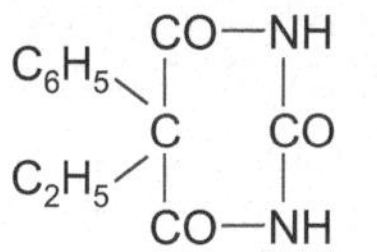

Phenobarbitone

4.15 URIC ACID

A. Introduction

Uric acid is a product of animal metabolism. It is present in the excreta of birds and reptiles. Guano, the excrement of birds found on islands near the Western Coast of South America, contains up to 25 per cent uric acid. About 90 per cent of snakes' excrement is ammonium urate. It also occurs in human urine in small amounts. Normal blood of human beings contains only small amount (0.2 mg per cent) of uric acid, but in the patients suffering from gout, the amount of uric and rises considerably. Uric acid was first discovered by **Scheele** in urinary calculi.

B. Structure of Uric Acid

The structure of uric acid has been elucidated through the extensive research work of Liebig, Wőhler, Baeyer and Fischer.

1. Molecular formula of uric acid is $C_5H_4N_4O_3$
2. On exhaustive methylation it forms tetramethyluric acid, which on hydrolysis gives four molecules of methylamine but not ammonia.

$$\underset{\text{Uric acid}}{C_5H_4N_4O_3} \xrightarrow{CH_3I} \underset{\text{Tetramethyl uric acid}}{C_9H_{12}N_4O_3} \xrightarrow{H_2O} \underset{\text{Methylamine}}{4CH_3NH_2}$$

   This shows that uric acid contains four amino groups and each methyl group in tetramethyl uric acid is attached to a nitrogen atom.

3. Oxidation of uric acid with nitric acid gives alloxan and urea in equimolecular quantities

$$\text{Uric acid} + H_2O \xrightarrow{HNO_3} \underset{\text{Alloxan}}{C_4H_2N_2O_4} + NH_2{-}\overset{O}{\overset{\|}{C}}{-}NH_2$$

4. **Structure of Alloxan**

   The structure of alloxan was established as (21) on the basis of hydrolytic experiments

   (*i*) On hydrolysis with alkali alloxan gives one molecule of urea and one molecule of mesoxalic acid.

$$C_4H_2N_2O_4 + 2H_2O \longrightarrow NH_2{-}\overset{O}{\overset{\|}{C}}{-}NH_2 + CO\begin{matrix}\diagup COOH \\ \diagdown COOH\end{matrix}$$

   Alloxan does not contain any free amino or carboxyl groups. Alloxan is, therefore, a cyclic ureide *i.e.*, it is mesoxalyl urea.

   The structure of alloxan has been confirmed as (21) as it can be obtained by direct condensation of urea and mesoxalic acid.

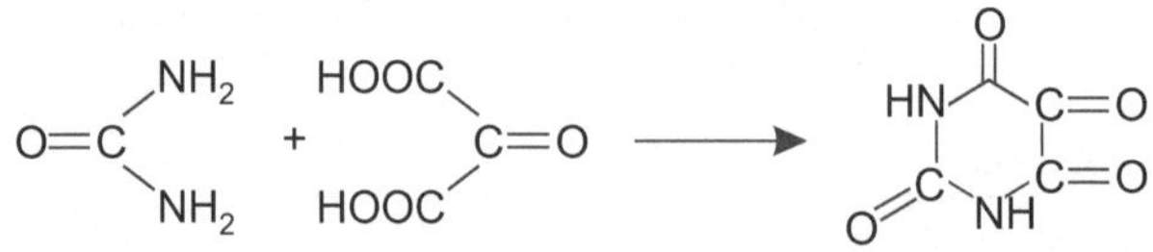

(21) Alloxan

   (*ii*) Barbituric acid can also be converted to alloxan monohydrate as given below :

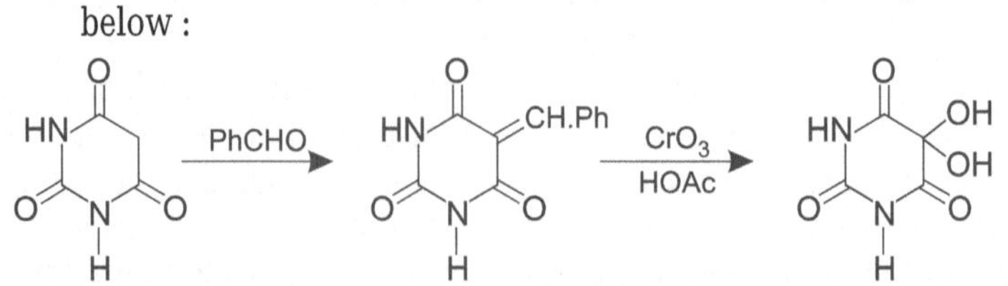

Barbituric acid — Alloxan monohydrate

5. Uric acid on oxidation with an aqueous suspension of lead dioxide gives allantoin (22) and carbon dioxide.

$$\text{Uric acid} \xrightarrow{(O)} \underset{\substack{(22)\\ \text{Allantoin}}}{C_4H_6N_4O_3} + CO_2$$

6. **Structure of Allantoin**

   (*i*) Allantoin (22) on hydrolysis with alkali forms two molecules of urea and one molecule of glyoxylic acid.

$$C_4H_6N_4O_3 \xrightarrow{H_2O} 2NH_2CONH_2 + \begin{array}{c} CHO \\ | \\ COOH \end{array}$$

   Thus, allantoin is the diureide of glyoxylic acid.

   (*ii*) On oxidation with nitric acid, allantoin forms urea and parabanic acid (23) in equal amounts.

$$\text{Allantoin} \xrightarrow{HNO_3} NH_2CONH_2 + \underset{\substack{(23)\\ \text{Parabanic acid}}}{C_3H_2N_2O_3}$$

   (*iii*) Parabanic acid is oxalyl urea as on hydrolysis it gives urea and oxalic acid.

$$\text{Parabanic acid} \xrightarrow{H_2O} NH_2CONH_2 + \begin{array}{c} COOH \\ | \\ COOH \end{array}$$

   The structure of parabanic has been confirmed by synthesis

Parabanic acid

   Thus, allantoin contains parabanic acid nucleus joined to a molecule of urea. The point of attachment is deduced from the following experimental evidence.

   (*iv*) When reduced with concentrated hydriodic acid at 100°C, allantoin forms urea and hydantoin (24).

$$\text{Allantoin} \xrightarrow[100°C]{HI} NH_2CONH_2 + \underset{\substack{(24)\\ \text{Hydantoin}}}{C_3H_4N_2O_2}$$

   (*v*) Hydantoin on controlled hydrolysis gives hydantoic acid (25) which on further hydrolysis gives glycine, ammonia and carbon dioxide. This indicates that hydantoin is glycollylurea. Thus, we have

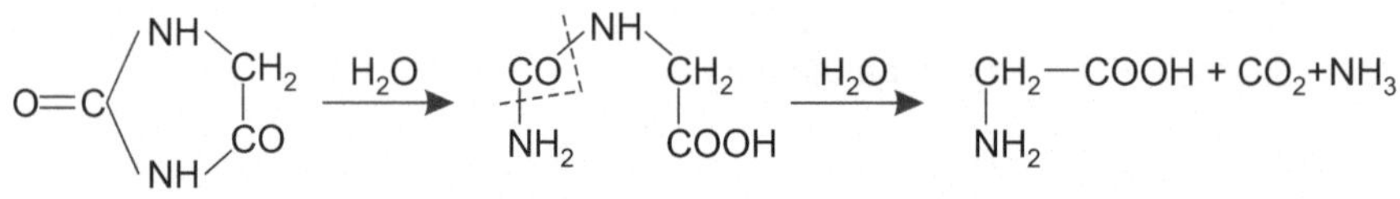

(24) (25)

This structure (24) for hydantoin has been confirmed by synthesis

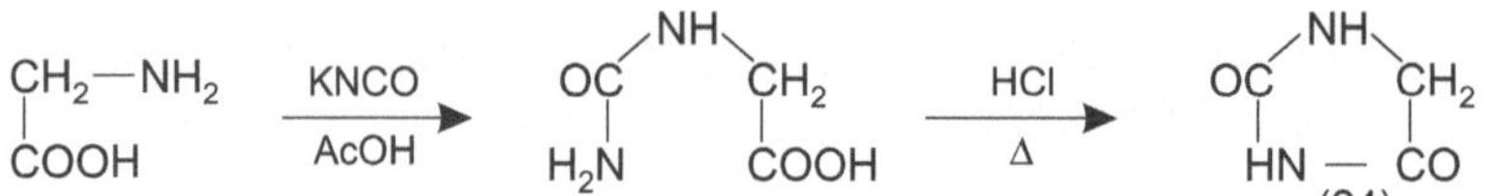

(24)

Thus, the structure of allantoin is (22). All the above reactions can be shown as below:

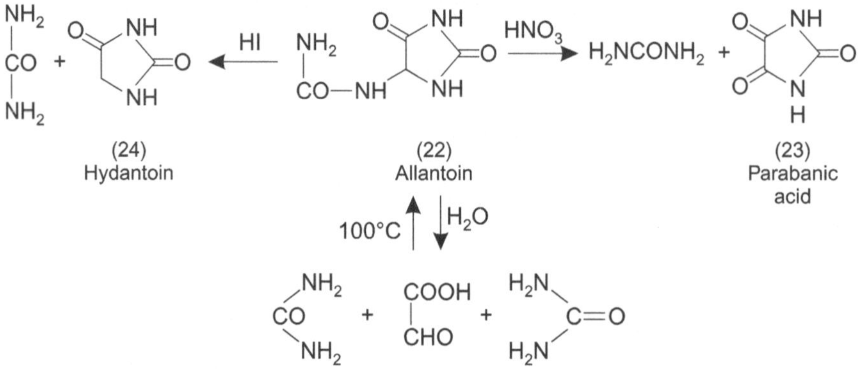

(24) Hydantoin; (22) Allantoin; (23) Parabanic acid

The structure for allantoin has been confirmed by synthesis. It can be obtained by heating urea with glyoxylic acid at 100°C. This structure is also confirmed by the following synthesis:

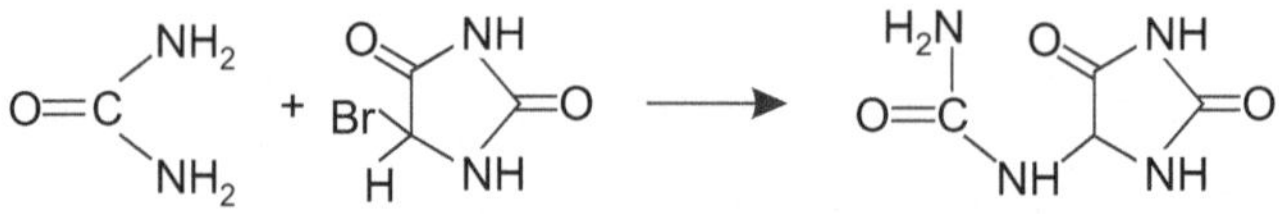

Allantoin

7. Now uric acid on oxidation loses one carbon atom and gives allantoin and at the same time it gives alloxan on oxidation with nitric acid. The formation of these two products can only be explained from the following structure for uric acid (26)

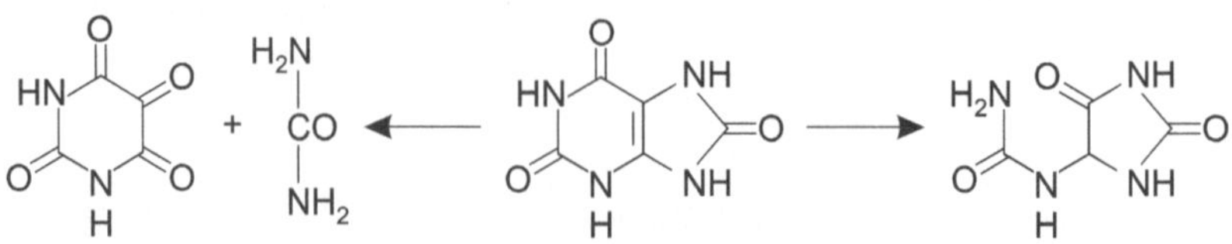

Alloxan (26) Uric acid Allantoin

8. The structure was confirmed by Fischer who prepared two isomeric mono methyl uric acids. One gave methyl alloxan and urea on oxidation with nitric acid and the other gave alloxan and methyl urea

   Four monomethyl derivatives of uric acid are possible, which were also prepared by Fischer.

C. Synthesis of Uric Acid

1. Synthesis by Behrend and Roosen

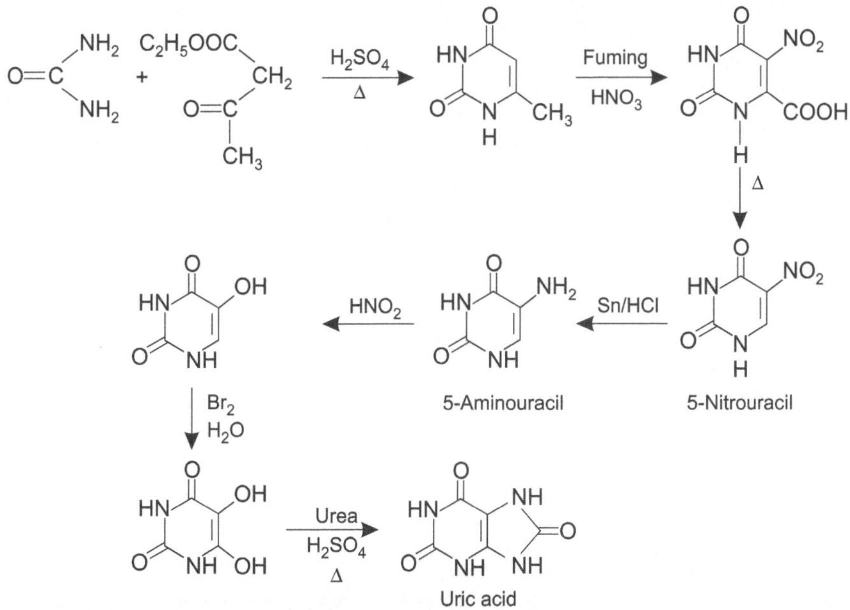

2. Baeyer Synthesis completed by Fisher

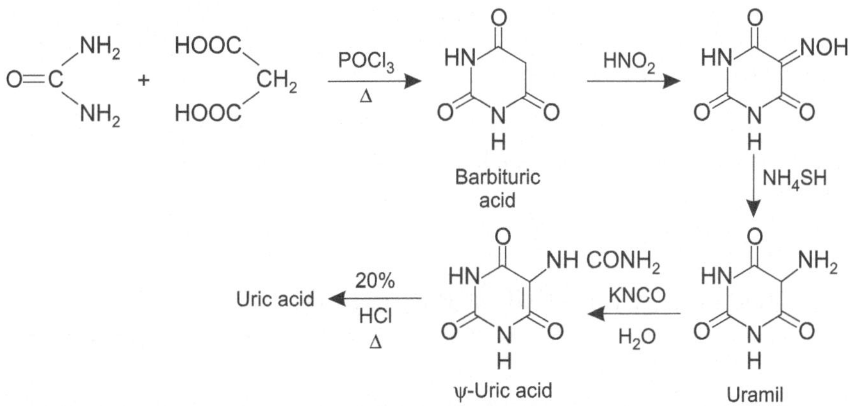

3. Traube's Synthesis

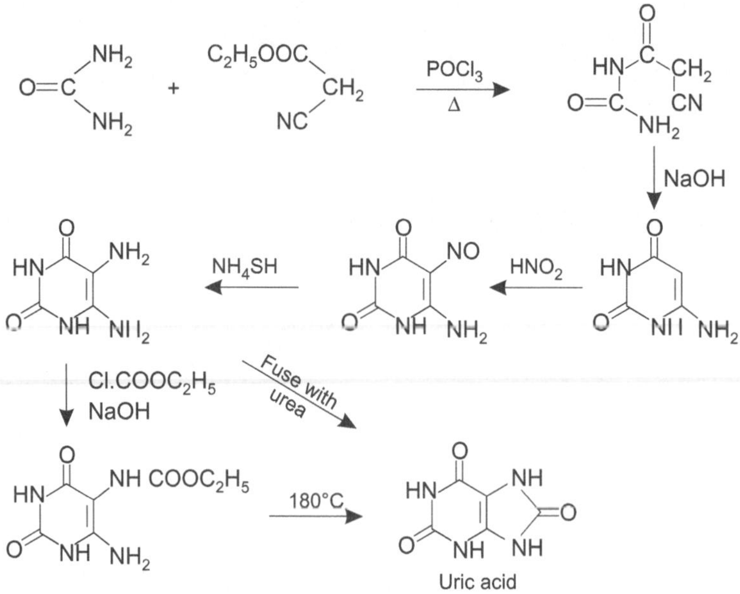

4.16 CAFFEINE

We all consume large quantity of caffeine in the form of tea and coffee. It is an important methylated xanthine that occurs naturally in tea leaves and coffee beans. Caffeine has stimulant action on the central nervous system.

A. Isolation from Tea Leaves

The dried tea leaves are digested with boiling water. The extract is treated with excess of lead acetate solution. Proteins and tannins get precipitated. They are filtered off and the filtrate is acidified with dilute sulphuric acid to remove excess of lead acetate as lead sulphate. The filtrate is concentrated and extracted with chloroform. Chloroform is distilled off and the residue is crystallised from water to give silky needles of caffeine.

Caffeine is a colourless crystalline solid, m.p. 236°C. It is fairly soluble in alcohol and chloroform but sparingly soluble in water and ether. It has a bitter taste and sublimes on heating.

B. Structure of Caffeine

1. Molecular formula of Caffeine is $C_8H_{10}N_4O_2$.
2. Caffeine and uric acid have the same skeleton structure. This is shown by the following facts:
   (*i*) On oxidation with potassium chlorate in hydrochloric acid, it gives dimethylalloxan (27) and methyl urea in equimolar proportions

(*ii*) The structure of dimethyl alloxan is established as (27) because on hydrolysis it gives *N, N′*-dimethyl urea and mesoxalic acid.

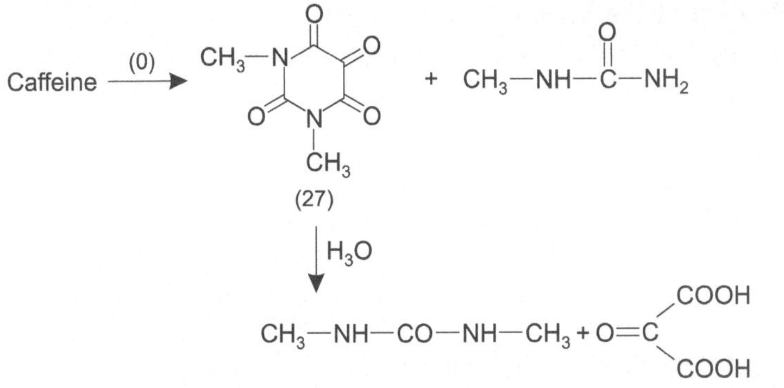

Dimethyl urea Mesoxalic acid

The formation of dimethylurea establishes the positions of two methyl groups and one oxygen atom in caffeine.

The problem now is to assign the position of the third methyl group and second oxygen atom.

The third methyl group may be either at position 7 or 9 and the oxygen atom at 6 or 8.

3. Position of the Third Methyl Group

We have seen the oxidation of caffeine to dimethyl alloxan and methyl urea. Fisher isolated another oxidation product which was shown to be *N*-methylhydantoin (28) because on hydrolysis it gives *N*-methylglycine, carbon dioxide and ammonia.

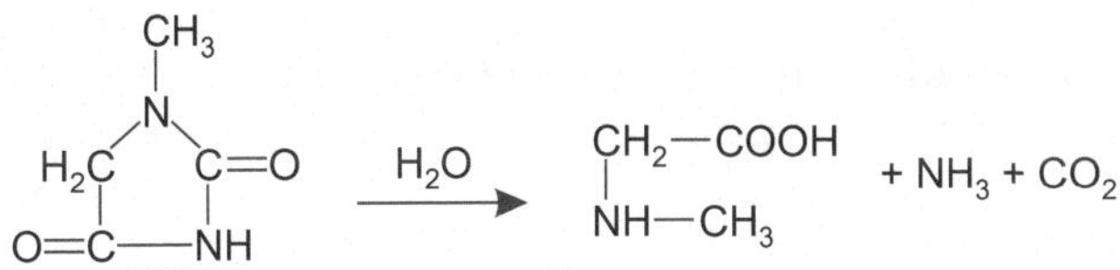

(28)

Therefore, two skeleton structures are possible (29 or 30) for caffeine, each could give the above oxidation products

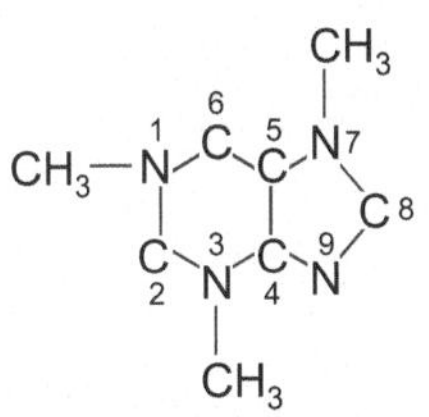

Methyl at 7 position
(29)

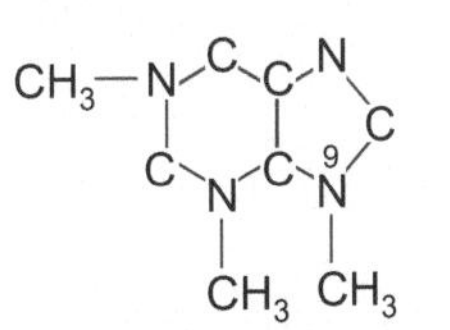

Methyl at 9 position
(30)

Fischer was able to isolate fourth oxidation product which was found to be *N, N′*-dimethyloxamide. Examination of the structures showed that oxamide can be obtained from structure (29) and not from structure (30). So structure (29) is the skeleton of caffeine.

4. Position of the Second Oxygen Atom

As mentioned above the second oxygen atom may be at either position 6 or 8. Therefore, two structures are possible for the caffeine molecule.

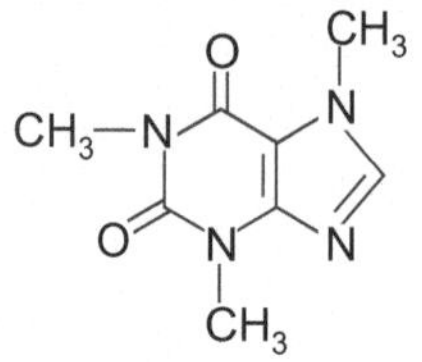
(31)

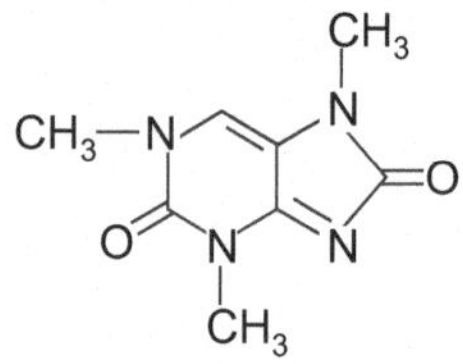
(32)

5. Final Structure of Caffeine

Structure (31) was shown to be caffeine by Fisher. He carried out the following reactions:

Caffeine $C_8H_{10}N_4O_2$ $\xrightarrow{Cl_2}$ Chlorocaffeine $C_8H_9ClN_4O_2$ $\xrightarrow[NaOH]{CH_3OH}$ Methoxy Caffeine $C_8H_9N_4O_2 - OCH_3$

$\downarrow$ dil HCl, Δ

Oxycaffeine $C_8H_{10}N_4O_3 + CH_3Cl$

Oxycaffeine was shown to be identical with trimethyl uric acid as it was converted into tetramethyl uric acid on methylation with methyl iodide in the presence of aqueous sodium hydroxide. Therefore, methoxycaffeine is either (33) or (34) and oxycaffeine (35) or (36).

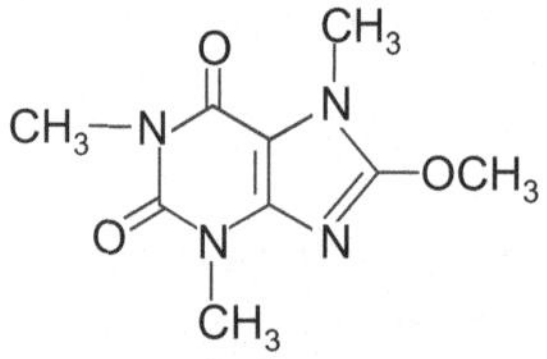
(33)

or

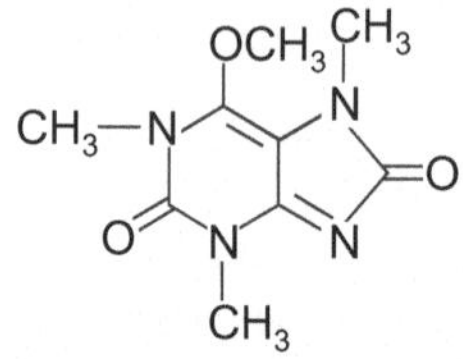
(34)

Methoxycaffeine

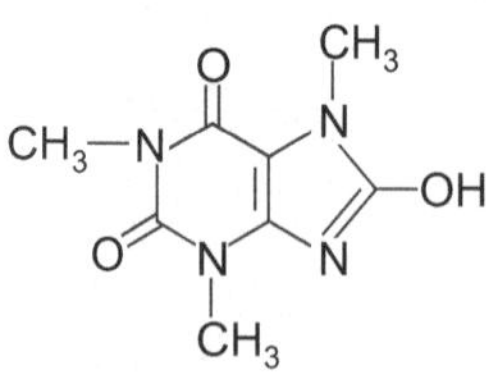
(35)

or

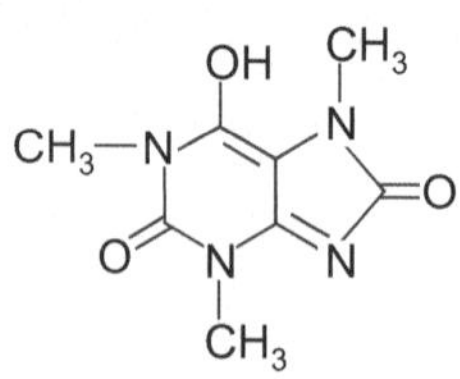
(36)

Oxycaffeine

Further, when oxycaffeine is heated with methyl iodide, it is converted into a mixture of tetramethyluric acid and methoxycaffeine. The formation of these two products simultaneously, suggests that oxycaffeine is a tautomeric substance *i.e.*, it contains amido-imidol system

$$-NH-\overset{|}{C}=O \rightleftharpoons -N=\overset{|}{C}-OH$$

Such a system is possible only when caffeine has the structure (31).

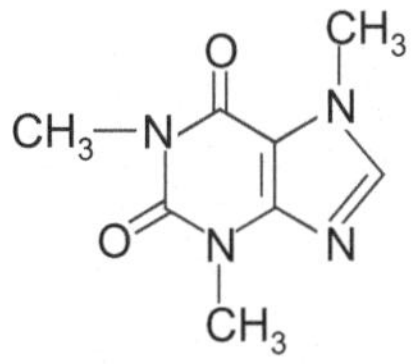

caffeine

The formation of oxycaffeine from caffeine can then be shown in the following manner:

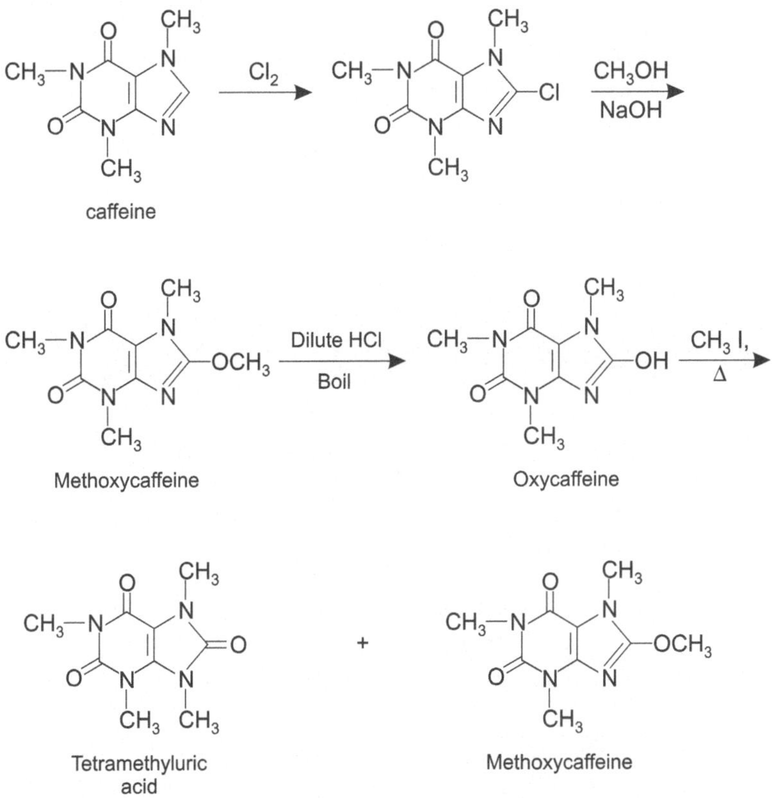

The above structure for caffeine has been confirmed by synthesis.

C. Synthesis of Caffeine

1. Fischer's Synthesis

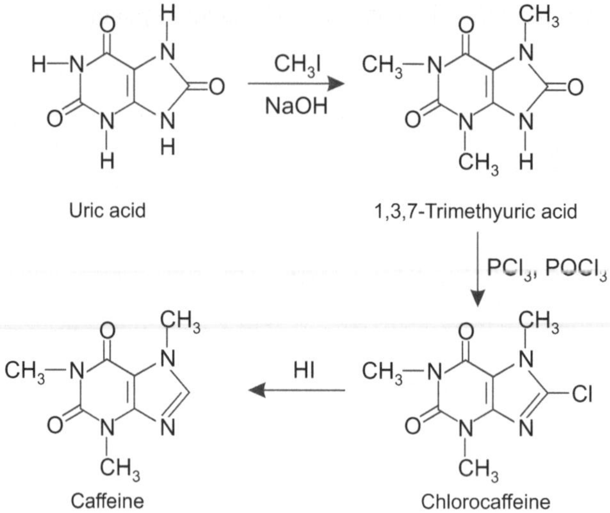

2. Synthesis based on Traube's Method.

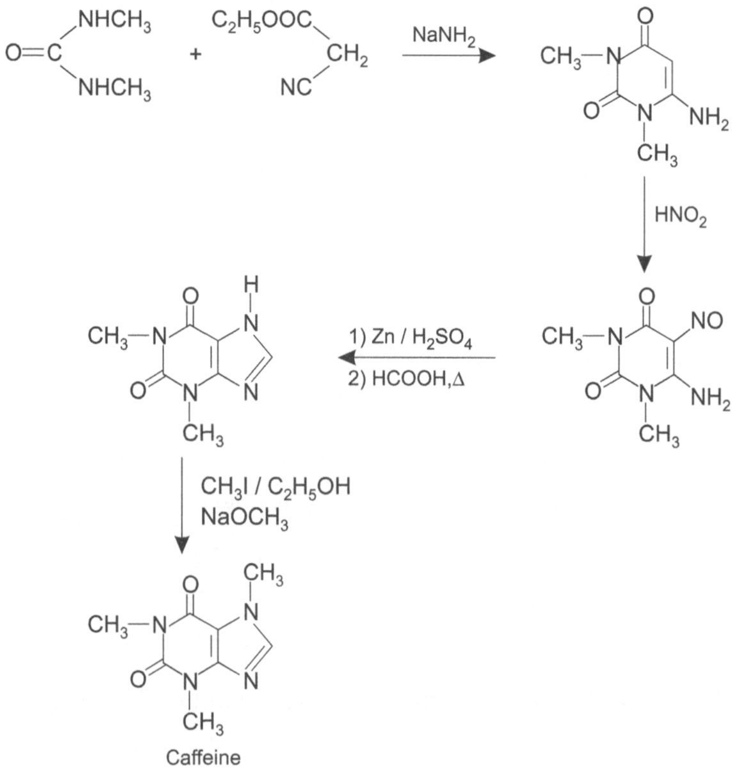

4.17 NUCLEOSIDES

We have already seen pyrimidine and purine bases are present in nucleic acids. Pyrimidine bases commonly found in nucleic acids are cytosine, uracil and thymine. The most impotant purine bases are adenine and guanine. When a purine or a pyrimidine base combines with a sugar, a nucleoside is formed. The sugar may be ribose or 2-deoxyribose. Thus, the combination of adenine and ribose forms the nucleoside, adenosine (19) and guanosine (20) is obtained from guanine and ribose. Similarly, we get cytidine from cytosine and ribose, uridine from uracil and ribose; thymidine from thymine and deoxyribose.

Nucleosides formed with ribose are termed ribonucleosides, while those with deoxyribose are termed deoxyribonucleosides.

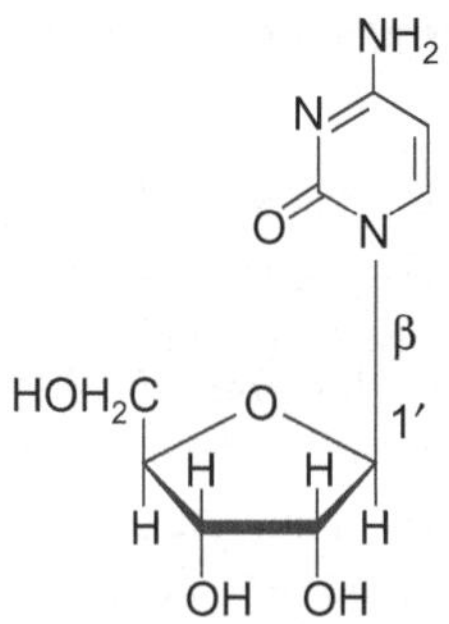

Cytidine
A ribonucleoside

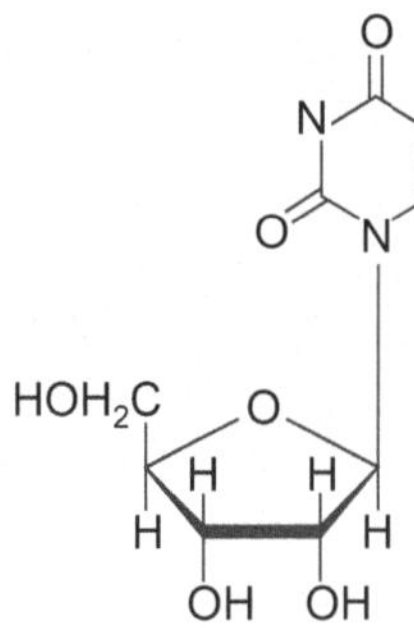

Uridine

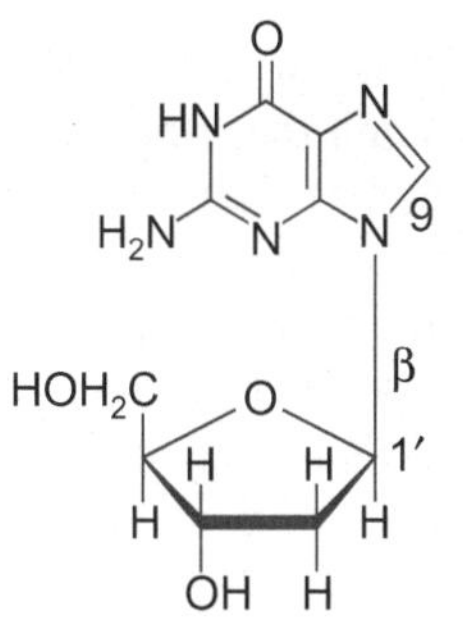

Deoxy guanosine
A deoxy ribonucleoside

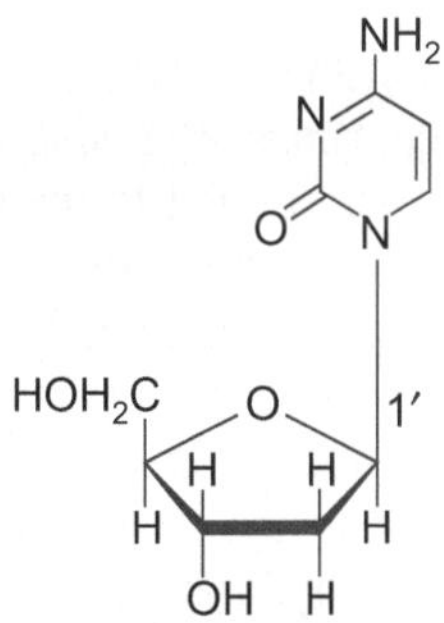

Deoxy cytidine

In a nucleoside, a base forms a glycosidic linkage with the sugar. Since nucleosides are non-reducing the aldehyde group of the sugar cannot be free. That means C–1 of the sugar is joined to the base. Now the problem is to decide which atom of the base is joined to C–1 of the sugar. This was decided by studying the reactions of cytidine and uridine.

- Cytidine on treatment with nitrous acid is converted into uridine. This means the sugar residue is linked in the same position in both of these nucleosides. The point of linkage cannot be 3 or 4 since cytidine has a free amino group at position 4 and there cannot be a hydrogen atom on N–3.
- When uridine is treated with an excess of bromine followed by the addition of phenylhydrazine, a uridine derivative is obtained which contains two phenylhydrazino groups at the positions 5 and 6. This type of compound can be obtained only if uracil is substituted in position 1 and positions 5 and 6 are free. Thus, the sugar is attached to N–1. In a similar way, it has been proved that the other pyrimidine nucleosides have the sugar residue linked at N–1. This linkage has also been confirmed by the X-ray analysis of cytidine.

In case of nucleosides derived from purine bases, it has been proved that C–1 of the sugar is attached to N–9 position of the purine base. This has been established by the following facts:

Adenosine has a free amino group at position 6. Therefore, the sugar cannot be at C–6 or N–1. Similarly guanosine has a free amino group at position-2, the sugar cannot be at C–2 or N–3. Therefore, only positions 7, 8 and 9 are possible points of attachment. Position 8 was excluded because this point would involve carbon-carbon bond which would be quite stable and not hydrolysed readily by dilute acids. Thus, positions 7 or 9 are free. Todd *et al.* had synthesised adenosine and guanosine in which the sugar is known to be in the 9-position. They showed that their synthetic compounds are identical with the natural products.

Nature of the ring in the sugars

Degradative experiments have shown that the sugar is present as the furanose form in the nucleosides. We have already studied that ring size of glucose and fructose can be determined by methylation studies. Here also methylation of a pyrimidine riboside followed by hydrolysis gives trimethylribose which on oxidation with nitric acid forms dimethylmesotartaric acid. The formation of this acid shows that the ribose ring is in the furanose form. Had it been pyranose, it would have given trimethoxyglutaric acid (Fig. 4.3). This is also confirmed by oxidation with periodic acid which consumes one molecule of the reagent with no loss of carbon and a dialdehyde is formed.

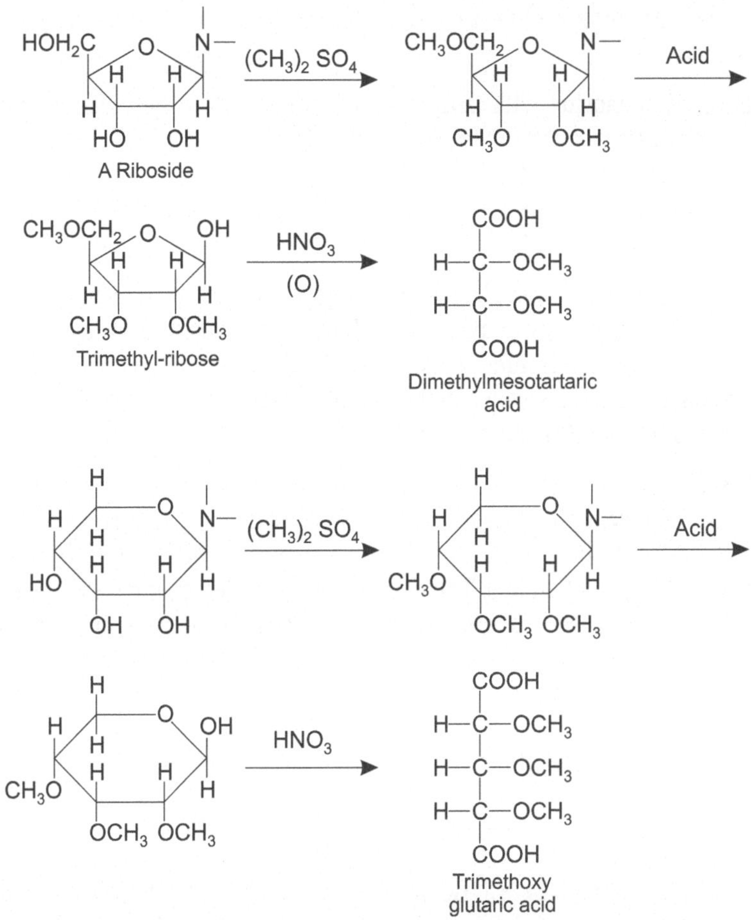

Fig. 4.3.

Deoxyribose has also been shown to have furanose structure. It is found that pyrimidine deoxyribosides are not degraded by periodic acid. This agrees with the 2-deoxyribofuranose structure since the molecule does not contain two adjacent hydroxyl groups.

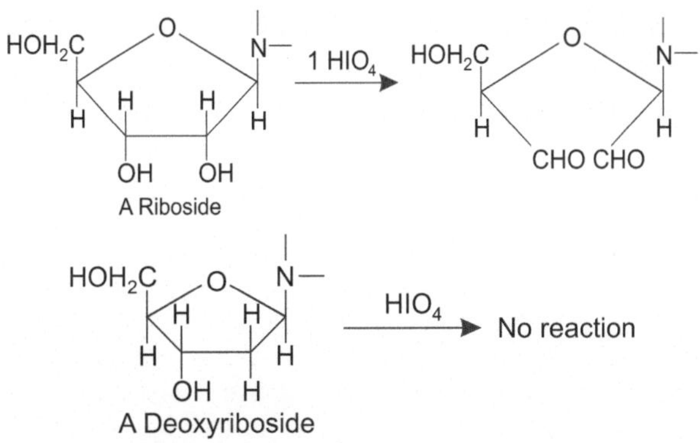

Lastly, the configuration of the furanoside link has been shown to be β- by different experiments.

Oxidation of adenosine with periodic acid was shown to be identical with the product obtained from the oxidation of 9-β-D-mannopyranosidyladenine. This proves that the sugar residue is at position 9, has the furanose structure and that the linkage is β. Similar experiments with other ribonucleosides suggest that all these compounds have a β-configuration.

In case of cytidine, it was shown by X-ray analysis that the sugar residue is attached to N–1 and is β-ribofuranoside. Since other ribonucleosides exhibit the same general pattern, it is inferred that all are furanosides with the β-configuration. Deoxyribonucleosides also exist in the β-configuration as shown by Manson *et al.* from absorption spectra measurements.

Nucleosides are usually stable towards alkaline hydrolysis but are readily hydrolysed by acids. Deoxynucleosides undergo acid hydrolysis more readily than ribonucleosides. In general, the order of hydrolysis is guanosine > adenosine > cytidine > uridine > thymidine. The reasons for this order are not clear. We already know that the furanose structure occurs only when the sugar is in the form of a glycoside. On hydrolysis, the furanose liberated immediately changes into the more stable pyranose form.

4.18 NUCLEOTIDES

When phosphoric acid is esterified to one of the hydroxyl groups of the sugar portion of the nucleoside, we get a nucleotide. Thus, a nucleotide contains a purine or a primidine base, ribose or 2-deoxy-ribose and a phosphoric acid unit. The point of attachment may be 2′, 3′ or 5′ in the ribose molecule and 3′ or 5′ in the deoxy-ribose molecule.

A nucleotide is named for the parent nucleoside with the suffix "monophosphate" or 'diphosphate' depending upon the number of phosphate units added. For example, adenosine can react with phosphoric acid to form the ribonucleotide, adenosine mono-phosphate (AMP). The position of the phosphate ester is specified by the number of the carbon at the hydroxyl group to which it is esterified. Thus, we have adenosine 3′-monophosphate or deoxycytidine 5′-monophosphate. The 5′-nucleotides are most commonly encountered in nature. Some examples are given in Fig. 4.4.

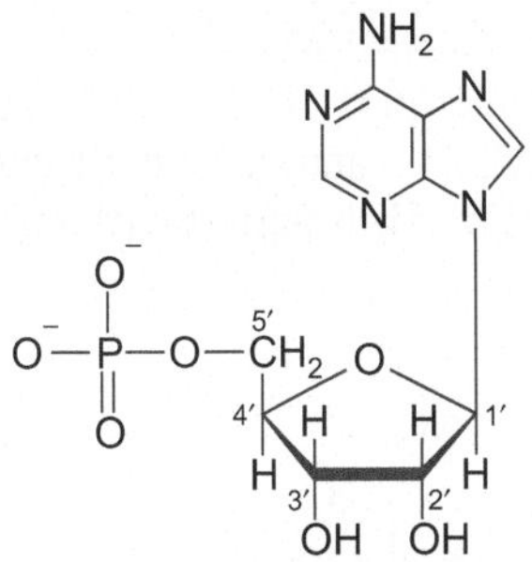

Adenosine 5′-mono-phosphate

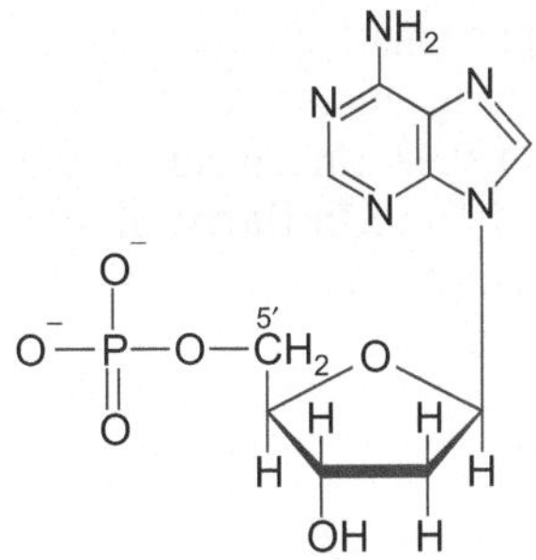

Deoxyadenosine-5′-monophosphate

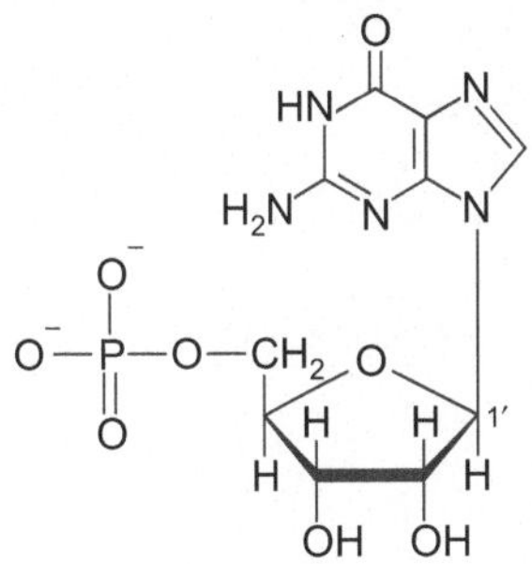

Guanosine 5′-mono-phosphate

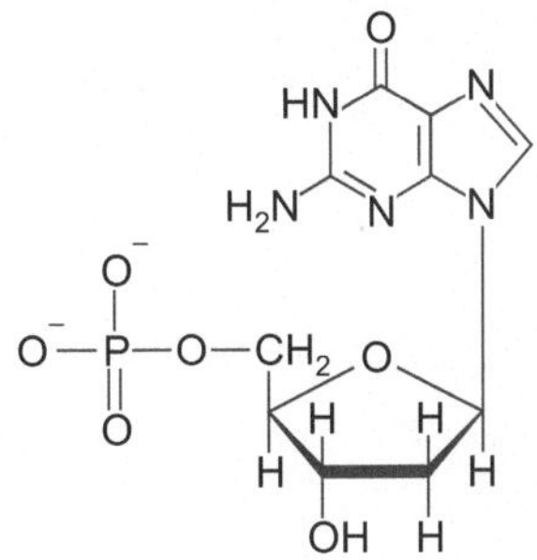

Deoxy-guanosine-5′-monophosphate

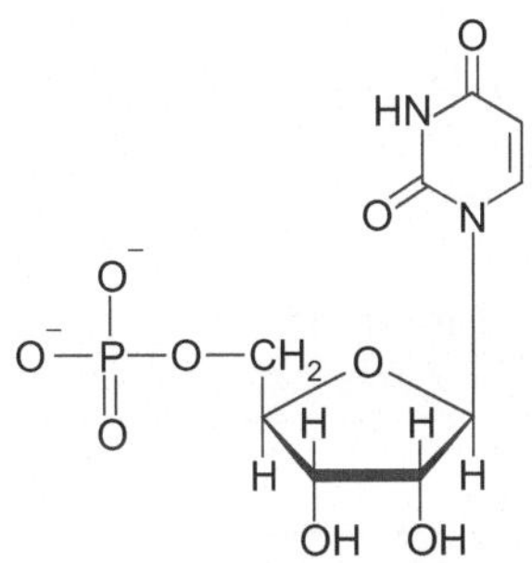

Uridine 5′-mono-phosphate

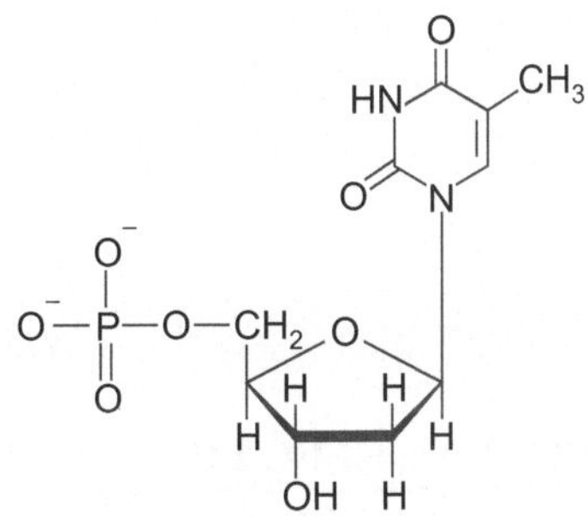

Deoxy-thymidine-5′-monophosphate

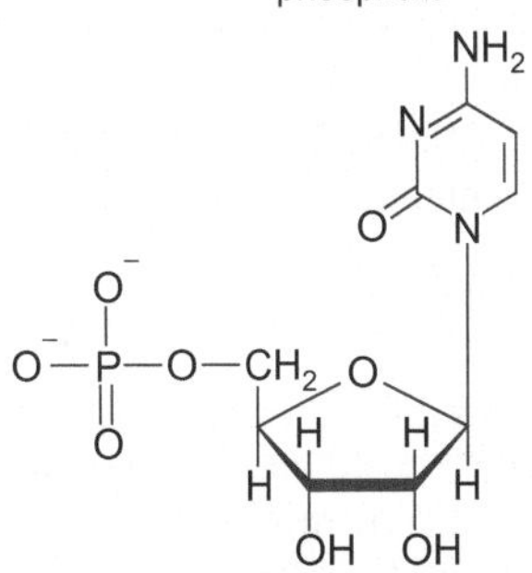

Cytidine 5′-mono-phosphate

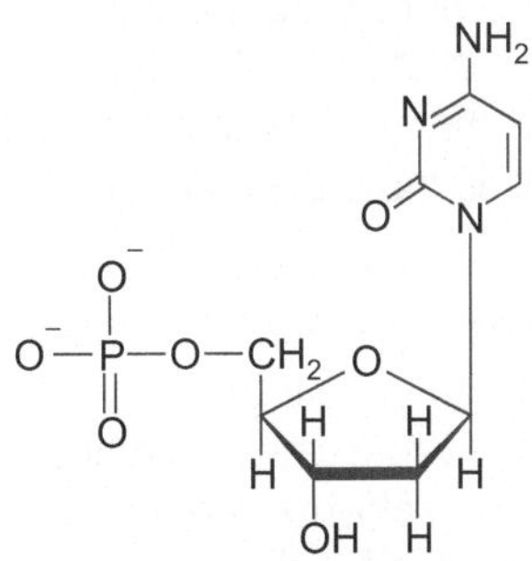

Deoxycytidine-5′-monophosphate

Fig. 4.4.

If additional phosphate groups form anhydride linkages with the first phosphate, the corresponding nucleoside diphosphates and triphosphates are formed. For example, adenosine forms adenosine diphosphate (ADP) and adenosine triphosphate (ATP) (Fig. 4.5). Similarly, deoxyadenosine diphosphate (d ADP) and deoxyadenosine triphosphate (d ATP) are formed. Nucleoside triphosphates serve directly as precursors for DNA and RNA synthesis.

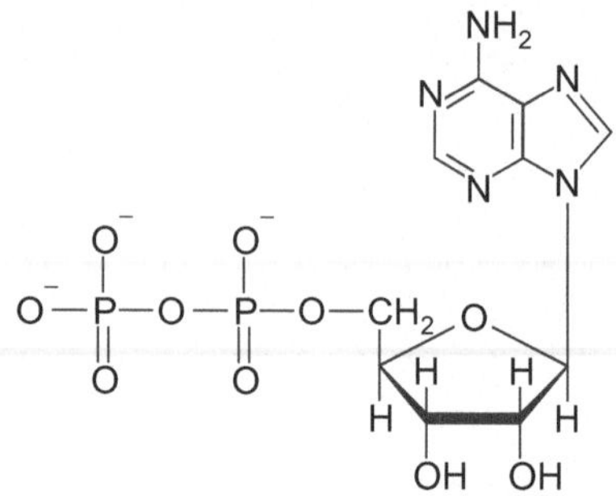
Adenosine 5′-diphosphate (ADP)

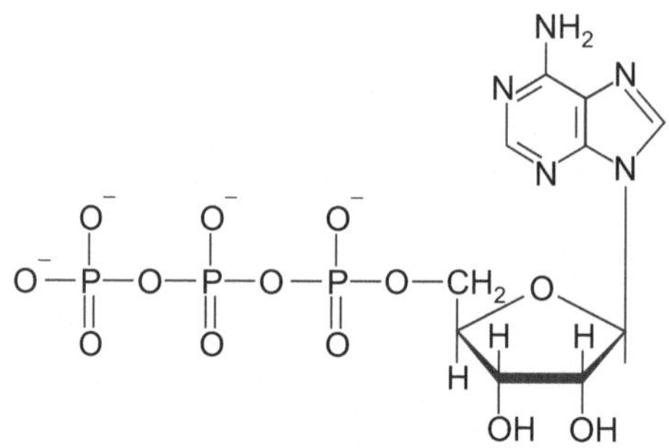
Adenosine 5′-triphosphate (ATP)

Fig. 4.5.

4.19 POLYNUCLEOTIDES

Nucleotides polymerise to form long chain molecules known as polynucleotides which actually constitute the nucleic acids. When two nucleotides are joined together the resultant molecule is called a dinucleotide while three form a trinucleotide. Oligonucleotides may contain nucleotides up to ten and if the number of nucleotides is more than ten they form a polynucleotide.

Thus, nucleic acids are obtained as a result of polymerisation of nucleotides. The linkage between monomers in nucleic acids involves formation of two ester bonds by phosphoric acid. The hydroxyls to which the phosphoric acid is esterified are those bonded to 3′ and 5′ carbons on adjacent residues. In other words, the polynucleotide chains are composed of monomer units linked by 3′,5′-phosphodiester bonds. The nucleotide residues of nucleic acids are numbered from the 5′-end, which normally carries a phosphate group, to the 3′-end which normally has a free hydroxyl group.

4.20 THE NUCLEIC ACIDS

We already know that there are two nucleic acids. In deoxyribonucleic acid (DNA), the monomers are deoxyribonucleotides and the chief bases are adenine (A), guanine (G), cytosine (C) and thymine (T). In RNA (ribonucleic acid) the monomers are ribonucleotides and the chief bases are A, G, C and uracil (U).

The basic differences between DNA and RNA are:

(*i*) Ribose is the sugar moiety of the RNA molecule whereas it is deoxyribose in DNA.

(*ii*) The pyrimidine base present in RNA is uracil whereas it is thymine in DNA.

(*iii*) RNA exists as a single strand whereas DNA exists as a double strand.

The backbone of a nucleic acid consists of alternating phosphates and pentoses. The bases are attached to the sugars of this backbone.

The structure of a fragment of an RNA chain is shown in Fig. 4.6. In this repeated linkage is a 3′, 5′-phosphodiester bond. As mentioned earlier the nucleotide residues are numbered from the 5′ end. Thus, the bases in RNA are A, C, G and U. There is a β-glycosidic bond between ribose and each base.

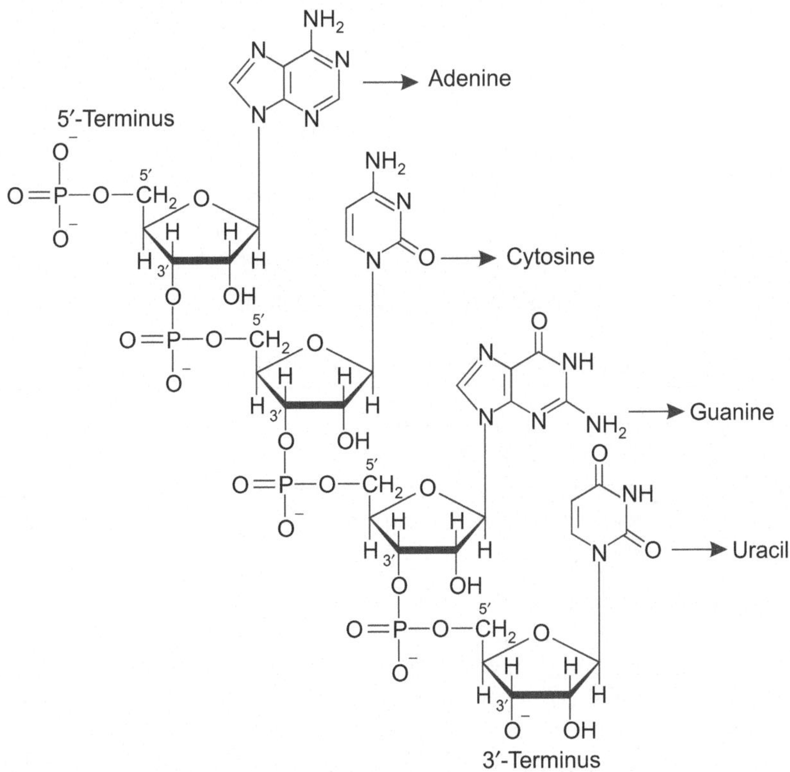

Fig. 4.6 A fragment of an RNA chain.

A portion of the DNA chain is shown in Fig. 4.7. It differs from the RNA chain only in the fact that here the sugar is 2′-deoxyribose rather than ribose. The bases from the 5′-terminus are thymine, guanine, cytosine and adenine. Each base forms a β-glycosidic bond with 2′-deoxyribose.

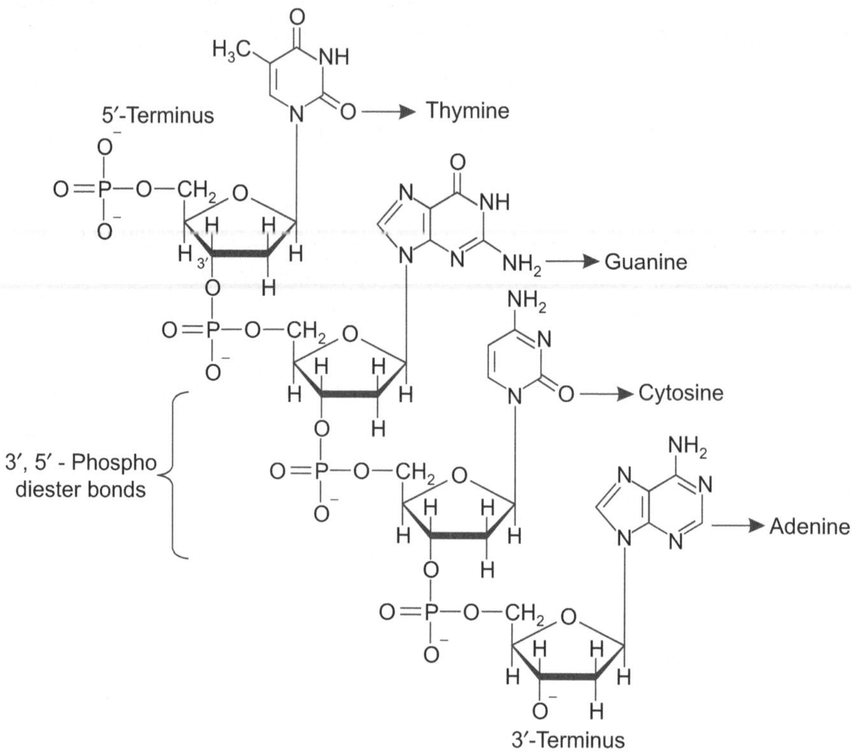

Fig 4.7 A portion of a DNA chain.

4.21 STRUCTURE OF DNA (THE DOUBLE HELIX)

Fig. 4.7 represents the primary structure of the DNA molecule. But it was suspected that most naturally occurring DNAs did not exist as single chains. Instead two or more chains appeared to interact with one another in some way. The resulting macromolecules were known to be long, thin and rigid such that they formed highly viscous solutions in water.

The double helix structure of DNA was proposed by **James Watson** and **Francis Crick** in 1953. It was based primarily on model building and X-ray diffraction pattern. Information from X-ray patterns was added to information from chemical analysis that showed that the amount of **A** was always the same as the amount of **T** and that the amount of **G** always equaled the amount of **C**. Both of these lines of evidence were used to conclude that DNA consists of polynucleotide chains wrapped around each other to form a helix. Hydrogen bonds between bases on opposite chains determine the alignment of the helix. The paired bases lie in planes perpendicular to the helix axis. The sugar-phosphate backbone is the outer part of the helix. The chains run in antiparallel directions one 3′ to 5′ and the other 5′ to 3′ (Fig. 4.8).

Other important features of the **Watson** and **Crick** Model are:

1. The base pairing is complementary, meaning that adenine pairs with thymine and that guanine pairs with cytosine. This complementary base pairing occurs along the entire double helix, the two chains are also referred to as complementary chains. An adenine-thymine (A-T) base pair has two hydrogen bonds between them whereas a guanine-cytosine (G-C) base pair has three.

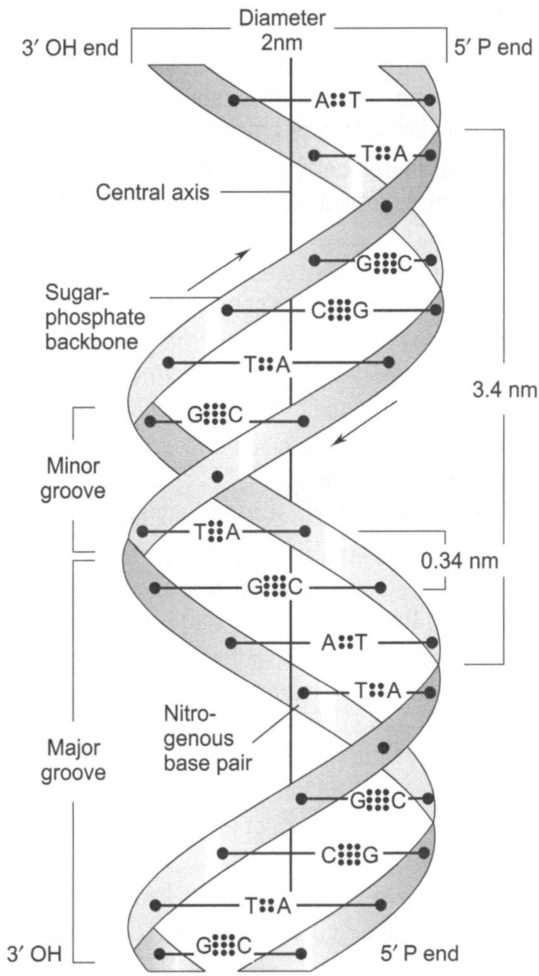

Fig. 4.8 Double helical structure of a part of DNA molecule.

2. The inside diameter of the sugar phosphate backbone of the double helix is about 11 Å (1.1 nm). The distance between the points of attachment of the bases to the two strands of the sugar-phosphate backbone is the same for the two base pairs (A-T and G-C), about 11 Å (1.1 nm). This gives DNA an extremely ordered structure.
3. The outside diameter of the helix is 20 Å (2 nm). The length of one complete turn of the helix along its axis is 34 Å (3.4 nm) and contains ten base pairs.
4. The double helix has empty spaces known as grooves. There is a major groove and a smaller minor groove in the double helix. Both can be sites at which drugs or polypeptides bind to DNA.

We can summarise the important features of Watson and Crick's double helical model of DNA as follows:

- The double helix comprises of two complementary polynucleotide chains.
- The two polynucleotide chains are wrapped clockwise around a central axis to form a circular stair-case type structure.
- The two strands of double helix are antiparallel *i.e.,* they run in opposite direction. One chain is aligned in $5' \rightarrow 3'$ direction and the other is aligned in $3' \rightarrow 5'$ direction.
- Each polynucleotide chain has a sugar-phosphate backbone and nitrogenous bases are directed inside the helix *i.e.,* towards the central axis. The sugar-phosphate backbone is formed of covalently bonded deoxyribose and phosphate groups and therefore, occur on the outside of the helix.
- The phosphate groups give the molecule a negative charge.
- The nitrogenous bases are hydrophobic in nature. They avoid water and therefore, remain inside the helix.
- The diameter of the double-helix is 20 Å or 2 nm.
- The nitrogenous bases lie almost perpendicular to the axis of the molecule. Therefore, these are stacked one on top of another. The hydrophobic interactions between the bases provide stability to DNA.
- The nitrogenous bases of two anti-parallel polynucleotide strands are linked through hydrogen bonds. There are two-hydrogen bonds between A and T and three between G and C (Fig. 4.9). The hydrogen bonds are the only attractive force between the two polynucleotides of the double helix. These serve to hold the strands together.
- There are ten base pairs in one turn of the helix. Therefore, each turn of the helix is 34 Å or 3.4 nm.
- The double helix has two helical grooves – a major groove, which is deep and wide and a minor groove which is shallow and narrow.

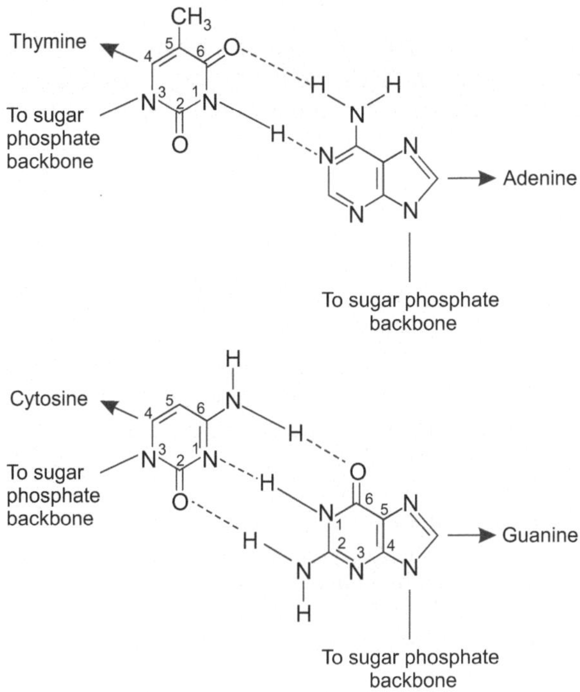

Fig. 4.9 Hydrogen bonds between A and T and between G and C.

4.22 CONFORMATIONAL VARIATIONS IN DNA (FORMS OF DNA)

We have discussed above the common form of DNA, which is called B-DNA. The double helical structure described by Watson and Crick was also of B-DNA. However, other secondary structures can occur depending on conditions such as nature of the positive ion associated with DNA and the specific sequence of bases. Thus, we have three important forms of DNA. These are A, B and Z-forms. B and Z-conformations occur as cellular DNA.

1. B-DNA

This form of DNA occurs in all living beings under normal conditions *i.e.*, this is the naturally occuring form. It occurs under low salt concentration and high degree of hydration. Its features have been described above. As given above its each turn measures 34 Å and each turn has ten base pairs. Each base pair occupies 3.4 Å. Its two strands are wounded in right-handed coils and each strand follows a clockwise path.

2. A-DNA

It is the dehydrated form of DNA. It occurs under conditions of decreased hydration and increased concentration of Na^+ ions. Its polynucleotide chains are coiled clockwise and have right-handed configuration. It has eleven base pairs per turn of the helix and a diameter of 23 Å. A-DNA is more compact and bulky as the distance between base pairs is 2.7 Å instead of 3.4 Å. The orientation of bases is also different. It is presumed that A-DNA is derived from B-DNA as a result of hydrophobic molecules under dry conditions.

3. Z-DNA

Another variant form of the double helix is Z-DNA, which is left-handed *i.e.*, it winds in the direction of the fingers of the left hand. It occurs in a small proportioin in cellular DNA. It is called Z-DNA because the phosphodiester backbone of its polynucleotide strands follows a zig-zag course. Z-DNA was discovered by Alexander Rich in 1979. It has 12 base pairs per turn. One complete helix is 45 Å as against 34 Å in B-DNA. The diameter of Z-DNA is 18 Å. It has only one groove. Z-DNA is less stable than B-DNA. It is formed only when purines and pyrimidines are present alternatively in the chain. Z-DNA is more compact. The Z-conformation is stabilised by high salt concentration.

4.23 DENATURATION AND RENATURATION OF DNA

We have already seen that the hydrogen bonds between the base pairs are responsible for holding the double helix together. The amount of stabilizing energy associated with the hydrogen bonds is not so great, but the hydrogen bonds hold the two polynucleotide chains in the proper alignment. Moreover, the stacking of the bases contribute the largest part of the stabilization energy.

Energy must be added to a sample of DNA to break the hydrogen bonds and to disrupt the stacking interactions. If we heat a DNA sample in solution, the two strands of DNA molecule separate. The phenomenon of separation of the two strands of DNA molecule by heating is known as **denaturation or melting**. The denatured DNA becomes single-stranded.

If native DNA molecules are heated to nearly 100°C, the hydrogen bonds between bases of two antiparallel polynucleotide strands break and two polynucleotide chains separate. This heat denaturation of DNA can be monitored experimentally by observing the absorption of ultraviolet light. There is a sharp increase in absorbance of UV light of 260 nm.

The strands of a DNA molecule separate over a temperature range. The mid-point of this range is called *melting temperature* or *transition temperature* and is represented by *Tm*. It depends upon base composition. DNA rich in G-C pairs melts at a higher temperature. In addition to the effect of base pairs, G-C pairs are more hydrophobic than A-T pairs, so they stack better, which also affects the transition temperature (Fig. 4.10).

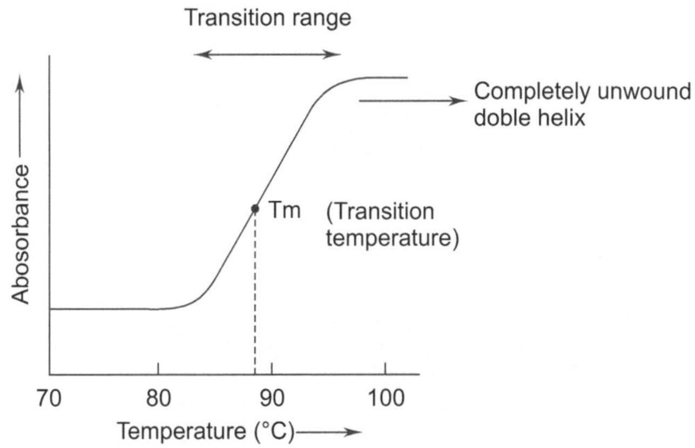

Fig. 4.10 Transition temparature.

Renaturation of denatured DNA is possible on slow cooling. When heated DNA solution is slowly cooled, the separated strands can recombine and form the same base pairs responsible for maintaining the double helix.

4.24 REPLICATION OF DNA

Whenever a cell divides to produce daughter cells, it is necessary to have duplication of DNA giving rise to a new DNA molecule with the same base sequence as the original. This duplication process is called **replication**. During DNA duplication, the two chains dissociate and each one serves as a template for the synthesis of a new complementary chain. In this way, two DNA molecules are produced, each having exactly the same molecular constitution. Approximately twenty or more enzymes and proteins are needed during replication. Replication occurs at the rate of between 50 nucleotides per second in mammals and 500 nucleotides per second in bacteria.

Watson and Crick suggested a simple mechanism of DNA replication.

The hydrogen bonds between the complementary bases on the two strands of the parent DNA molecule break so that the two polynucleotide strands of DNA separate. The strands thus separated are complementary to each other. Each separated strand acts as a template for the synthesis of a new complementary strand. Thus, two identical daughter DNA molecules are formed. As the base pairing is very specific, each nucleotide of the separated chains attracts its complementary nucleotide from the cell cytoplasm. After the nucleotides are attached by the hydrogen bonds, the sugar molecules join through their phosphate components, the formation of the new polypeptide is completed.

Thus, each strand of the double helix DNA serves as a template or model on which its complementary strand is synthesised. This method of DNA replication is described as **semi-conservative**, because each new DNA molecule contains one strand from the original DNA and one newly synthesised strand (Fig. 4.11).

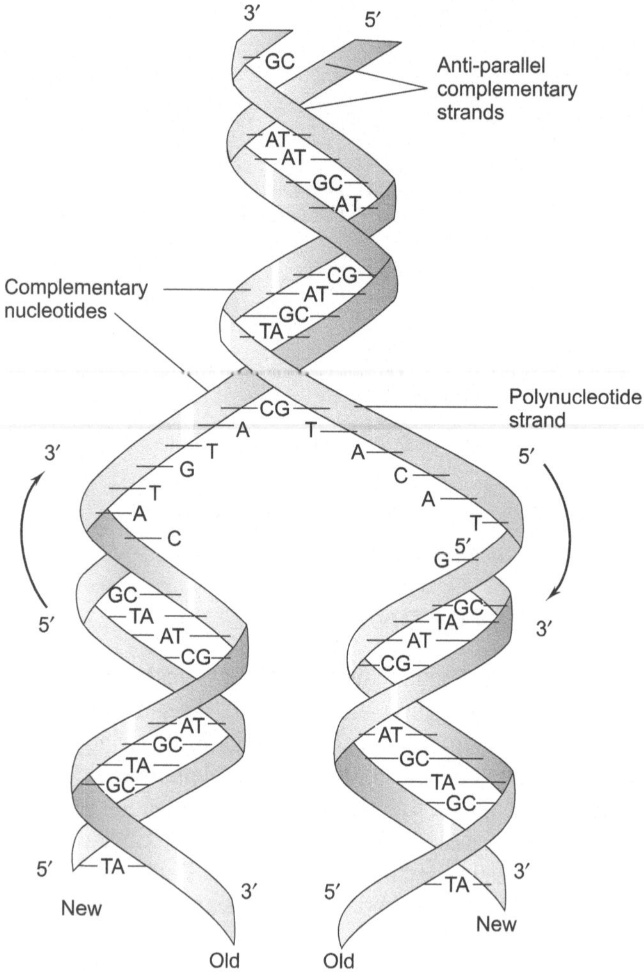

Fig. 4.11 Semi-conservative replication of DNA.

We can summarise the basic rules for DNA replication as below:

1. The base pairing is specific during DNA replication *i.e.*, adenine pairs with thymine and guanine with cytosine.
2. Nucleotide monomers are added one by one to the 3′ end of the growing strand by the enzyme DNA polymerase.
3. The sequence of bases in each daughter strand is complementary to the base sequence in the template strand.
4. 3′ Carbon of deoxyribose present on the 3′ end of the new polynucleotide chain of DNA has OH group and is free to bind to another nucleotide. 5′ Carbon of the deoxyribose on the 5′ end of polynucleotide chain has a phosphate. Therefore, the new polynucleotide chain is always synthesised in 5′ → 3′ direction.

4.25 RIBONUCLEIC ACID (RNA)

We have already seen that ribonucleic acid (RNA) is a single-stranded nucleic acid which is found in all living cells. RNA is chiefly found in the cytoplasm but also in the nucleolus. Inside the cytoplasm it occurs freely as well as in the ribosomes. The different forms of RNA are associated with the transmission of information from nucleus into the cytoplasm and with the synthesis of proteins for the regulation of cell activities. RNA is also the only macromolecule which also functions for the storage of information and also as a catalyst.

The primary structure of RNA has already been given in Fig. 4.6. More commonly RNA is a single-stranded molecule consisting of an unbranched polynucleotide chain, but it is often folded back on itself forming helices. However, DNA is a double stranded structure and its two polynucleotide chains are wounded spirally around the main axis.

Like DNA, in RNA also we have 3′–5′ phosphodiester bonds. As the sugar found in RNA is ribose, the nucleotides of RNA are called ribonucleotides. We know the four bases found in RNA are adenine, cytosine, guanine and uracil. In addition to these bases, some unusual nitrogenous bases are also found in RNA. However, here the base composition of RNA does not agree to the ratio,

$$A :: U \simeq G :: C = 1.$$

In DNA, nucleotides of the two polynucleotide strands pair through hydrogen bonds. In RNA, intramolecular pairing between the nucleotides of the single strand of RNA provides stability to the molecule.

4.26 PRINCIPAL KINDS OF RNAs AND THEIR STRUCTURES

There are six kinds of RNA-**transfer RNA**, **ribosomal RNA**, **messenger RNA**, **small nuclear RNA**, **micro RNA** and **small interfering RNA**. All these play an important role in the life processes of cells. DNA is the hereditary material, whereas RNAs perform different functions during protein synthesis. The various kinds of RNA participate in the syntehsis of proteins in a series of reactions ultimately directed by the base sequence of the cell's DNA. The base sequences of all types of RNA are determined by that of DNA. The process by which the order of bases is passed from DNA to RNA is called transcription. Transfer RNA, ribosomal RNA and messenger RNA are the three functional types of RNAs.

A. Transfer RNA (*t*-RNA)

It is the smallest of the three important kinds of RNA. Different types of *t*-RNA molecule can be found in every living cell because at least one *t*-RNA bonds specifically to each of the amino acids that commonly occur in proteins. A *t*-RNA is a single-stranded polynucleotide chain containing 75 to 93 nucleotides. Each *t*-RNA molecule has a 3′-OH terminus and a 5′-mono-phosphate terminus. Its polynucleotide chain undergoes secondary and tertiary foldings because of internal complementary base pairing. Therefore, a *t*-RNA molecule acquires a precise L-shaped three dimensional configuration. In 1965, R. W. Holley proposed a clover leaf model of *t*-RNA molecule. For this, he got Nobel Prize with Khorana and Nirenberg in 1968.

The transfer RNA plays a key role in protein synthesis. It picks up a specific amino acid from the cytoplasm, carries it to the site of protein synthesis and attaches itself to ribosome in accordance with the sequence specified by *m*-RNA. Finally, it transmits its amino acid to the polypeptide chain which is being synthesised.

B. Ribosomal RNA (*r*-RNA)

The ribosomal RNA constitutes bulk of the cellular RNA up to 85% of the total RNA. In contrast with *t*-RNA, *r*-RNA molecules tend to be quite large and only a few types of *r*-RNA exist in a cell. Cells contain millions of ribosomes and each ribosome contains several molecules of *r*-RNA. The RNA portion of a ribosome accounts for 60%–65% of the total weight and the protein portion constitutes the remaining 35%–40% of the weight. Ribosomes are the structures where the proteins are synthesised.

All ribosomes are constructed from two subunits, the larger subunit is twice the size of smaller one. Both subunits contain RNA and protein. Ribosomal RNA cannot be removed from the ribosomes without destroying them. All RNAs are single stranded, however many of these are hydrogen bonded.

r-RNA differs 'in base content from *t*-RNA and *m*-RNA. It is relatively rich in guanine and cytosine. The base components in **E. Coli** have a molar ratio of adenine 21 : uracil 19 : guanine 37 : cytosine 23.

C. Messenger RNA (*m*-RNA)

It is the least abundant of the main types of RNA. It constitutes about 5%–10% of the total cellular RNA. Messenger RNA carries genetic information from chromosomal DNA to the cytoplasm, where it acts as a template for protein synthesis. It is complementary to DNA and carries the copy of the same base sequence as found in that part of DNA from which it is copied. The only difference is that thymine is substituted by uracil. The sequences of bases in *m*-RNA specify the order of amino acids in proteins. In rapidly growing cells, many different proteins are needed within a short time interval. Therefore, fast turnover in protein synthesis becomes essential. It is logical that *m*-RNA is formed when it is needed, directs the synthesis of proteins and then it is degraded so that the nucleotides can be recycled. Of the main types of RNA, *m*-RNA is the one that usually turns over most rapidly in the cell. Both *t*-RNA and *r*-RNA can be recycled intact for many rounds of protein synthesis.

We can consider here some characteristic features of *m*-RNA:

- It is formed as a complementary strand to one of the two strands of DNA.
- As it carries the same sequence of base arrangement as found in that part of DNA from which it is copied, *m*-RNA, therefore, contains the same information as coded in that part of DNA.
- The molecules of *m*-RNA are linear and longest of all the three types of RNAs.
- There is one *m*-RNA for each polypeptide chain. *m*-RNA acts as a template for protein synthesis.
- It has a high turnover. However, it has a short life span and withers away after a few translations.
- Messenger RNA molecules are heterogeneous in size as are the proteins whose sequences they specify.

❑❑❑

CHAPTER 5

TERPENOIDS

Learning Objectives

In this chapter, we will study:

- The essential oils-occurrence in nature
- Isoprene rule
- Classification of Terpenes-their isolation from natural sources
- Citral-its isolation, structure and synthesis
- Dipentene-isolation, structure and synthesis
- Isolation and structure of geraniol
- Structure of α-terpineol
- Structure and synthesis of menthol
- α-Pinene and camphor

5.1 INTRODUCTION – THE ESSENTIAL OILS

We are quite familiar with the characteristic odour of flowers, fruits, leaves and even wood of several plants. The flowers like rose and jasmine have pleasant odours. Lemon oil and clove oil have been known to us since long. Sandalwood oil has been in use as a perfume since ancient times. The pleasant smell of these is actually due to the presence of certain volatile oils known as **essential oils**. The essential oils obtained by steam distillation or solvent extraction of these plants or their parts have been widely used in perfumery, food flavourings and medicine from the earliest times. The simple mono-and sequiterpenes are the chief constituents of these essential oils.

The term 'terpene' is usually restricted to hydrocarbons derived from $C_{10}H_{16}$ while the term terpenoids represents the hydrocarbons as well as the oxygenated derivatives. This is due to fact that since the suffix *'ene'* signifies unsaturated hydrocarbons, the name terpene is inappropriate to include compounds such as alcohols, aldehydes, ketones *etc.*

5.2 OCCURRENCE IN NATURE

Mono-and sesquiterpenoids are the chief constituents of essential oils. The essential oils are complex mixtures, however, some may consist mainly of other constituents. For example, oil of bitter almonds contains benzaldehyde and oil of winter green has methylsalicylate.

Some of the important essential oils with their terpenoid constituents are given below:

Essential Oil	*Terpenoids*
Turpentine	Pinene
Caraway	Carvone, limonene
Citronella	Geraniol, citronellal
Eucalyptus	Cineole
Jasmine	Linalool
Lavender	Linalool
Lemon	Limonene, citral
Sweet orange	Limonene, linalool
Rose	Geraniol
Cardamom	Terpineol
Camphor	Camphor
Ginger	Zinziberene

Mono - and sesquiterpenoids are of considerable commercial importance. Due to their pleasant smelling nature they are particularly used in the perfumery industry. Various other terpenoids show biological activity and are useful in pharmacy.

The di - and triterpenoids are not steam volatile. They are obtained from plant and tree gums by solvent extraction. The tetraterpenoids form a group of compounds known as **carotenoids**. Rubber is the most important polyterpenoid.

5.3 ISOPRENE RULE

On thermal decomposition almost all terpenoids yield isoprene as one of the products.

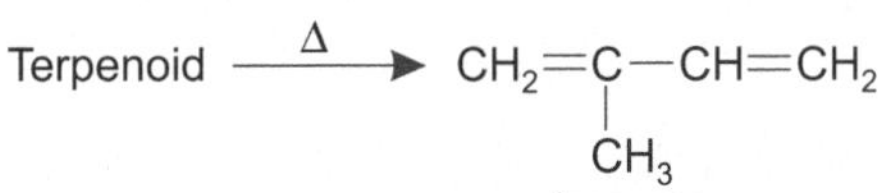

Isoprene
(2-Methyl-1,3 butadiene)

This observation led to the suggestion that all naturally occuring terpenoids can be built up of 'isoprene units'. This is known as **'isoprene rule'** and was first pointed by **Wallach**. This means that all terpenoids can be divided into isoprene units linked together in certain fashion.

It was pointed out by **Ingold** that the isoprene units in natural terpenoids were linked with one another through C_1 to C_4 positions *i.e.*, head to tail union. Head is the C_1 position or the branched end of isoprene. This divisibility into isoprene units and their 'head to tail' union is referred to as **'special isoprene rule'**. For example, in a terpenoid of the formula $C_{10}H_{16}$ head to tail linkage can be shown in the following manner:

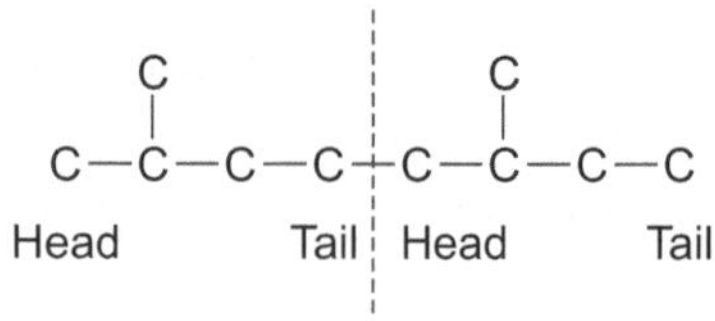

However, the above rule cannot be taken as a fixed rule as several exceptions occur. There are also some terpenoids whose carbon content is not a multiple of five and some whose C content is a multiple of five but cannot be divided into isoprene units.

The isoprene units making up some naturally occurring terpenoids are shown below:

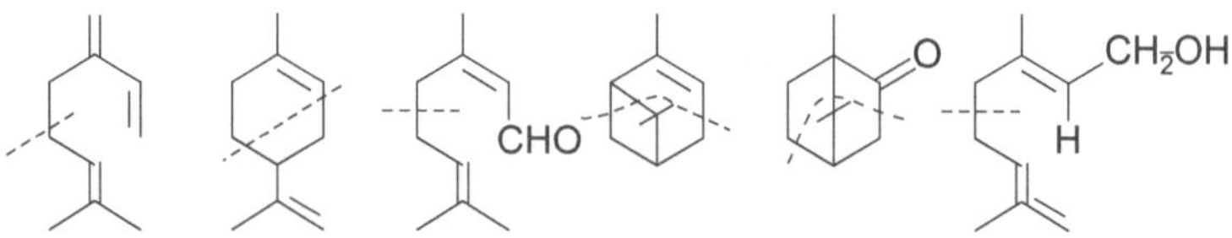

Myrcene Limonene Citral Pinene Camphor Geraniol

5.4 CLASSIFICATION OF TERPENOIDS

Terpenoids are classified on the basis of number of isoprene units present in them. Natural terpenoid hydrocarbons have the molecular formula as $(C_5H_8)_n$ where value of n is used as a basis for classification. For monoterpenoids $n = 2$ and $n = 4$ for diterpenoids and so on. Thus, we have

1. **Monoterpenoids** with 2 isoprene units, Mol. formula is $C_{10}H_{16}$.
2. **Sesquiterpenoids** ($C_{15}H_{24}$) with 3 isoprene units.
3. **Diterpenoids** ($C_{20}H_{32}$) with four isoprene units.
4. **Sesterterpenoids** ($C_{25}H_{40}$) with five isoprene units.
5. **Triterpenoids** ($C_{30}H_{48}$) with six isoprene units.
6. **Tetraterpenoids** ($C_{40}H_{64}$) with eight isoprene units.
7. **Polyterpenoids** $(C_5H_8)_n$ where $n > 40$ with several isoprene units.

The sesquiterpenoids have been discovered recently and so far only very few are known. In addition to the hydrocarbons there are oxygenated molecules of each class which also occur naturally. These are mainly alcohols, aldehydes and ketones. The terpenoids are further classified as acyclic and cyclic terpenoids depending upon whether they possess an open chain or a cyclic structure.

5.5 ISOLATION FROM NATURAL SOURCES

The essential oils obtained from certain plants contain mostly mono- and sesquiterpenoids. Essential oils are extracted from plants by several methods as outlined below:

1. Steam distillation

This is the most widely used method. The plant material is macerated and then steam distilled. The essential oils go into the distillate. The distillate is extracted by pure volatile solvents like petroleum ether. The solvent extracts are dried and the solvent evaporated or distilled under reduced pressure. However, the method should be used with great care since some essential oils decompose under these conditions.

2. Extraction by solvents

As described above some essential oils are sensitive to heat and hence decompose during distillation. In such cases, the plant material is powdered and extracted directly with light petroleum ether at 50°C. The solvent is then removed by distillation under reduced pressure to give the essential oil.

3. Adsorption in purified fats (Enfleurage method)

The fat taken in glass plates is warmed to about 50°C. The flower petals are spread on the surface of the fat for several days until the fat is saturated with the essential oils. The petals are then removed and the fat is digested with ethyl alcohol when the essential oils present in fat are dissolved in ethyl alcohol. If some fat is also dissolved during digestion, it is removed by cooling to about 20°C. The solvent is distilled off to get the essential oils.

The essential oils so obtained contain a number of terpenoids and these are separated by fractional distillation. The terpenoid hydrocarbons distil first and these are followed by oxygenated derivatives. Distillation of the residue under reduced pressure gives the sesquiterpenoids and these are separated by fractional distillation.

More recently, various forms of chromatography have been used for the isolation and separation of terpenoids. Adsorption chromatography, using alumina or silica gel as adsorbents, has provided the most important method for the isolation and separation of terpenoids particularly triterpenoids. Gas chromatography is also being used to a great extent for this purpose.

5.6 CITRAL

A. Introduction

Citral is the most important member of acyclic monoterpenoids and finds many industrial applications. Citral is widely distributed in nature and occurs in lemon grass oil to the extent of 60–80 per cent. It also occurs in citrous fruits-in oils of lime, lemon, citronella *etc.* Citral is a yellow liquid, b.p. 225°–228°C and has the smell of lemons. It is optically inactive.

Two geometrical isomers of citral are possible. Both these exist in natural citral. The trans form is called **citral-*a***, also named as **geranial**. It is more stable and present upto 90 per cent in natural citral. The *cis*-isomer is known as **citral-*b*** or **neral**, which is present to the extent of 10 per cent.

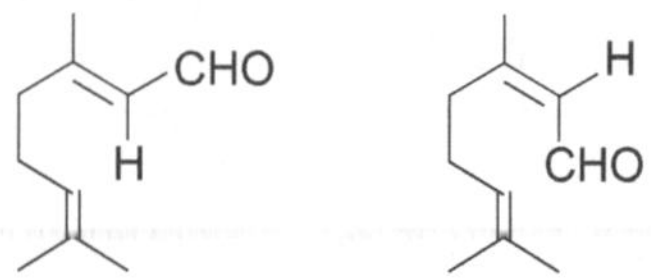

Citral-*a* (*trans* form) Citral-*b* (*cis* form)

Both the forms have been obtained. Citral is obtained as its crystalline bisulphite product which on hydrolysis gives back citral.

B. Isolation of Citral

Citral can be isolated from lemons. Lemons are crushed and subjected to steam distillation. Lemon oil is obtained which contains citral. The lemon oil is subjected to fractional distillation. The fraction collected between 225°–230°C contains citral, which is treated with sodium bisulphite. The bisulphite addition product of citral is separated and hydrolysed with sodium carbonate solution to give citral.

Citral is also obtained from lemon grass oil by fractional distillation in vacuum and purified *via* the crystalline bisulphite adddition product.

Citral is used for the preparation of β-ionone which is a special perfume.

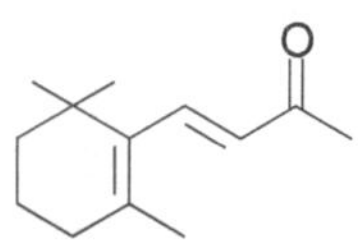

β-Ionone

C. Structure of Citral

The structure of citral has been derived on the basis of the following observations:

1. Molecular formula of citral is $C_{10}H_{16}O$.
2. **Presence of double bonds:** Citral is found to have two double bonds, as two molecules of bromine can add to citral to give tetrabromo derivative.
3. **Presence of aldehyde group:** Citral forms bisulphite addition product and oxime and 2,4-dinitrophenylhydrazone indicating the presence of a carbonyl group (*oxo* group). It reduces Fehling solution and Tollen's reagent.

Citral is reduced by sodium - amalgam to 1° alcohol, geraniol, $C_{10}H_{18}O$. On oxidation with silver oxide it forms geranic acid, $C_{10}H_{16}O_2$ containing the same number of carbon atoms.

These reactions indicate that citral has got an aldehyde group.

4. **U. V. absorption spectrum** of citral has got λ_{max} = 238nm (ε_{max} 13500), showing the presence of α, β – unsaturated system in the citral molecule.

5. **Oxidative degradation:**

   (*a*) Oxidation of citral with alkaline potassium permanganate followed by chromic acid gives acetone, oxalic acid and laevulic acid.

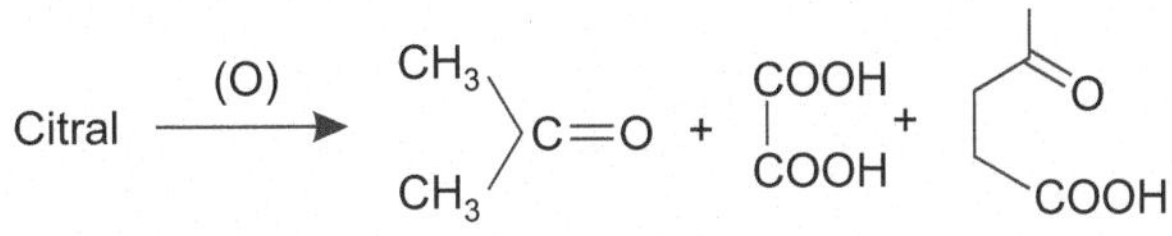

   (*b*) Ozonolysis of citral gives, acetone, glyoxal and laevulic aldehyde.

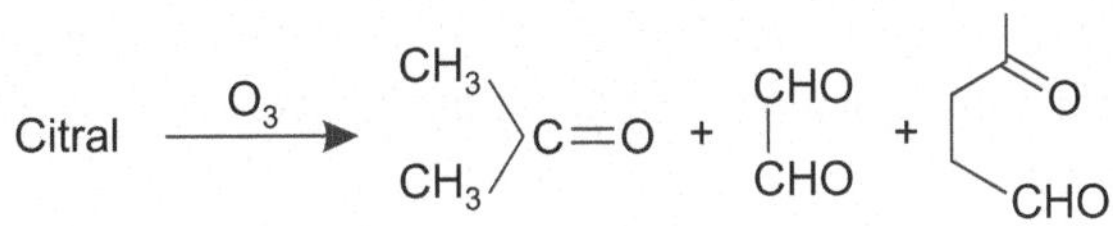

6. The carbon-skeleton of citral is shown by cyclisation of citral. On heating with $KHSO_4$ citral forms *p*-cymene. This reaction was used by Semmler to determine the position of methyl and isopropyl groups in citral.

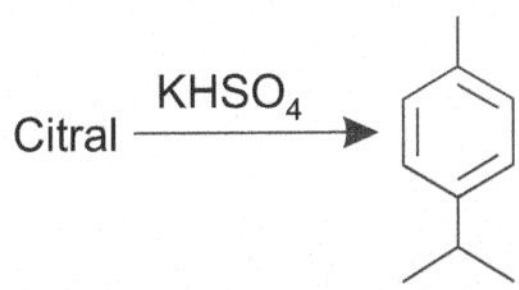

p-Cymene

Therefore, citral has got the following skeleton structure (1) in which two isoprene units are joined in head and tail fashion.

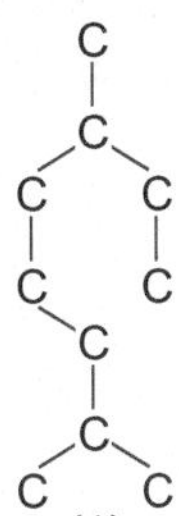

(1)

The formation of the above oxidation products can be explained by assuming structure (2) for citral.

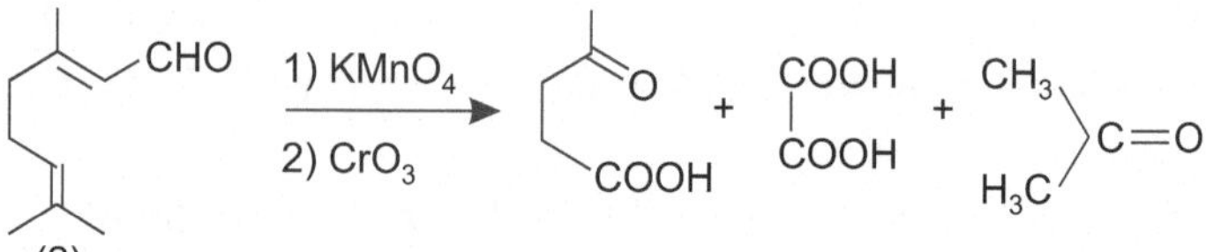

(2)

7. Hydrolysis of citral to methylheptenone:

The above structure of citral is supported by the hydrolysis of citral to methylheptenone (3). Citral on treatment with aqueous potassium carbonate is converted into 6-methylhept-5-en-2-one and acetaldehyde. Thus, we have

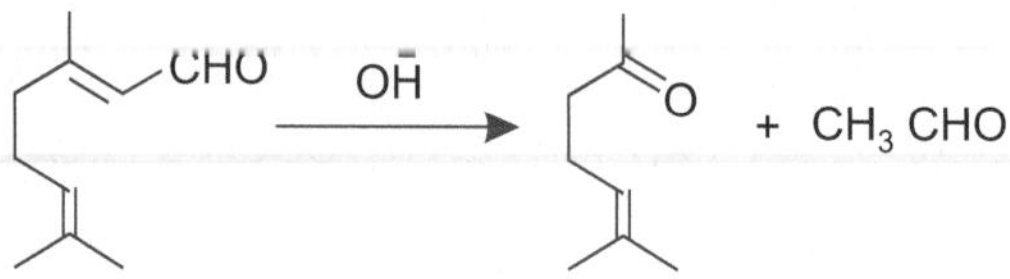

(2) (3)

Further, heptenone is itself oxidised to acetone and laevulic acid and on ozonolysis it gives acetone and laevulic aldehyde.

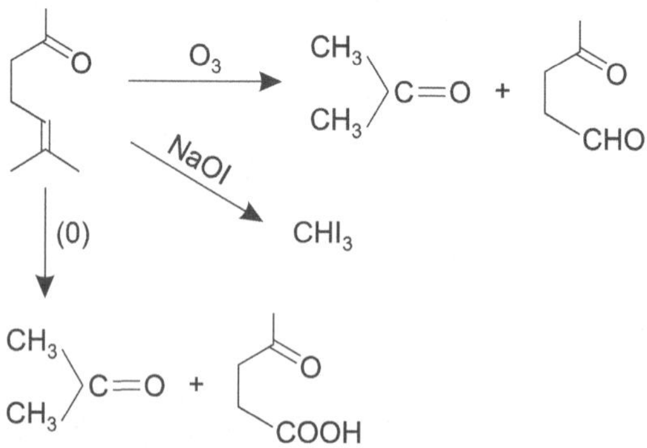

The structure of citral is finally confirmed by the synthesis of heptenone which is then converted into citral.

D. Synthesis of Citral

(a) Synthesis of methylheptenone:

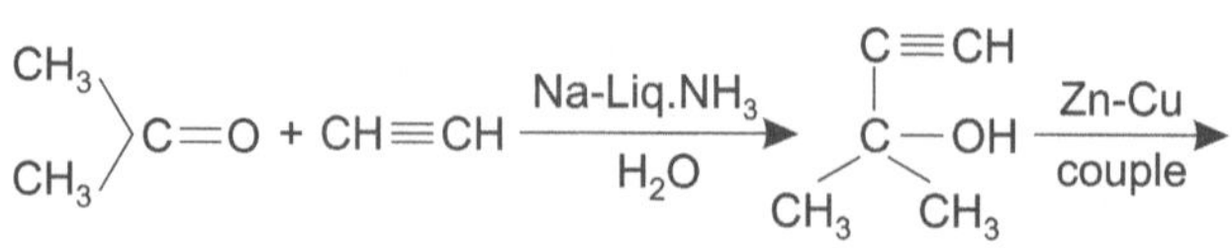

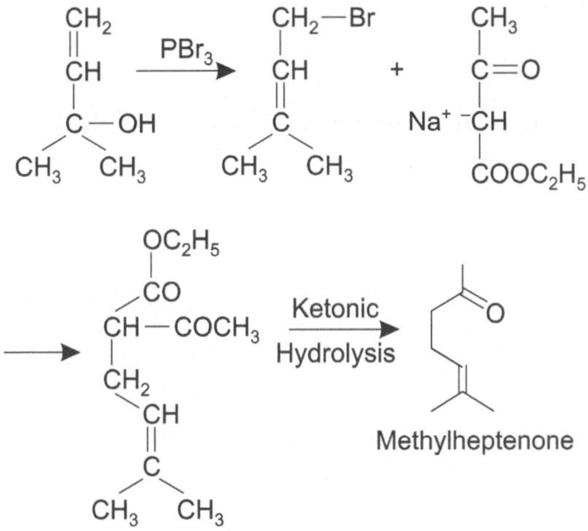

(*b*) Conversion of methylheptenone to Citral:

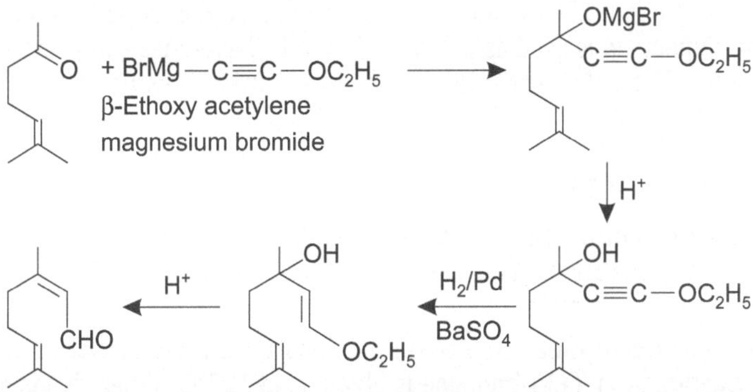

Methylheptenone can also be converted into geranic ester by the Reformatsky reaction, which is then converted into citral by heating a mixture of the calcium salt of geranic and formic acids.

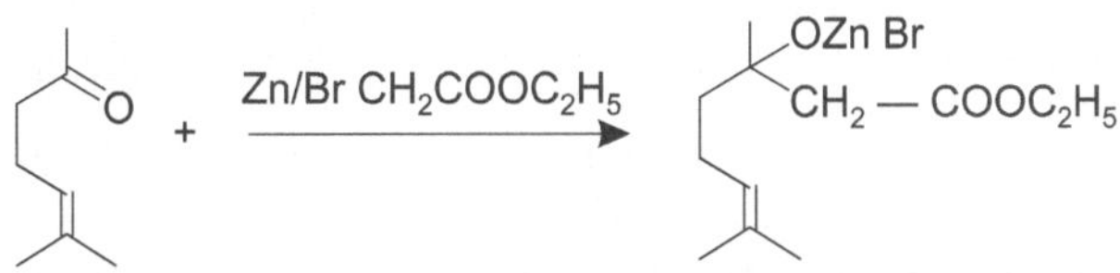

Methylheptenone

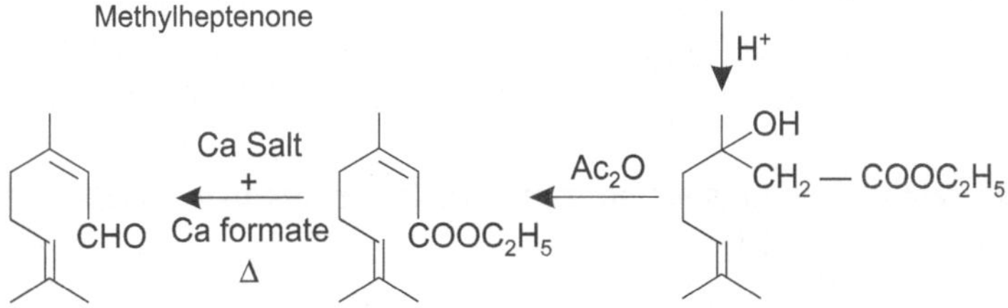

Citral

5.7 DIPENTENE

A. Introduction

The racemic modification of limonene is known as dipentene. It is a cyclic monoterpenoid. Limonene is optically active. The (+) form of limonene occurs in lemon and orange oils.

The (–) limonene occurs in peppermint oil and along with (+) pinene in pine needle oil. The (±) limonene, known as dipentene is present in turpentine oil along with *α*-pinene and terpinolene.

α-Pinene

Terpinolene

The racemic modification is also produced by racemisation of the optically active forms at about 250° C. Of the three optical isomers dipentene is the most important. Dipentene is used as a flavour in foods, beverages and pharmaceutical preparations. It is also used in the manufacture of resins, *p*–cymene and *p*–menthane.

B. Isolation of Dipentene

(*a*) **From turpentine oil:** The turpentine oil is subjected to fractionation. The steam volatile fractions contain monoterpenoids. On fractional distillation (±) form is separated.

(*b*) **From bitter orange oil:** The oil obtained from bitter oranges is subjected to fractionation. (+) Limonene is obtained at 117°C. When this is heated to 250°– 300°C, we get dipentene.

C. Structure of Dipentene (or Limonene)

The structure of dipentene has been arrived at by the following observations:

1. Molecular weight determinations show that the molecular formula of dipentene is $C_{10}H_{16}$.
2. **Presence of two double bonds:** Dipentene or limonene adds on two molecules of bromine to form a crystalline tetrabromo product. It also reacts with two molecules of hydrogen bromide to give the corresponding dihydrobromide.

   On catalytic hydrogenation using H_2/Ni it gives *p*–menthane consuming two molecules of H_2,

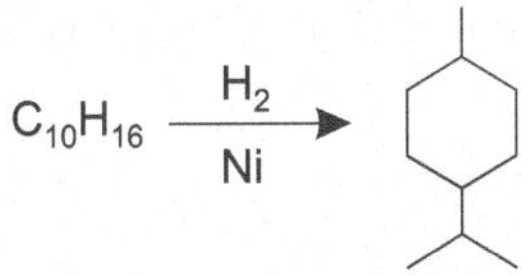

p-Menthane

p-Menthane is a completely saturated compound and its corresponding alkane is $C_{10}H_{22}$. p-Menthane has got cyclic structure, therefore, dipentene has a cyclic structure.

3. **Carbon skeleton :** Carbon skeleton of dipentene is further shown by the conversion of limonene to p-Cymene when it is heated with sulphur or selenium.

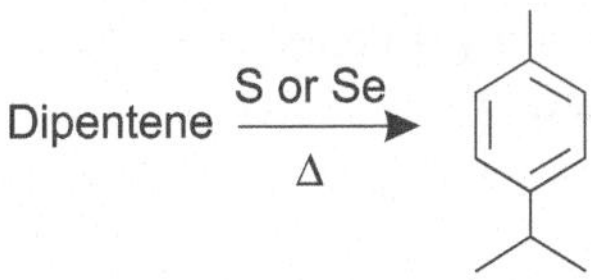

p-Cymene

These reactions show that carbon skeleton of dipentene is

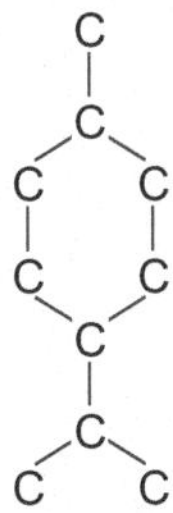

4. **Position of the double bonds :**

   (a) Dipentene when treated with NOCl consumes only one mole of NOCl. The nitrosochloride derivative obtained is colourless. This shows that one double bond is present in the ring. As the product is colourless the C-atoms of the double bond are not tertiary.

   (b) Dipentene does not undergo Diels-Alder reaction. That means double bonds are not conjugated.

   (c) Dipentene nitrosochloride on treatment with alcoholic KOH gives carvoxime

   $$\text{Dipentene} \xrightarrow{\text{NOCl}} \underset{\text{nitrosochloride}}{\text{Dipentene}} \xrightarrow[\text{EtOH}]{\text{KOH}} \text{Carvoxime}$$

   The structure of carvoxime is known. It, therefore, follows that dipentene must have a double bond in position 8 of the molecule.

   (d) When dipentene is treated with 5% H_2SO_4, we get α-terpineol (4). α-Terpineol has a double bond at 1, 2-position.

(*e*) (+)-α-Terpineol on dehydration with $KHSO_4$ gives limonene or dipentene.

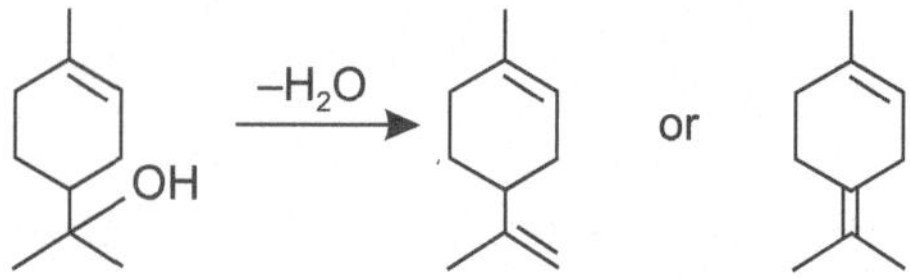

(4) (5) (6)

α-Terpineol Limonene

The position of one double bond in the ring is decided *i.e.,* it is at 1, 2-position. The position of the second double bond comes from the reaction with NOCl.

Moreover, structure (5) contains a chiral centre C-4 and hence can exhibit optical activity.

The reactions with NOCl can be shown as below:

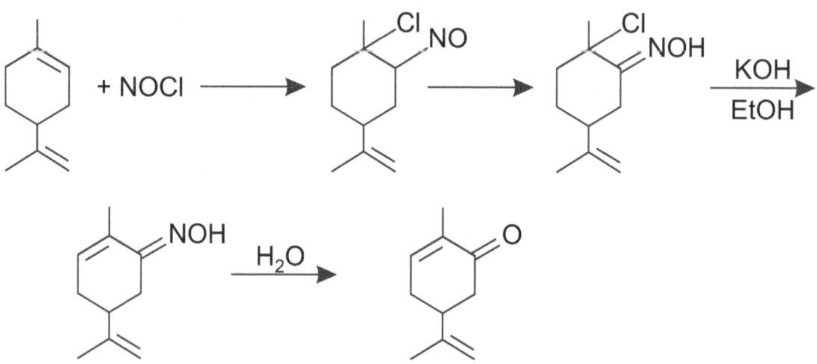

Carvoxime Carvone

(*f*) Presence of a double bond at 8 (9) position can also be shown by the following reactions:

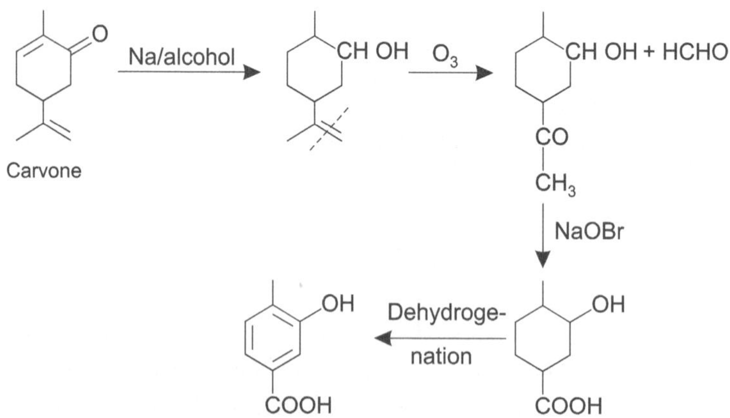

m-Hydroxy-*p*-Toluic acid

Formation of formaldehyde shows the presence of a double bond at position 8 (9).

(*g*) Dipentene on oxidation with $KMnO_4$ gives terpenylic acid and terebic acid which are also obtained by oxidation of α-terpineol.

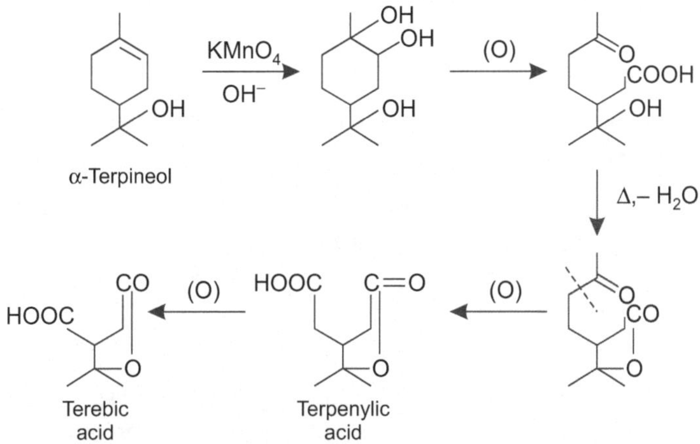

(*h*) The correlation between limonene and dipentene can be shown as follows:

(+) or (–) Limonene adds on two molecules of hydrogen chloride in the presence of moisture to form limonene dihydrochloride and this is identical with dipentene dihydrochloride

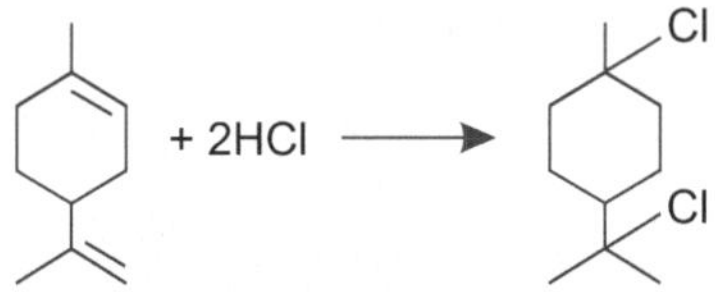

(+) or (–)-Limonene

Limonene dihydrochloride no longer contains a chiral centre. It is, therefore, optically inactive.

Dipentene can be regenerated by heating the dihydrochloride with sodium acetate in acetic acid or boiling with aniline.

D. Synthesis of Dipentene

1. **From Isoprene :** By dimerisation of isoprene at 280° C, we get dipentene.

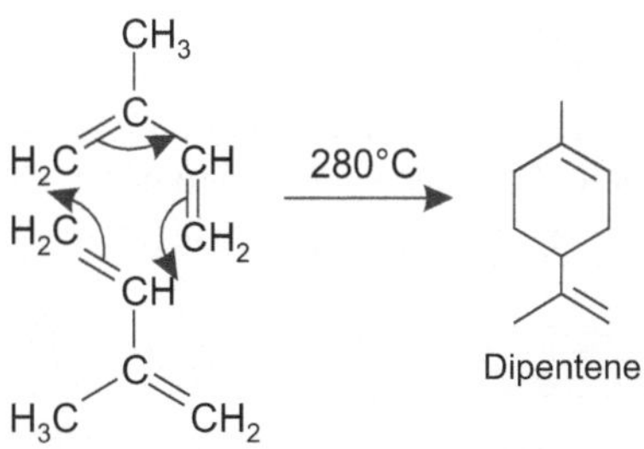

2. From α-terpineol by dehydration with $KHSO_4$:

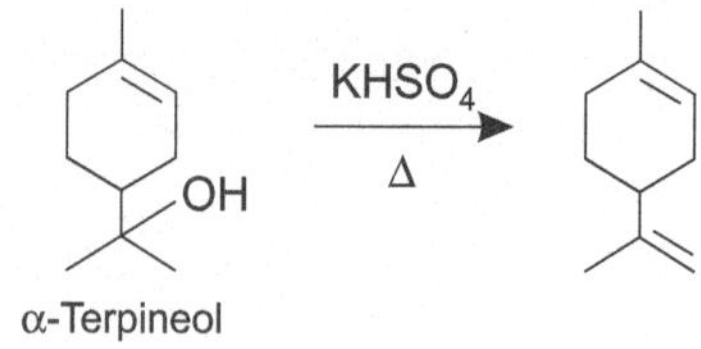

E. Synthesis of α-Terpineol can be carried out by the following reactions:

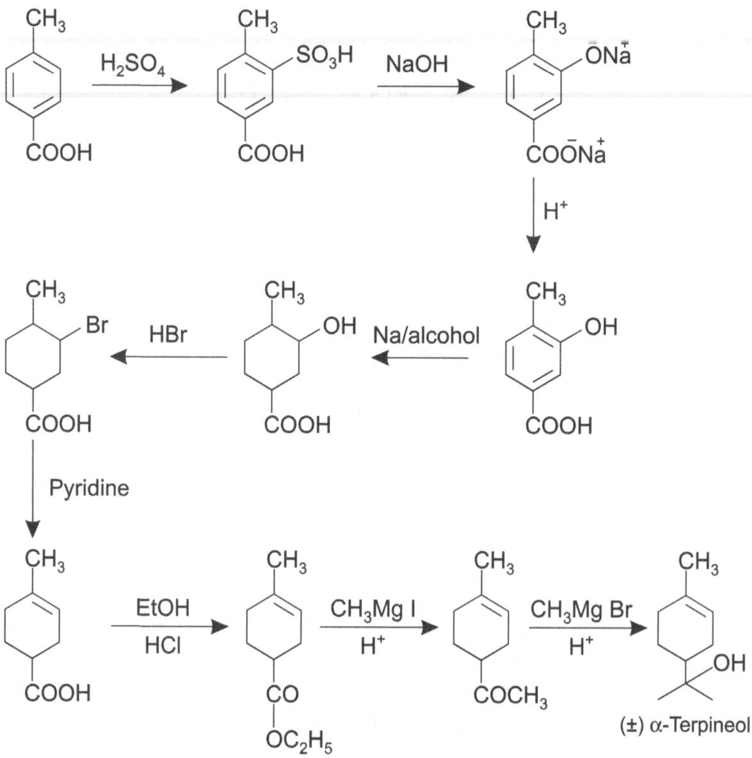

5.8 GERANIOL

A. Introduction

While discussing the structure of citral, we have seen that citral-*a* is also known as geranial and citral-*b* is known as neral. Geraniol is the alcohol corresponding to geranial and nerol is the alcohol corresponding to neral. Geraniol is found in many essential oils, particularly rose oil. It also occurs in lemon-grass, lavender, citronella and palmrosa oils.

B. Isolation

Geraniol is isolated from the oil of palmrosa by treating it with anhydrous calcium chloride and decomposing the crystalline derivative with water.

Geraniol is obtained as a colourless liquid (b. p. 229°–230°C). It has a pleasant rose like odour. It is insoluble in water but dissolves in ethanol. Geraniol is used in the preparation of artificial rose scents. It is widely used in perfume, cosmetic and flavour industries.

C. Structure of Geraniol

1. Molecule formula of Geraniol has been found to be $C_{10}H_{18}O$.
2. It adds on 2 molecules of bromine to form a tetrabromide and on catalytic hydrogenation it consumes two molecules of hydrogen. Geraniol, therefore, contains two double bonds.
3. Geraniol does not form an oxime indicating the absence of an aldehyde or a ketone group.
4. On oxidation geraniol gives successively citral-*a* (an aldehyde) and geranic acid containing the same number of carbon atoms as geraniol. This means geraniol is a primary alcohol and the positions of the double bonds in geraniol are the same as in citral-*a*. Geraniol, therefore, has the following structure as given below:

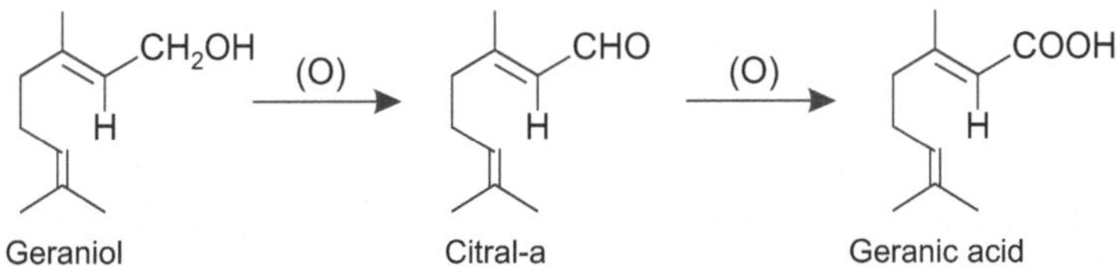

5. Reduction of citral produces geraniol and some nerol is also formed.

   We know that geraniol and nerol are structurally related. This can be shown by following facts:

   * Both, geraniol and nerol, add on two molecules of hydrogen during catalytic hydrogenation and give the same saturated alcohol, $C_{10}H_{22}O$.
   * On oxidation geraniol and nerol give the same oxidation products. Hence, geraniol and nerol are geometrical isomers.
6. Geraniol and nerol both can be cyclised to give α-terpineol by means of dilute sulphuric acid. This cyclisation is 9 times as fast with nerol as it does with geraniol. The faster rate of cyclisation indicates that the alcoholic group is very near to the carbon which is involved in the ring formation. This means the configuration of nerol is ***cis*** and geraniol has been assigned the ***trans*** configuration.

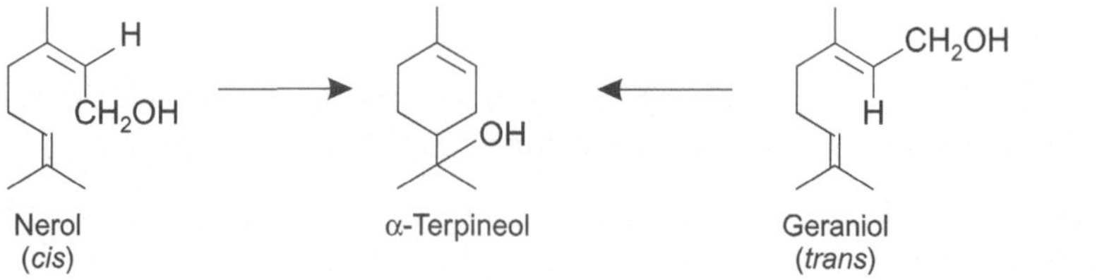

7. The hydration of geraniol and nerol to α-terpineol involves the formation of same intermediate as given below:

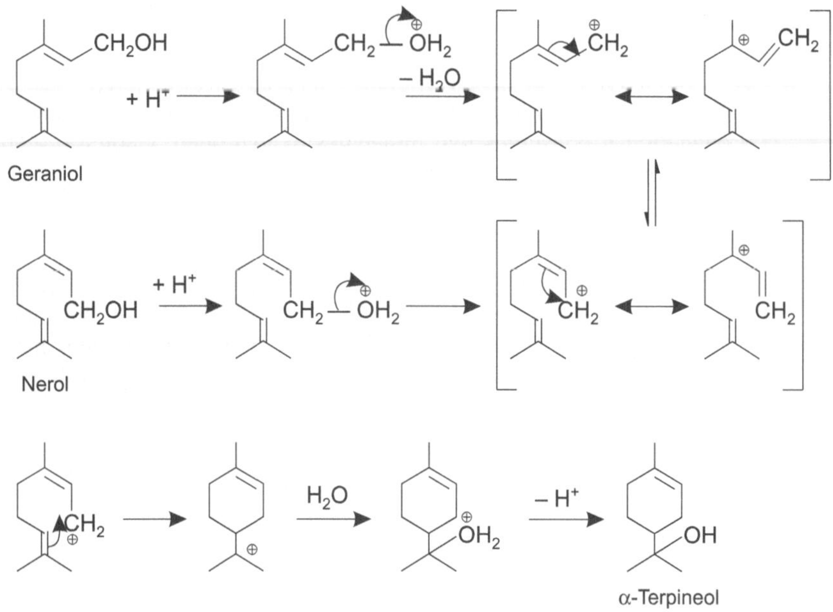

5.9 α-TERPINEOL

A. Introduction

α-Terpineol is the most important monocyclic terpenoid. Its structure confirms the structures of several other monoterpenoids because α-terpineol can be converted into many other terpenoids and *vice versa*.

α-Terpineol is an optically active monoterpenoid. It occurs naturally in the (+), (-) and (±) forms. The racemic modification is a crystalline solid, m.p. 35°C. α-Terpineol is extensively used in the preparation of certain perfumes.

B. Structure of α-Terpineol

1. The molecular formula of α-terpineol is $C_{10}H_{18}O$.
2. It adds on only one molecule of bromine to form a dibromide. It, therefore, contains one double bond.

3. Oxygen atom is present as a tertiary alcoholic group.
4. The saturated hydrocarbon of α-terpineol has the molecular formula $C_{10}H_{20}$. This corresponds to the general formula of cycloalkanes (C_nH_{2n}). It means that α-terpineol is a monocyclic terpenoid.
5. On heating with sulphuric acid α-terpineol forms *p*-cymene indicating that α-terpineol contains *p*-cymene skeleton

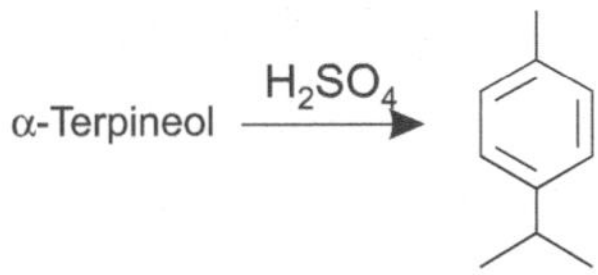

p-Cymene

It follows, therefore, that α-terpineol is a *p*-menthane derivative with one double bond and a tertiary alcoholic group.

6. Oxidation of α-terpineol with dilute (1%) alkaline potassium permanganate gives a trihydroxy compound (8), $C_{10}H_{20}O_3$. This on oxidation with chromic acid produces a compound with molecular formula, $C_{10}H_{16}O_3$ (9). Compound (9) was shown to be a lactone of monocarboxylic acid. There is no loss of carbon atoms in its formation, the double bond must, therefore be in the ring of α-terpineol.

   Further, on warming with alkaline permanganate the lactone (9) gave acetic acid and a compound $C_8H_{12}O_4$ (10). The formation of acetic acid suggests that (9) contains a CH_3 CO – group. Therefore, (9) is a methyl ketone and a lactone.

   The acid (10) was named as terpenylic acid. Its reactions showed that it was the lactone of a monohydroxy dicarboxylic acid. Oxidation of terpenylic acid gives terebic acid, $C_7H_{10}O_4$ (11), which is the lactone of a monohydroxy dicarboxylic acid. The reactions are shown below:

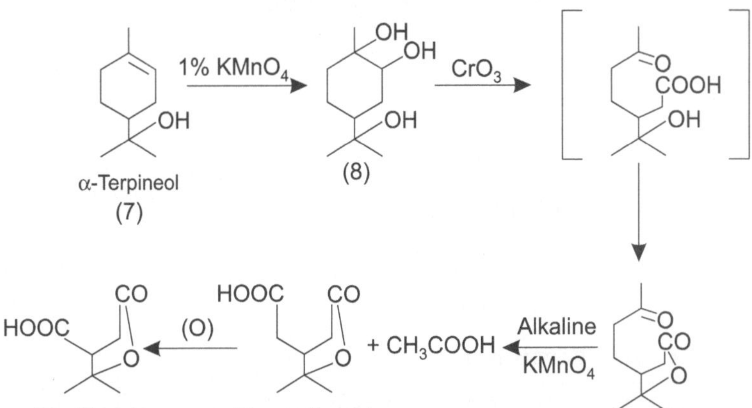

α-Terpineol
(7)

(8)

(9)

Terebic acid
(11)

Terpenylic acid
(10)

The above reactions confirm structure (7) for α-Terpineol.

The structures of terpenylic acid (10) and terebic acid (11) are confirmed by synthesis.

C. Synthesis of Terpenylic acid

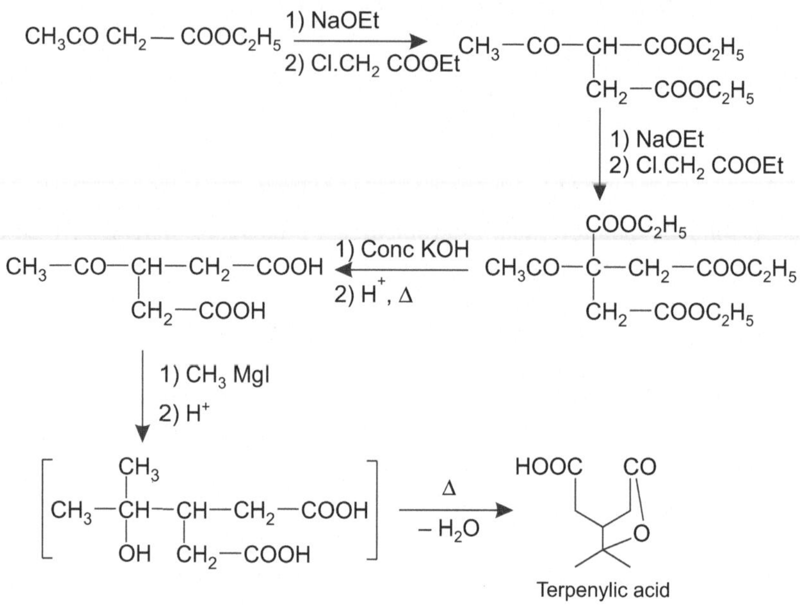

D. Synthesis of Terebic acid

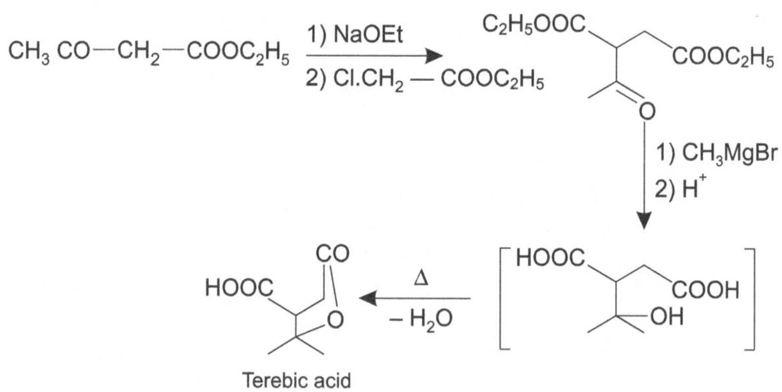

E. Synthesis of α-Terpineol

1. The structure of *α*-terpineol is finally confirmed by its synthesis which starts with *p*-toluic acid.

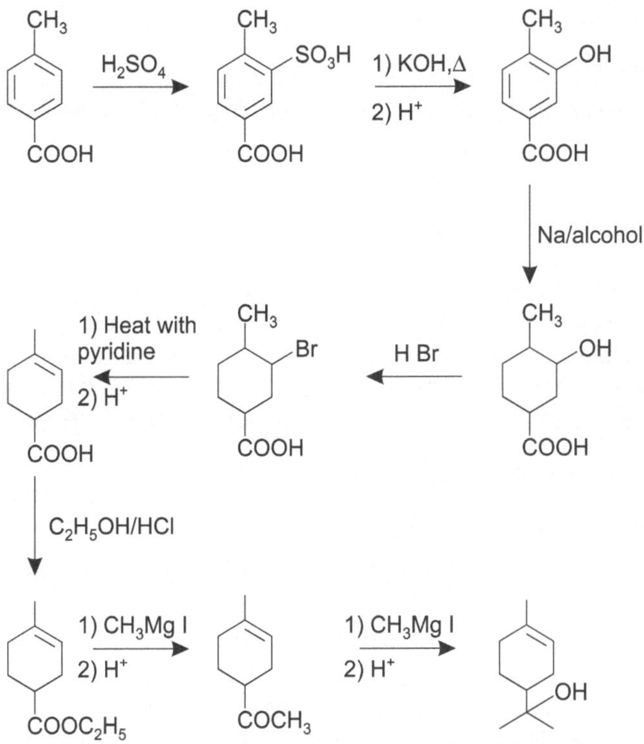

(±)-Terpineol

2. A much simpler synthesis of α-terpineol has been carried out by Alder and Vogt. In this isoprene and methyl vinyl acetate condense by Diels-Alder reaction to give the ketone which is converted to α-terpineol.

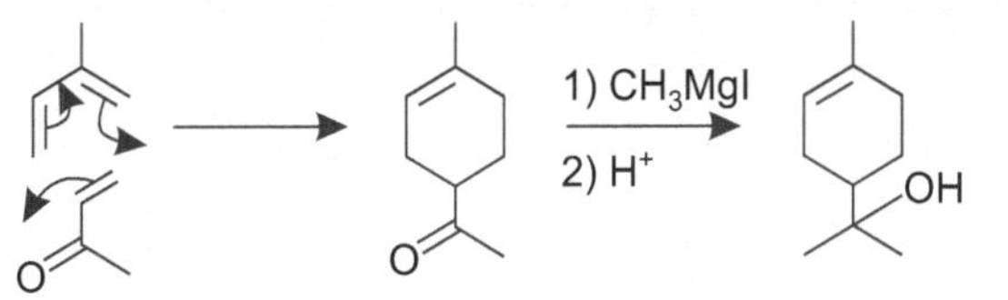

Two other terpineols-β-terpineol and γ-terpineol also occur naturally.

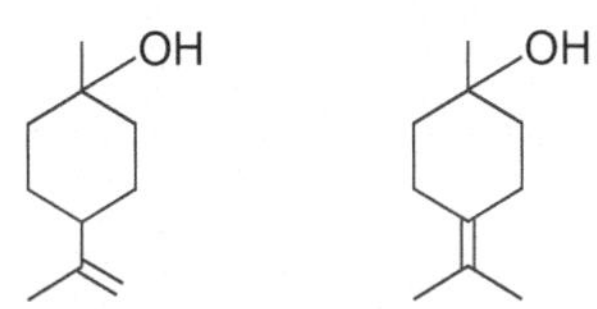

β-Terpineol γ-Terpineol

5.10 MENTHOL

A. Introduction

Menthol occurs in the oil of peppermint to the extent of 80 per cent. When this oil is cooled, crystals of menthol are obtained. Menthone obtained by the oxidation of menthol also occurs naturally in the oils of peppermint and geranium.

Menthol is an optically active compound. Only the (–) form of menthol occurs naturally. (–)-Menthol is a colourless solid, m.p. 43° C. Menthol has got anaesthetic and antiseptic action. It is widely used in many pharmaceutical preparations, in face creams, shaving creams, tooth pastes and in ointments because it imparts cooling sensation to the skin. It is used in Vicks inhalers because its vapours have a cooling and mildly anaesthetic action on the inflamed nasal tissues. It is also used in chewing gums and certain brands of cigarettes.

B. Structure of Menthol

1. Menthol has molecular formula as $C_{10}H_{20}O$.
2. Menthol is a saturated compound. No addition of bromine and hydrogen is there.
3. Menthol forms esters with acids. Therefore, it must possess an alcoholic group.
4. On oxidation with chromic acid menthol is converted into menthone, which is a ketone. The alcoholic group in menthol is, therefore, secondary.
5. Reduction of menthol with hydrogen iodide gives *p*-menthane. Menthol has, therefore, a *p*-menthane skeleton.
6. Finally, menthol is obtained by reduction of (+) pulegone. As the structure of pulegone is known, it therefore follows that menthol must be (12).

   All the above reactions can be shown as given below:

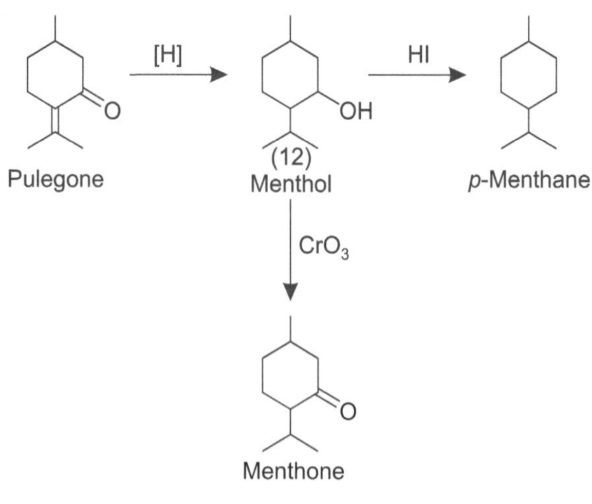

7. The structure of menthol is confirmed by the oxidation products of menthone.

   Menthone on oxidation with $KMnO_4$ gives a keto acid, $C_{10}H_{18}O_3$ (13), which on further oxidation forms acetone and *β*-methyladipic acid. The reactions can be explained on the basis of structure given above for menthone.

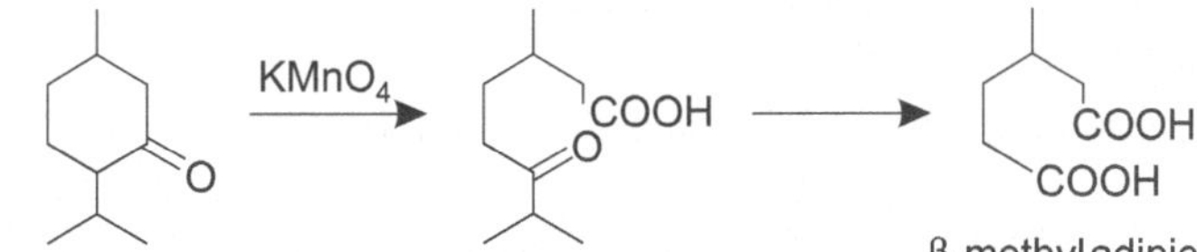

Menthone (13) β-methyl adipic acid

C. Synthesis of Menthol

The structure of menthone and menthol have been confirmed by their synthesis as given below:

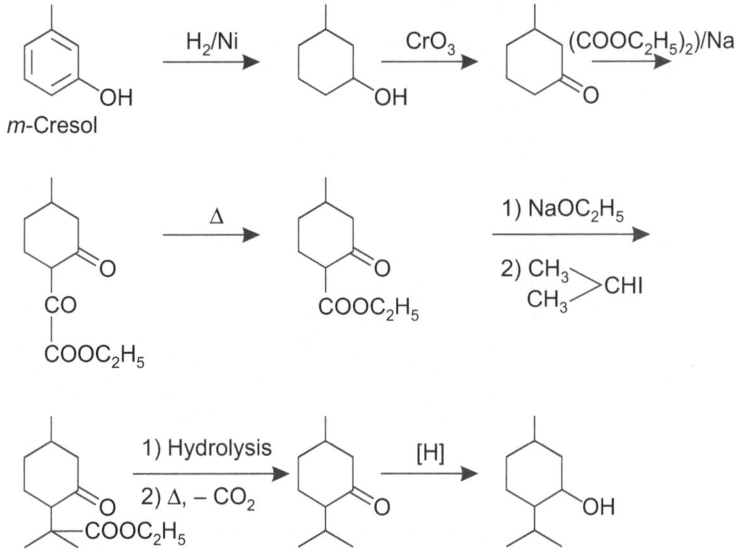

Menthone Menthol

5.11 α-PINENE

α-Pinene is the most widely distributed terpenoid. It is present in many essential oils. The most common source of α-pinene is turpentine oil, in which it occurs in both (+) and (–) forms. α-Pinene is a liquid, b.p. 156°C. Turpentine oil, which contains large amount of pinene, is used as a solvent for resins and as a thinner for paints and varnishes.

A. Structure of α-Pinene

1. The molecular formula of α-pinene is $C_{10}H_{16}$.
2. It adds on two bromine atoms indicating the presence of one double bond. Thus the molecular formula of the parent hydrocarbon is $C_{10}H_{18}$. This corresponds to the general formula of a bicyclic compound, α-pinene is, therefore, bicyclic.
3. α-Pinene on treatment with alcoholic sulphuric acid gives α-terpineol.

α-Pinene $\xrightarrow{C_2H_5OH/H_2SO_4}$

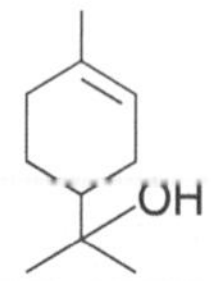

α-Terpineol

The structure of α-terpineol is well-known. Its formation from α-pinene suggests the following points:

(*i*) α-Pinene possesses six-membered ring. The position of the double bond is same as in α-terpineol.

(*ii*) Again α-pinene, which is bicyclic, takes up one molecule of water to give α-terpineol. The presence of hydroxyl group at C-6 in α-terpineol suggests that the C-6 of α-terpineol is involved in the formation of the second ring in α-pinene.

There are three possibilities as shown by the following structures for the C-6 to form other ring of α-pinene. The fourth (17) possibility is ruled out on the ground that a double bond cannot be formed by a carbon atom occupying the bridge-head of the bicyclic system (Brett's rule).

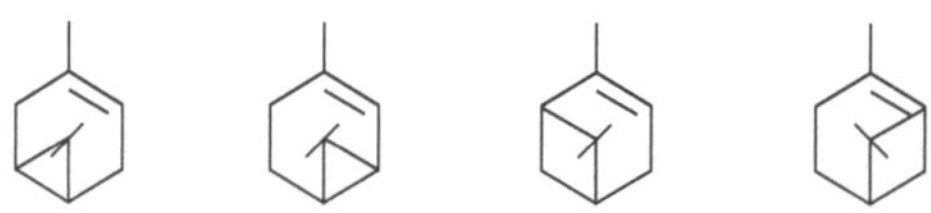

(14) (15) (16) (17)

(*iii*) Moreover, the gem-dimethyl group of α-terpineol is not present in the six-membered ring of α-pinene and hence, it must be in the other ring.

4. The second ring was shown to be four-membered by Baeyer by carrying out the following oxidative degradation of α-pinene.

α-Pinene $\xrightarrow[KMnO_4]{1\%\ \text{alkaline}}$ Pinene glycol (18) $\xrightarrow[KMnO_4,\ \Delta]{\text{Alkaline}}$ Pinonic acid (19) $\xrightarrow{NaOBr}$ Pinic acid (20) + $CHBr_3$

Pinic acid (20) $\xrightarrow[2)\ Ba(OH)_2\ \ 3)\ CrO_3]{1)\ Br_2}$ *Cis*-Norpinic acid (21)

The formation of bromoform suggests that the pinonic acid has an acetyl group. The final product of oxidation, *cis*-norpinic acid, $C_8H_{12}O_4$, was shown to be a saturated dicarboxylic acid. Therefore, the gem – dimethyl group must be present in *cis*-norpinic acid. Thus, the formula for this acid may be written as $(CH_3)_2$–C_4H_4–$(COOH)_2$, which indicates that the saturated parent hydrocarbon of this acid is C_4H_8 and hence, the acid must be a cyclobutane derivative. So, norpinic acid is probably a dimethyl cyclobutane dicarboxylic acid.

Working backwards, α-pinene must have structure (16) which explains all the above reactions as given below:

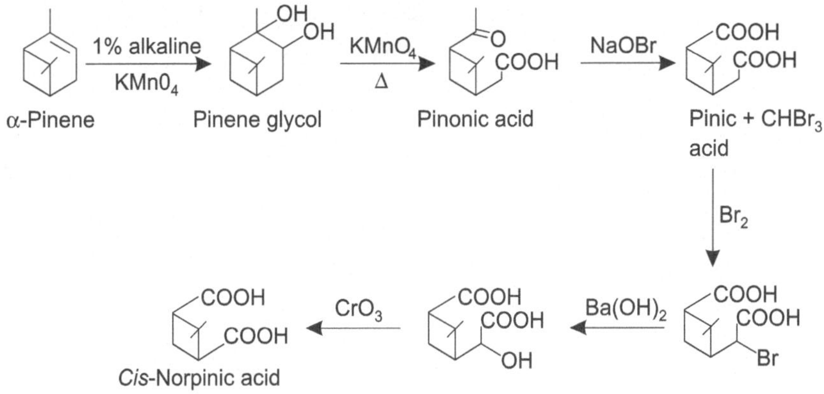

5. **Synthesis of nor–pinic acid:** The synthesis of nor–pinic acid confirms its structure.

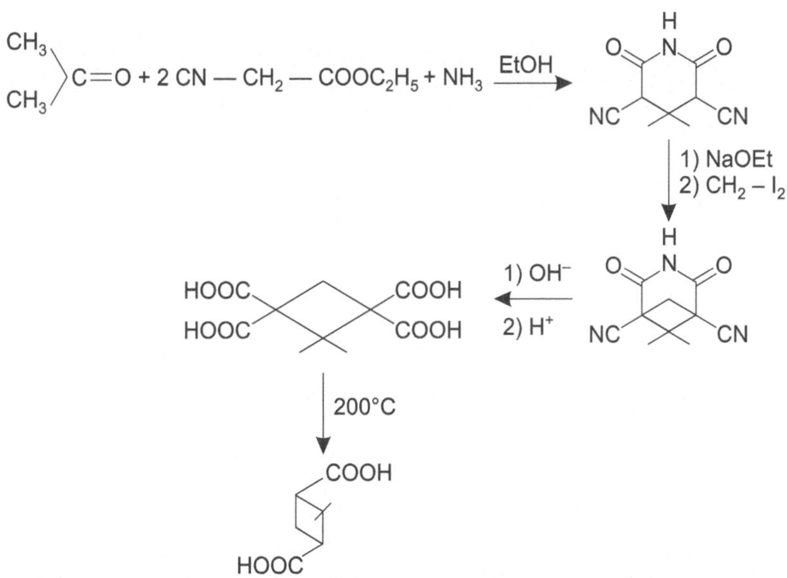

The above method gives *trans*-isomer of nor–pinic acid. This is readily converted into the *cis*-isomer by heating the *trans*-isomer with acetic anhydride when *cis*-anhydride is formed. On hydrolysis, this gives the *cis* acid

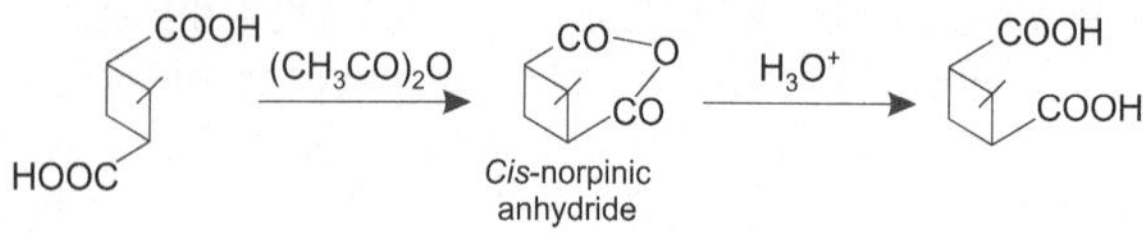

B. Synthesis of α-Pinene

Finally, the structure of α-pinene is confirmed by its synthesis from pinonic acid, which is synthesised from nor–pinic acid as given below:

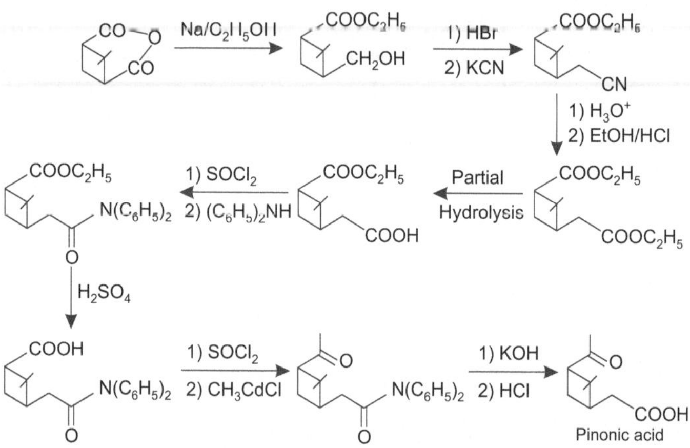

Pinonic acid is converted into α-pinene by the following sequence of reactions.

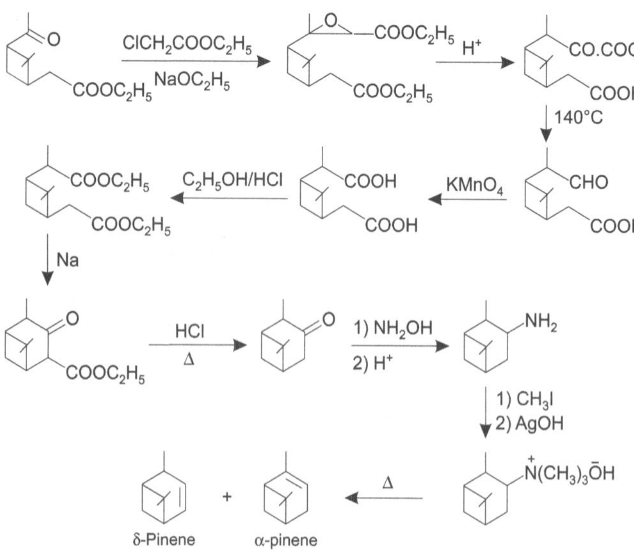

The mixture of α-pinene and δ-pinene is separated and identified.

5.12 CAMPHOR

A. Introduction

Camphor is the most important constituent of the oil of camphor. It occurs naturally in the camphor tree of Formosa and Japan. In this tree, although camphor is present in all the parts, the highest proportion is in the trunk.

Camphor is a solid, m.p. 180°C and is optically active. Both the (+) and (–) forms occur naturally. The racemic form is usually a synthetic product. Camphor is mainly used for smokeless powders and as explosive. It is also used for medicinal purposes as a mild disinfectant and stimulant for the heart muscles.

B. Structure of Camphor

1. The molecular formula of camphor is $C_{10}H_{16}O$.
2. The general reactions of camphor show that it is a saturated molecule.
3. The oxygen atom present is in the form of a keto group. This is shown by the fact that camphor forms an oxime. However, oxidation of camphor gives a dicarboxylic acid containing 10 carbon atoms. If camphor contained an aldehyde group, oxidation would have given a mono carboxylic acid containing 10 carbon atoms.

   This means that the parent hydrocarbon of camphor has the molecular formula, $C_{10}H_{18}$ corresponding to the general formula, C_nH_{2n-2}. Camphor is, therefore, bicylic.
4. On treatment with nitrous acid, camphor forms an oxime indicating the presence of $-CH_2CO-$ group.
5. Camphor has got *p*-cymene skeleton is shown by the fact that on distillation with zinc chloride or phosphorus pentoxide it gives *p*-cymene.

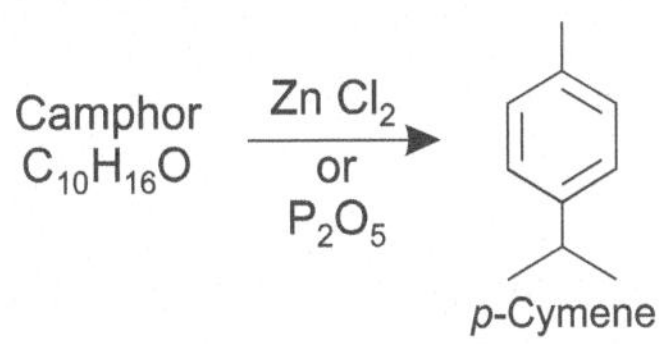

6. Oxidation of camphor with nitric acid gives camphoric acid ($C_{10}H_{16}O_4$) (22). Further oxidation of camphoric acid gives camphoronic acid, $C_9H_{14}O_6$ (23).
7. **Structure of camphoric acid and Camphoronic acid:**
   (*i*) Various reactions of camphoric acid show that it is a dicarboxylic acid and it is saturated. As camphoric acid is formed from camphor, it means the ring containing the keto group is opened and therefore, camphoric acid must be a monocyclic compound.

(*ii*) Camphoronic acid (23) was shown to be a saturated tricarboxylic acid. On distillation at atmospheric pressure it gives isobutyric acid, trimethyl succinic acid, carbon dioxide and carbon. The formation of these products can only be explained from the following structure (23) for camphoronic acid.

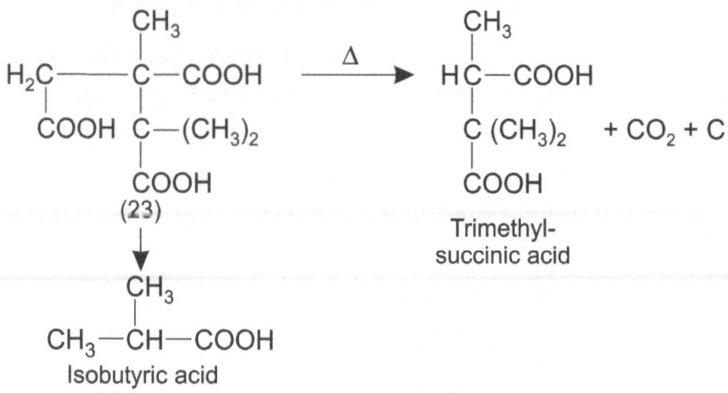

(*iii*) If camphoronic acid has structure (23), then camphoric acid must contain three methyl groups *i.e.*, it has the molecular formula as $(CH_3)_3$–C_6H_5–(COOH), which corresponds to C_5H_{10} as the parent hydrocarbon. Therefore, camphoric acid is a cyclopentane derivative. We have already seen that camphoric acid is monocyclic. Thus, oxidation of camphoric acid to camphoronic acid may be written as

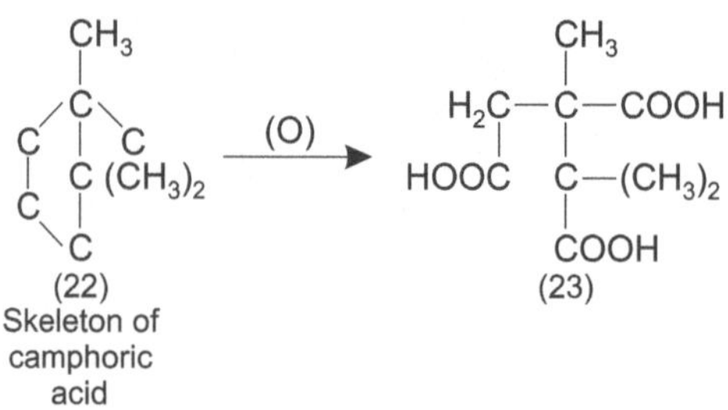

(*iv*) As camphoric anhydride forms only one mono bromo derivative, there is only one α-hydrogen atom in camphoric acid. Therefore, carbon of one carboxyl group must be C_1 because this is the only C atom attached to a tertiary carbon atom.

(*v*) Two –COOH groups are produced as a result of oxidation of camphor. One –COOH group comes from –CO, the other must come from $-CH_2$ group. Therefore, C_1 carbon must be the carbon of the keto or methylene group in camphor.

Thus, two structures (22) and (24) are possible for camphoric acid. Structure (24) suggests that camphor will contain a six-membered ring with a gem dimethyl group. The conversion of camphor into *p*-cymene cannot be explained by this structure. If we assume structure (22) for camphoric acid, then camphor would have structure (25). We can explain the above reactions *i.e.*, conversion of camphor to camphoric acid and camphoronic acid

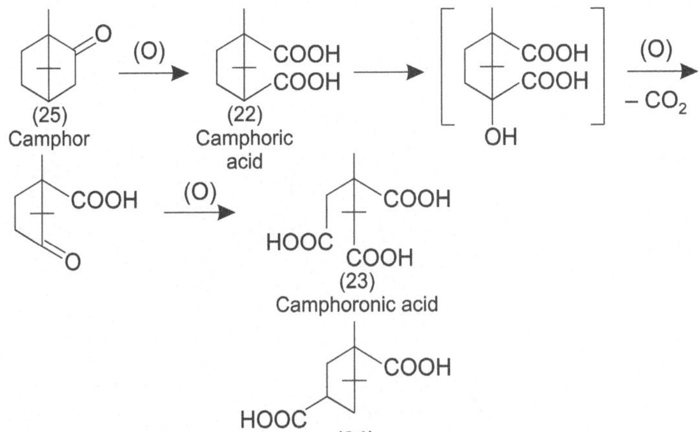

(25) Camphor; (22) Camphoric acid; (23) Camphoronic acid; (24)

On the basis of above observations Brett suggested two structures for camphor

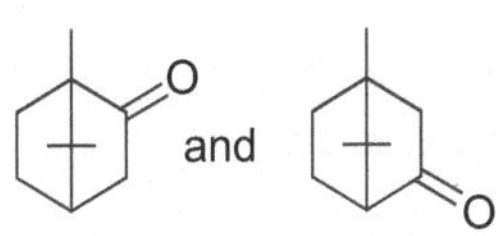

(25) (26)

He rejected structure (26) in favour of (25) because camphor gives carvacrol (27) when distilled with iodine. The formation of this compound can be expected from (25) but not from (26)

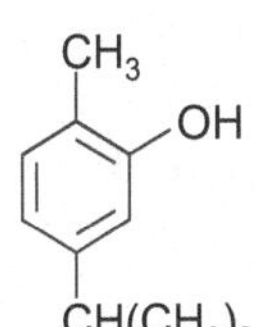

(27)

The above structure (25) for camphor was finally confirmed by the synthesis of camphoronic acid, camphoric acid and camphor.

C. Synthesis of (±) Camphoronic Acid

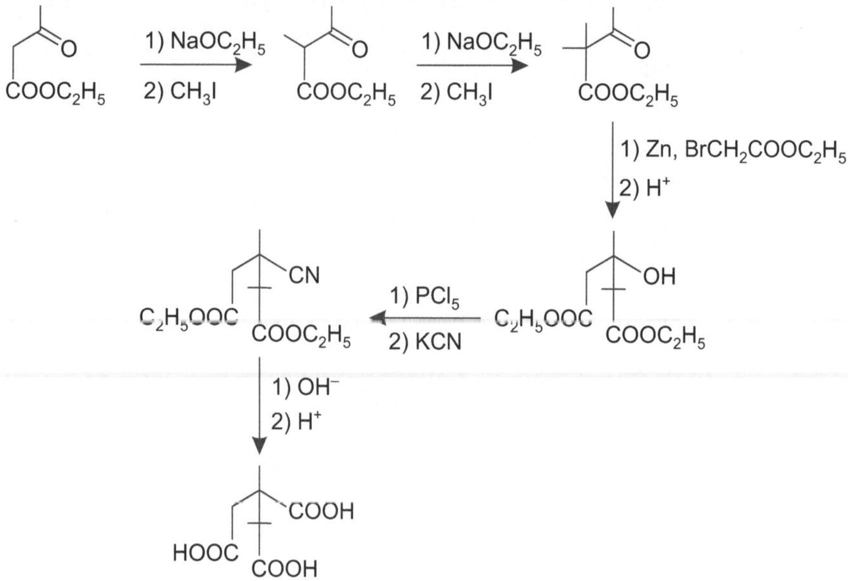

Camphoronic acid

D. Synthesis of (±) Camphoric Acid

3, 3 – Dimethyl glutaric ester is synthesised first starting with mesityl oxide and diethylmalonate. The complete reactions are as follows:

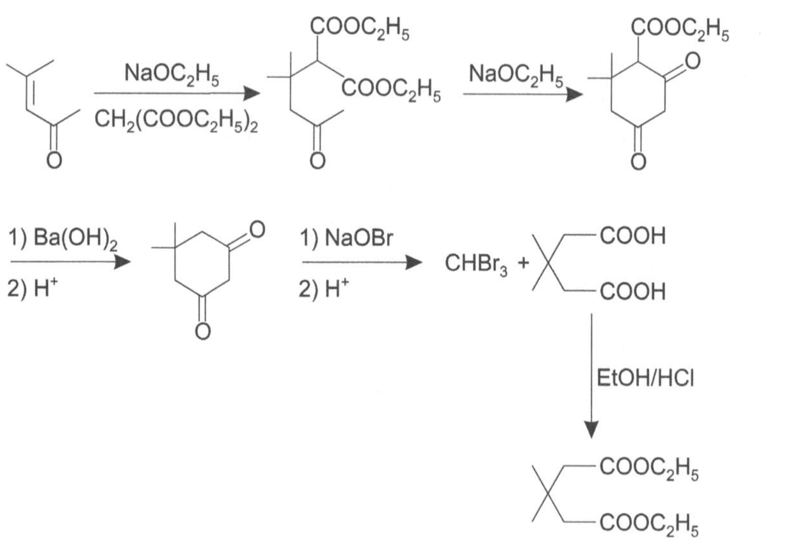

The conversion of dimethyl glutaric ester to camphoric acid involves the following reactions:

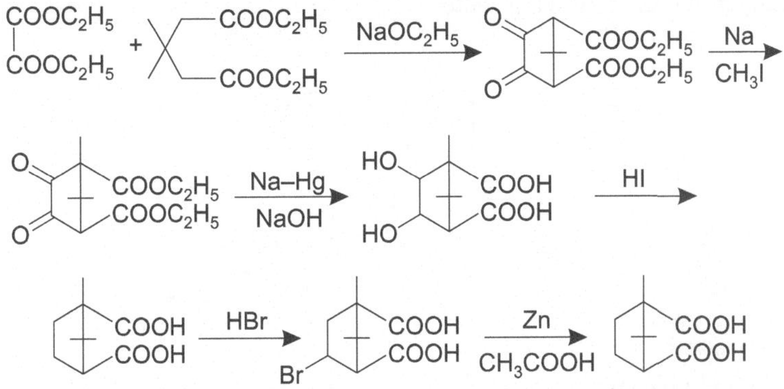

Camphoric acid

E. Synthesis of Camphor

Camphoric acid obtained above can be converted to camphor by the following reactions:

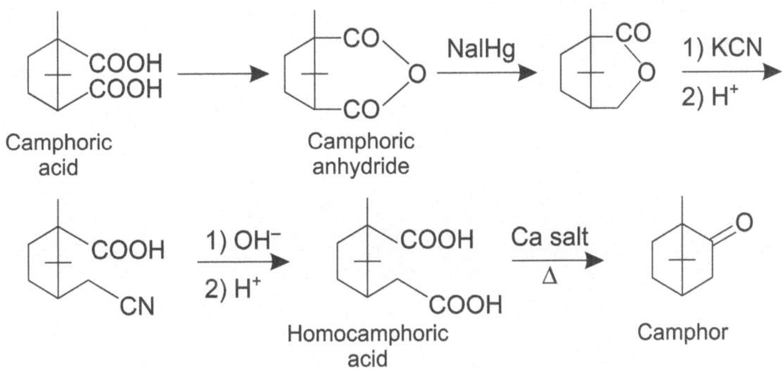

Money *et al.* synthesised (±) camphor from dihydrocarvone in just two steps.

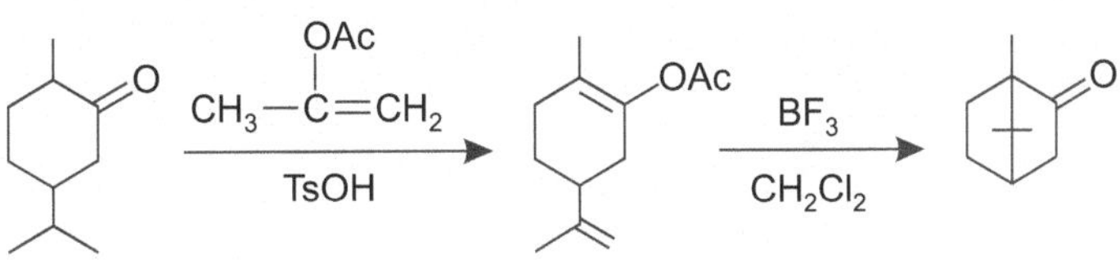

Dihydrocarvone

F. Commercial Preparation of Camphor

Camphor is prepared industrially from α-pinene which in turn is obtained from turpentine oil by the following route:

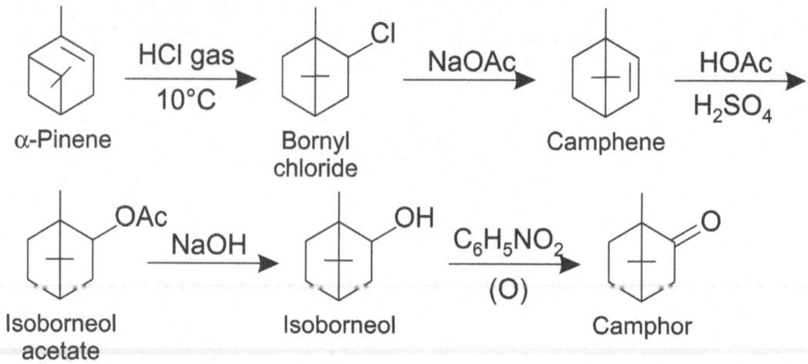

α-Pinene first gives pinene hydrochloride on treatment with HCl, which then gets converted to bornyl chloride

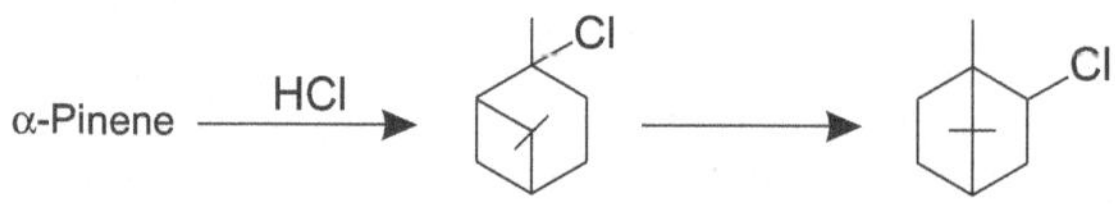

If we replace acetic acid by formic acid, we get isobornyl formate which gives isoborneol when treated with sodium hydroxide. Isoborneol is oxidised by atmospheric oxygen to camphor.

$$\text{Camphene} \xrightarrow{\text{HCOOH}} \text{Isobornyl formate} \xrightarrow{\text{NaOH}} \text{Isoborneol}$$

$$\xrightarrow{O_2} \text{Camphor}$$

❑❑❑

CHAPTER 6

LIPIDS-OILS AND FATS

Learning Objectives

In this chapter, we will study:

- What are lipids ?
- Classification of lipids
- Biological functions of lipids
- Oils and fats as simple lipids
- Fatty Acids and their composition in some fats and oils
- Omega-3 and omega-6 fatty acids
- Trans fats and their effect on the heart and brain
- Effect of fats on blood cholesterol levels
- Extraction of oils and fats
- Refining of oils and fats
- Hydrogenation
- Identification of fats and oils
- Flavour changes in oils and fats-rancidity

6.1 WHAT ARE LIPIDS ?

Lipids are broadly defined as any fat-soluble naturally occurring molecules. They are insoluble in water but soluble in ether, chloroform and hot alcohol. Lipids include many types of compounds containing a wide variety of functional groups such as fats, oils, waxes, cholesterol, sterols, fat-soluble vitamins (*A*, *D*, *E* and *K*), phospholipids and others. Lipids constitute major part of stored food material in animals and human beings and a major structural component. The main biological functions of lipids include energy storage, acting as structural components of cell-membranes and participating as important signalling molecules.

Although the term lipid is sometimes used as a synonym for fats, fats are a subgroup of lipids called triglycerides and should not be confused with the term fatty acid. Lipids also include molecules such as fatty acids and their derivatives, cholesterol and all other compounds obtained from cholesterol, Terpenoids and carotenoids also come under lipids. Lipids are hydrophobic in nature. They originate entirely or in part from two distinct types of biochemical subunits or "building blocks"-ketoacyl and isoprene groups.

We consume oils and fats in our daily diet. Dietary fats supply energy, carry fat-soluble vitamins-*A*, *D*, *E*, *K* and are a source of antioxidants and bioactive compounds. Fats are also incorporated as structural components of the brain and cell membranes.

6.2 CLASSIFICATION OF LIPIDS—TYPES OF LIPIDS

Lipids may be classified into four major groups :

A. Simple Lipids
   1. Fats and oils
   2. Waxes

B. Complex Lipids
   1. Phospholipids
   2. Glycolipids
   3. Sphingolipids

C. Derived Lipids
   1. Sterol lipids — Steroids and sterols
   2. Prenol lipids – Isoprenoids or Terpenes

D. Unclassified Lipids

   Tocopherols (vitamin E) and vitamin K

A. Simple Lipids

The simple lipids are esters of monocarboxylic fatty acids and aliphatic alcohols. Simple lipids are of the following types :

1. Fats and Oils

Fats and oils are triglycerides formed of glycerol and fatty acids. Glycerol contains three hydroxyl groups. When all the three alcoholic groups form ester linkages with fatty acids, the resulting compound is a triacylglycerol (1). The older name of this type of compound is triglyceride. The three ester groups are the polar parts of the molecule, whereas the tails of the fatty acids are non-polar. Three different fatty acids are usually esterified to the alcoholic groups of the glycerol molecule.

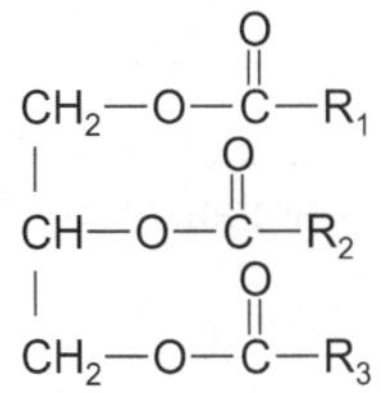

(1)

Triacylglycerols do not occur as components of the membranes as do other types of lipids. They accumulate in adipose tissue and provide a means of storing fatty acids particularly in animals. They serve as concentrated stores of metabolic energy. Complete oxidation of fat yields about 9.4 K cal/g in contrast to 4 K cal/g for carbohydrates and proteins.

Triglycerides are neutral fats. They are lighter than water with a specific gravity of about 0.86. They occur both as solids and liquids.

* **Fats** are solid at room temperature. They have saturated fatty acids.
* **Oils** are fluid at room temperature. They have large number of unsaturated fatty acids.

2. Waxes

Waxes are complex mixtures of esters of long chain carboxylic acids and long chain monohydroxy alcohols. For example, beeswax is composed of mainly palmitic acid esterified with either hexacosanol or triacontanol.

$$\underset{\text{Palmitic acid}}{CH_3(CH_2)_{14}COOH} + \underset{\text{Hexacosanol}}{HOCH_2-(CH_2)_{24}-CH_3} \longrightarrow$$

$$CH_3-(CH_2)_{14}-\overset{O}{\overset{\|}{C}}-OCH_2-(CH_2)_{24}-CH_3$$

Waxes are chemically inert and resistant to atmospheric oxygen. They are insoluble in water. Waxes frequently serve as protective coatings for both plants and animals. In plants, they coat stems, leaves and fruits. In animals, they are found on fur, feathers, and skin. Myricyl cerotate is the principal component of carnauba wax. It is extensively used in floor wax and automobile wax. The principal component of the wax produced by whales is cetyl palmitate. This wax is an important component of cosmetics.

$$CH_3-(CH_2)_{14}-\overset{O}{\overset{\|}{C}}-O-(CH_2)_{15}-CH_3$$

Cetyl palmitate

$$CH_3-(CH_2)_{24}-\overset{O}{\overset{\|}{C}}-O-(CH_2)_{29}-CH_3$$

Myricyl cerotate

B. Complex Lipids or Conjugated Lipids

Complex lipids or conjugated lipids are esters of fatty acids and alcohol but also have a phosphate, carbohydrate or protein group. They are of the following types:

1. Phosphoacylglycerols (Phospholipids)

If one of the alcoholic groups of glycerol is esterified by a phosphoric acid molecule and two fatty acids are esterified to the glycerol molecule, the resulting compound is called a phosphatidic acid (2). One molecule of phosphoric acid can form ester bonds both to glycerol and some other alcohol creating a phosphatidyl ester (3). Phosphatidyl esters are classed as phosphoacylglycerols or phospholipids. Fatty acids present may be widely different.

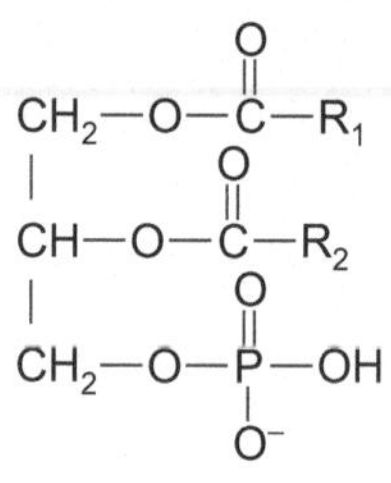

(2) Phosphatidic acid

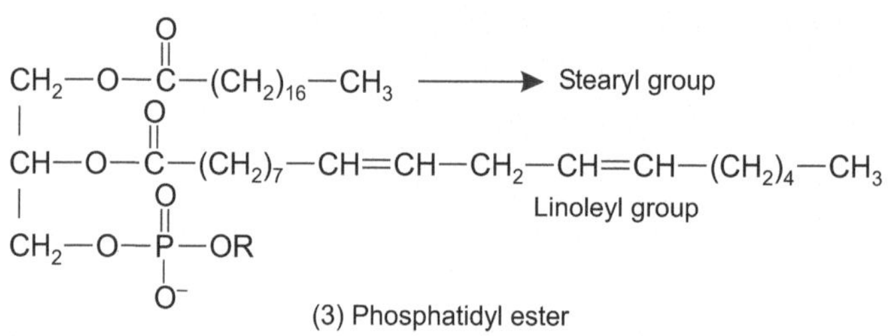

(3) Phosphatidyl ester

Phospholipids are most important lipids. They are important components of biological membranes and are essential for their proper functioning. They regulate cell permeability, control transport and metabolism of synthesised and dietary fat and participate in blood coagulation.

Phospholipids include the following types depending upon the nature of the second alcohol group esterified to the phosphoric acid.

(*a*) **Lecithins:** Lecithins are monoamino-monophospholipids. An example is that of phosphatidyl choline (4). They usually occur along with cholestrol in the nervous tissue and in egg yolk. The fatty acids most commonly found in lecithins are palmitic acid, stearic acid, oleic acid, linoleic acid and linolenic acid. But only two of these acids are present at a time in one molecule of lecithin. Lecithins play an important role in permeability, osmotic tension and surface conditions of cells.

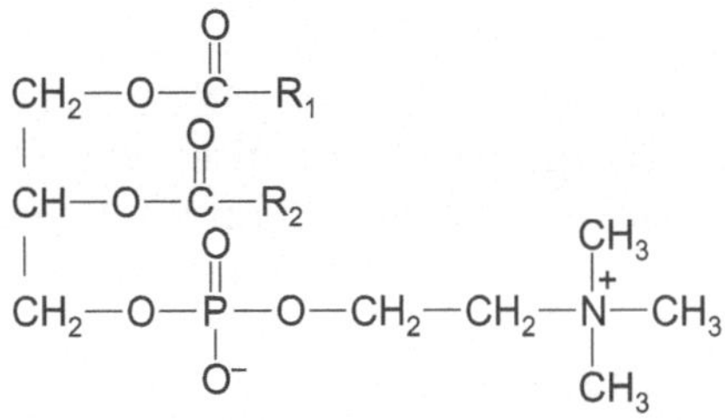

(4) Phosphatidylcholine, a Lecithin

(*b*) Cephalins: The composition of cephalins is very similar to lecithins but the choline radical of lecithin is replaced by aminoethanol group (5). The fatty acids present in cephalins are also different. Cephalins form components of certain lipoproteins and form an important factor in blood coagulation.

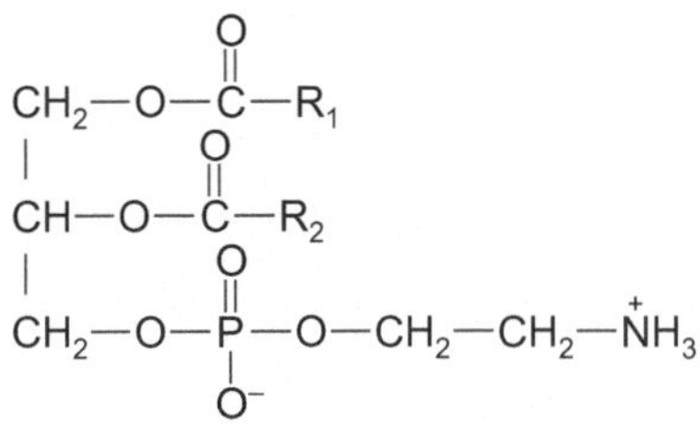

(5) Phosphatidylethanolamine

Other important phospholipids are phosphatidylserine (6) and phosphatidylinositol (7). Phosphatidylinositols and phosphatidic acids are either precursorrs of or are themselves, membrane-derived second messengers.

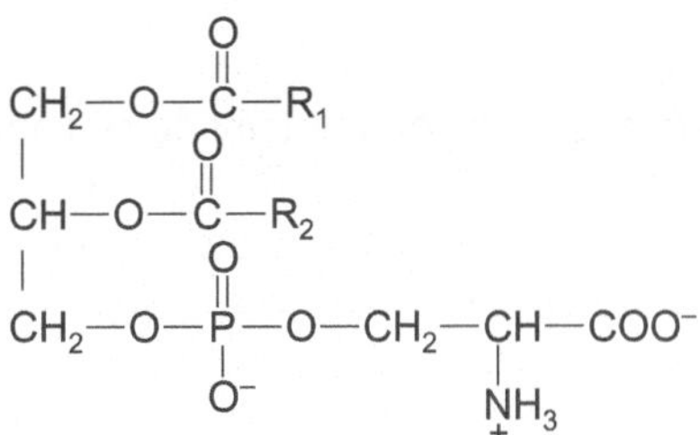

(6) Phosphatidylserine

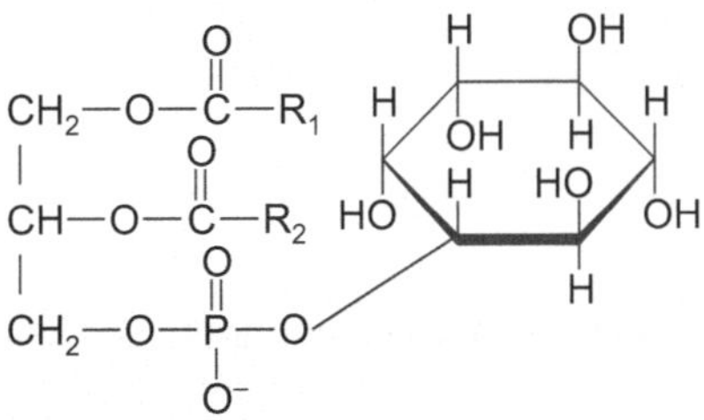

(7) Phosphatidylinositol

2. Glycolipids

If a carbohydrate is bound to an alcohol group of a lipid by a glycosidic linkage, the resulting compound is a glycolipid. Ceramides (8) are the parent compounds for glycolipids and the glycosidic bond is formed between the primary alcohol group of the ceramide and a sugar residue. The resulting compound is called a cerebroside. In most cases, the sugar is glucose or galactose. For example, glucocerebroside (9) contains glucose.

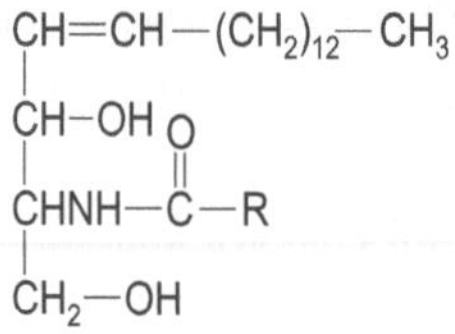

(8) A Ceramide

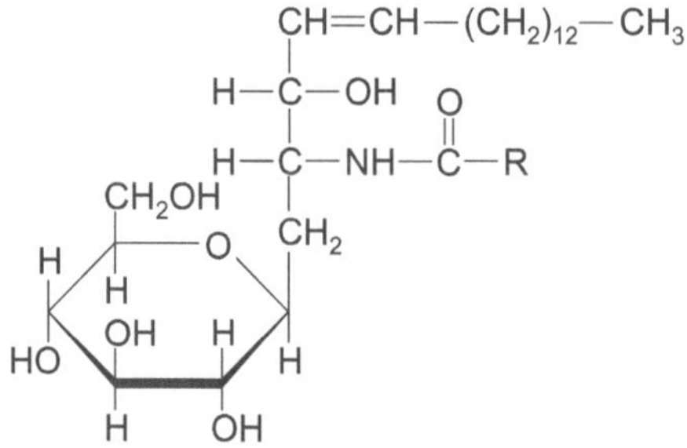

(9) A Glucocerebroside

Cerebrosides are found in nerve and brain cells, primarily in cell membranes.

3. Sphingolipids

Sphingolipids do not contain glycerol, but they do contain the long-chain amino alcohol sphingosine (10). Sphingolipids are found in both plants and animals, they are particularly abundant in the nervous system. As shown earlier, the simplest compounds of this class are the ceramides, which consist of one fatty acid linked to the amino group of sphingosine by an amide bond.

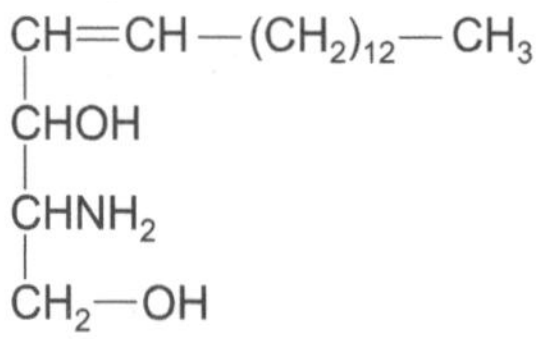

(10) Sphingosine

In sphingomyelins (11), the primary alcohol group of sphingosine is esterified to phosphoric acid, which in turn, is esterified to another amino alcohol, choline. Sphingomyelins occur in cell membranes in the nervous system.

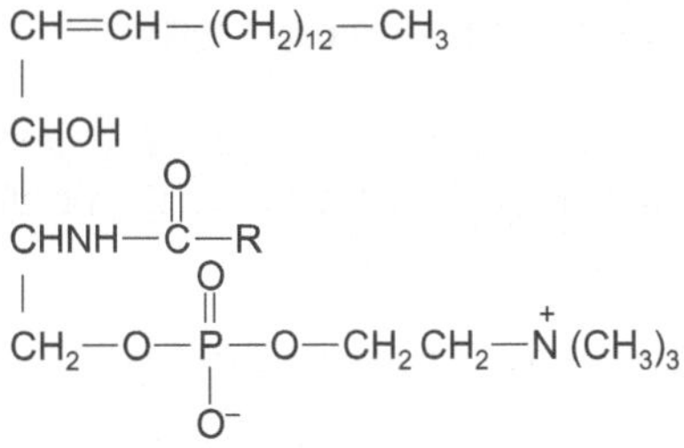

(11) A sphingomyelin

C. Derived Lipids

1. Sterol Lipids – Steroids and Sterols

Sterol lipids, such as cholesterol and its derivatives, are an important component of membrane lipids along with the glycerophospholipids and sphingomyelins. Cholesterol (12) is widespread in biological membranes especially in animals. The presence of cholesterol in membranes can modify the role of membrane-bound proteins. Cholesterol has a number of important functions including its role as a precursor of other steroids and of vitamin D. Steroids contain the same fused four-ring core structure. They contain three six-membered rings and one five-membered ring. There are many important steroids including sex hormones. All steroid hormones are derived from cholesterol. However, cholesterol is best known for its harmful effects on health when it is present in excess in the blood. Its excess in the blood develops a condition in which lipid deposits block the blood vessels and lead to heart disease.

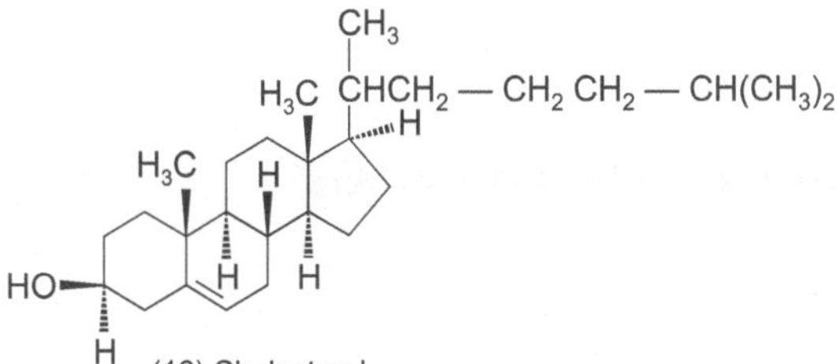

(12) Cholesterol

2. Prenol Lipids (Terpenes or Isoprenoids)

Prenol lipids are synthesised from the 5-carbon precursors. The simplest unit is isoprene (13). Two or more isoprene molecules take part in the synthesis of isoprenoids. Examples of isoprenoids are various terpenes (discussed in Chapter 5 of this book), fat soluble vitamins, bile salts, different types of pigments like β-carotene. Terpenoids are classified according to the number of C5 units present. Structures containing greater than 40 carbons are known as polyterpenes. Carotenoids are important for their function as anti-oxidants and as precursors of vitamin A.

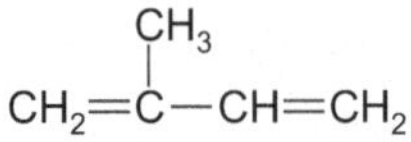

(13)

Sterols and steroid hormones mentioned above are also supposed to be derived from isoprenoids.

6.3 BIOLOGICAL FUNCTIONS OF LIPIDS

The main biological functions of lipids include energy storage, acting as structural components of cell-membranes. They play an important role in metabolism, nutrition and health.

1. Energy Storage and Rich Source of Energy

Triglycerides or triacylglycerols are a major form of energy storage in animals because of their high calorific value (9.4 K cal/g). Fats are stored in the adipose tissue and act as reserve food material. Animals use triglycerides for energy storage whereas plants, which do not require energy for movement, can afford to store food for energy in the form of starch.

2. Structural Constituents

Phospholipids and glycolipids are the structural components of all the membrane systems of the cell. The glycerophospholipids are the main structural component of biological membranes, such as the cellular plasma membrane and the intracellular membranes of organelles. In animal cells, the plasma membrane physically separates the intracellular components from the extracellular environment.

3. Solvents and Fat Transport

Lipids act as solvent for fat-soluble vitamins *A*, *D*, *E* and *K*. The fat-soluble vitamins, which are isoprene-based lipids are essential nutrients stored in the liver and fatty tissues. Phospholipids play an important role in the transport and metabolism of fatty acids.

4. Heat Insulators and Shock Absorbers

Fats deposited in the subcutaneous tissue act as insulators conserving body heat. The fat deposited around the visceral organs and underneath the skin acts as cushion and absorbs mechanical shocks.

Lipids form a protective waxy covering on the aerial parts of plants to check loss of water by evaporation.

5. Metabolism

Lipids play an important role in the metabolism. Triglycerides and phospholipids are broken down into free fatty acids by the action of lipases. Beta oxidation is the process by which fatty acids in the form of acyl-CoA molecules are broken down to generate acetyl-CoA. The acetyl-CoA is then ultimately converted into ATP, CO_2 and H_2O using citric acid cycle.

6. Nutrition and Health

Lipids play diverse and important roles in nutrition and health. Many lipids are absolutely essential for life. Adrenocorticoids, sex hormones and vitamin *D* are synthesised from cholesterol. However, there is also considerable awareness that abnormal levels of cholesterol and trans fatty acids are risk factors for heart disease amongst others.

Humans have a requirement for essential fatty acids, such as linoleic acid and alpha linolenic acid in the diet because they cannot be synthesised from simple precursors in the diet.

We have seen earlier that most vegetable oils are rich in linoleic acid (safflower, sunflower and corn oils). Alpha-linolenic acid is found in the green leaves of plants and in selected seeds, nuts and legumes. Most of the lipid found in food is in the form of triacyl glycerol, cholesterol and phospholipids.

Most of the saturated fatty acids in the diet are incorporated into adipose tissue stores because we get higher energy per carbon than is obtained from oxidation of unsaturated fatty acids. Dietary fat provides an average energy intake which is apprroximately twice that of carbohydrate or protein. A minimum amount of dietary fat is necessary to facilitate absorption of fat-soluble vitamins (*A, D, E* and *K*) and carotenoids. A minimal amount of body fat is also necessary to provide insulation that prevents heat loss and protects vital organs from shock due to ordinary activities.

High fat intake contributes to increased risk of obesity, diabetes and coronary and cardiovascular diseases. These diseases are due to the build up of plaque on the inside walls of arteries. Plaque is made up of cholesterol-rich low density lipoprotein (LDL), macrophages, smooth muscle cells, platelets and other substances. Saturated fats have a profound effect in increasing blood cholesterol levels and tend to increase LDL. They are found predominantly in animal products, butter, cheese and meat, but coconut oil and palm oil are common vegetable sources. Intake of monosaturated fats in oils such as olive oil is thought to be preferable to consumption of polysaturated fats in oils such as corn oil because monosaturated fats apparently do not lower high density lipoprotein (HDL) cholesterol levels. Keeping cholesterol in the normal range not only helps prevent heart attacks and strokes but may also prevent the progression of other coronary cardiovascular diseases. "Statins" are a class of drugs that lower the level of cholesterol in the blood by inhibiting the enzyme which is responsible for the biosynthesis of cholesterol in the liver.

6.4 OILS AND FATS AS SIMPLE LIPIDS

A. Introduction

We have already seen that oils and fats are simple lipids. They are triglycerides of fatty acids. Glycerol is a trihydric alcohol *i.e.*, it contains three —OH groups that can combine with up to three fatty acids to form monoglycerides, diglycerides and triglycerides. Fatty acids may combine with any of the three hydroxyl groups to create a wide diversity of compounds. The most common glycerides are triglycerides, also known as triacylglycerols, in which all three —OH groups are esterified by fatty acids. For example, **tristearin** is a component of beef fat in which the three —OH groups of glycerol are esterified by stearic acid.

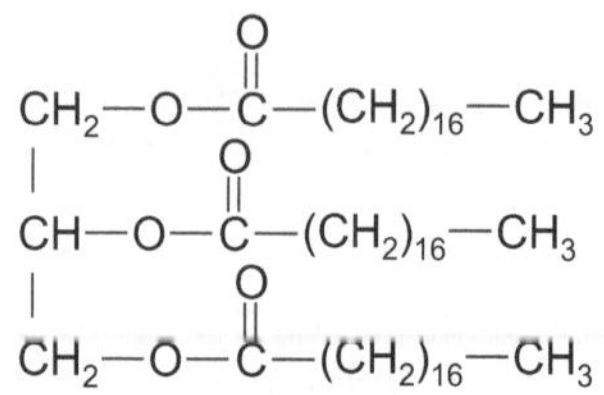

Tristearin

Triglycerides are commonly called **fats** if they are solid at room temperature and **oils** if they are liquid at room temperature. Most triglycerides derived from mammals are fats such as beef fat. These fats are solid at room temperature, the body temperature of the living animal keeps them somewhat fluid allowing for body movement. In plants and cold-blooded animals, triglycerides are generally oils, such as corn oil, peanut oil or fish oil. A fish requires liquid oils rather than solid fat because it would have difficulty in moving if its triglycerides solidified whenever it swam in cold water.

Oils and fats are commonly used for long-term energy storage in plants and animals. Fat is more efficient source of energy than carbohydrates because 1 gram of fat releases over twice as much energy as 1 gram of sugar or starch.

Most of the triglycerides which occur in nature are mixed triglycerides *i.e.*, they contain two or three different fatty acids in one molecule. Triglycerides have lower densities than water. That's why they float on water.

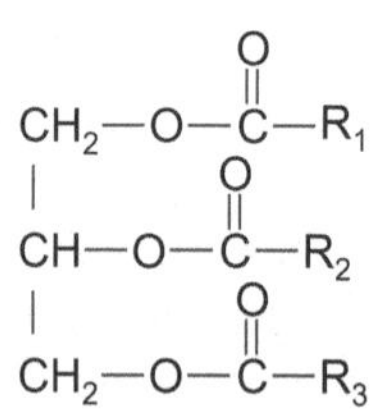

A Mixed Triglyceride

B. Animal and Vegetable Sources

Animal fats and oils are derived both from terestrial and marine animals. Animal fats are located in the adipose tissue cells. We have tallow from cattle, sheep and goat. More or less of fat occurs in the muscles, in the connective tissue, under the skin and in the bones. The amounts found in different parts of the body vary from species to species and in any given species with the age of the individual animal and its condition. The fats from the different parts and organs of a given animal differ somewhat in their properties. The fat from the interior of the animal is somewhat firmer than the fat from near the body surface *i.e.*, it melts at a higher temperature.

Marine oils include liver oils and fish oils. The different types of marine fats have been of great importance in the past and still possess considerable significance. For example, codliver oil is used for some special purposes - some are used as food stuffs and others serve industrial uses.

Vegetable fats and oils are found in greatest abundance in fruits and seeds. They are also present in small amounts in other parts of the plants *i.e.*, roots, branches and leaves. In some seeds and fruits, however, the fat content is very high, in some cases it is as high as 35 per cent. For example, in dried coconuts the oil is present upto 65 per cent. These are the commercial sources of vegetable oils. Oils are stored in seeds to serve as nourishment for the germination of embryo. Other important sources of vegetable oils are soya bean, groundnut, palm kernel, mustard, sesame and sunflower *etc*. All these give edible oils. However, cotton seed, linseed, castor seed and mowrah give non-edible oils.

The olive contains a large amount of fat in the pulp surrounding the kernel and only a small amount in the kernel itself, while in the oil palm both the pulp and the kernel contain large amounts. The fat from the pulp may have characteristics quite different from those of the fat in the kernel.

C. Chemical and Physical Characteristics

We know oils and fats are simple triglycerides of fatty acids. The reactions of oils and fats are the reactions of the triglycerides. Thus, they can undergo hydrolysis at all the three ester groups giving glycerol and three molecules of fatty acids. The fatty acids generated may be same or different. Triglycerides are easily hydrolysed by enzymes called lipases, which are present in the digestive tracts of human beings and animals. Three molecules of water are used up for each molecule of fat when one molecule of glycerol and three molecules of fatty acid are obtained. Conversely, the fat may be reconstituted from glycerol and fatty acids, in which event three molecules of water are set free for each molecule of fat synthesised.

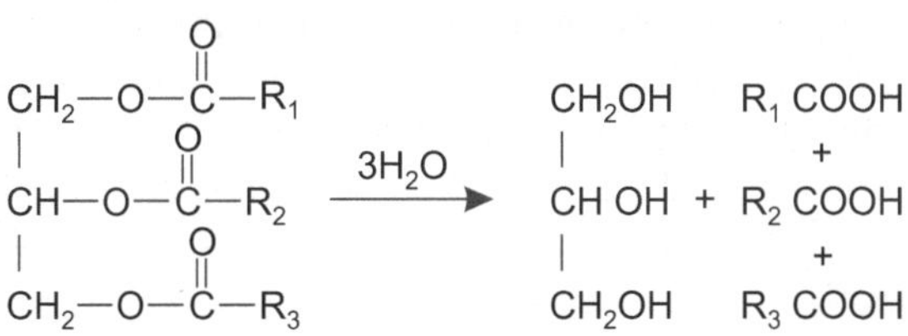

A Triglyceride Glycerol Fatty acid

The common physical properties of oils and fats are:

- They float on water but are not soluble in it.
- When pure they are colourless, odourless and tasteless.
- Triglycerides have lower densities than water.
- They are greasy to the touch and have lubricating properties.
- They are easily soluble in organic solvents like diethylether, acetone, chloroform and carbon tetrachloride.
- They are not readily volatile and may be burned without leaving any residue.

D. Saponification

When triglycerides are hydrolysed by alkalis, glycerol and salts of the fatty acids are obtained. This process is called *saponification*. Generally, sodium hydroxide or potassium hydroxide is used for hydrolysis. Sodium salts or potassium salts of these fatty acids are produced, which are called soaps. Sodium salts are hard soaps whereas potassium salts are soft soaps. The fatty acid salts of ammonium are also sometimes used for cleansing. Only a few other soaps are of practical importance, for example lead soaps are used as medicinal plasters, zinc soaps are used in ointments and aluminium soaps are used in waterproofing. Salts of lower fatty acids are soluble in water and have no cleansing action.

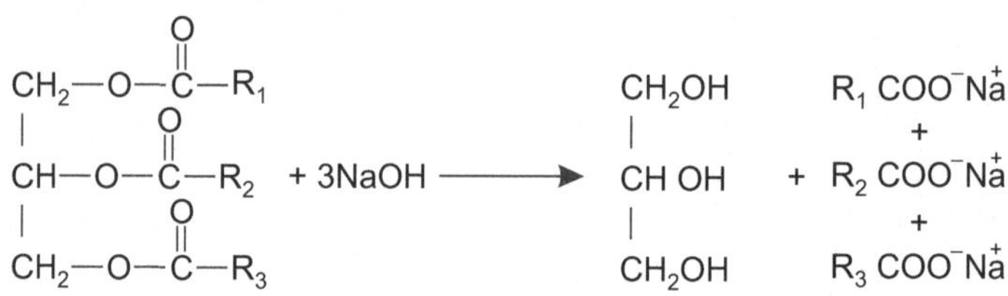

Mixed Triglyceride

Mixture of salts (soap)

6.5 FATTY ACIDS IN FATS AND OILS

A fatty acid has a carboxyl group at the polar end and a hydrocarbon chain at the non-polar end. Fatty acids are, therefore, amphipathic compounds because the carboxyl group is hydrophilic and the hydrocarbon tail is hydrophobic. Their carbon chain is long and unbranched. All fatty acids that occur in the living system contain an even number of carbon atoms because they are derived from two carbon acetyl units. If there are carbon-carbon double bonds in the chain, the fatty acid is unsaturated. If there are only single bonds, the fatty acid is saturated. The number of carbon atoms in the chain may vary from **4–30** but normally ranges from **14–18**.

Tables 6.1 and 6.2 give some examples of the saturated and unsaturated acids with their melting points. Butyric acid is one of the saturated short-chain fatty acids responsible for the characteristic flavour of butter.

TABLE 6.1 Typical Naturally Occurring Saturated Fatty Acids

Acid	*Number of Carbon Atoms*	*Formula*	*Melting Point (°C)*
Lauric	12	CH_3-$(CH_2)_{10}$-COOH	44
Myristic	14	CH_3-$(CH_2)_{12}$-COOH	58
Palmitic	16	CH_3-$(CH_2)_{14}$-COOH	63
Stearic	18	CH_3-$(CH_2)_{16}$-COOH	71
Arachidic	20	CH_3-$(CH_2)_{18}$-COOH	77

TABLE 6.2 Typical Naturally Occurring Unsaturated Fatty Acids

Acid	*Number of Carbon Atoms*	*Degree of Unsa-turation*	*Formula Point (°C)*	*Melting Point (°C)*
Palmitoleic	16	16 : 1	CH_3 $(CH_2)_5$ CH = CH $(CH_2)_7$-COOH	- 0.5
Oleic	18	18 : 1	CH_3 $(CH_2)_7$-CH = CH-$(CH_2)_7$-COOH	4
Linoleic	18	18 : 2	CH_3-$(CH_2)_4$-CH = CH-CH_2-CH = CH-$(CH_2)_7$-COOH	- 5
Linolenic	18	18 : 3	CH_3-$(CH_2$-CH = CH$)_3$-$(CH_2)_7$-COOH	- 11
Arachidonic	20	20 : 4	CH_3-$(CH_2)_4$-(CH = CH-$CH_2)_4$-$(CH_2)_2$-COOH	- 50

In unsaturated fatty acids, the stereochemistry at the double bond is usually ***cis*** rather than ***trans***. The difference between *cis* and *trans* acids is very important to their overall shape. A *cis* double bond puts a kink in the long chain hydrocarbon tail whereas the shape of a *trans* fatty acid is like that of a saturated fatty acid.

The notation used for fatty acids indicates the number of carbon atoms and the number of double bonds. Thus, 18 : 0 denotes an 18-carbon saturated fatty acid with no double bonds and 18 : 1 denotes an 18-carbon fatty acid with one double bond and so on. The double bond in oleic acid is at the ninth carbon atom, in linoleic acid there are two double bonds at the 9 and 12 carbon atoms whereas linolenic acid has three double bonds at 9, 12 and 15 carbon atoms.

Tables 6.1 and 6.2 show that saturated fatty acids have higher melting points than unsaturated fatty acids. Thus, stearic acid has a melting point of 71°C whereas oleic acid with one double bend has melting point of 4°C. This lowering of melting point is due to the 'kink' in the *cis* configuration of oleic acid (Fig. 6.1). In such a case, the molecule cannot pack tightly.

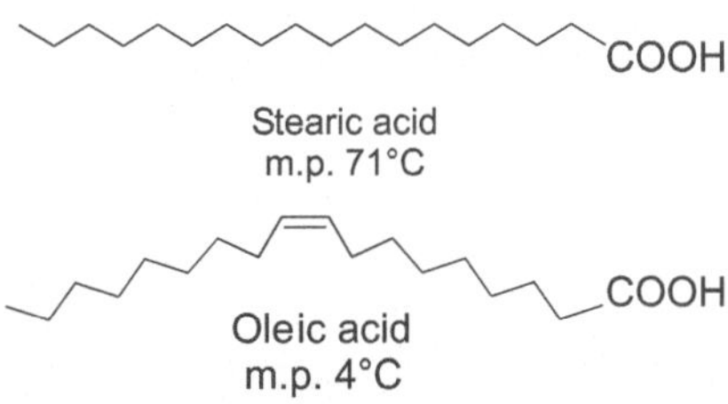

Fig. 6.1.

A second double bond lowers the melting point further (lenoleic acid, m.p. – 5°C) and a third double bond lowers it still further (m.p. – 11°C).

Fats and oils also have melting points that depend upon the degree of unsaturation in their fatty acids. Thus, tristearin has m.p. 72°C. It is a saturated fat that packs well in a solid lattice. Triolein (m.p. – 4°C) has the same number of carbon atoms but it has three double bonds whose kinked conformations prevent optimum packing in the solid.

6.6 FATTY ACID COMPOSITION IN SOME FATS AND OILS

Fatty acid compositions depend on the sources of the oils. While a large variety of fatty acids is found in natural fats and oils, only a few of them are of commercial importance. These are lauric acid, myristic acid, palmitic acid stearic acid, oleic acid, linoleic acid and linolenic acid. The triglycerides of these seven acids make up the great bulk of natural fats and oils. Fats and oils are practically always mixtures of triglycerides in varying proportions. In some fats one triglyceride predominates, in others another and in still others several of them are present. It appears that no natural fat or oil consists solely of a single triglyceride. They are always present as mixtures. The fats of different species of animals and plants vary widely. They largely depend upon the fatty acids that are present in them.

Table 6.3 gives the composition of various fatty acids in different oils and fats.

TABLE 6.3 Per cent of Weight of Total Fatty Acids

Oil or Fat	Saturated					Unsaturated		
	Capric	Lauric acid	Myristic acid	Palmitic acid	Stearic acid	Oleic acid	Linoleic acid	Linolenic acid
Almond oil	-	-	-	7	2	69	7	-
Beef Tallow	-	-	3	24	19	43	3	1
Butterfat (cow)	3	3	11	27	12	29	2	1
Coconut oil	6	47	18	9	3	6	2	-
Corn oil	-	-	-	11	2	28	58	1
Cotton - seed oil	-	-	1	22	3	19	54	1
Olive oil	-	-	-	13	3	71	10	1
Palm oil	-	-	1	45	4	40	10	-
Palm - kernel oil	4	48	16	8	3	15	2	-
Peanut oil	-	-	-	11	2	48	32	-
Safflower oil	-	-	-	7	2	13	78	-
Soyabean oil	-	-	-	11	4	24	54	7
Sunflower oil	-	-	-	7	5	19	68	1
Walnut oil	-	-	-	11	5	28	51	5

6.7 WHAT ARE OMEGA-3 AND OMEGA-6 FATTY ACIDS ?

Omega-3 (ω 3) and omega-6 (ω 6) fatty acids are unsaturated "Essential Fatty Acids" that need to be included in the diet because the human metbolism cannot create them from other fatty acids. These fatty acids use the Greek alphabet (α, β, γ, ..., ω) to identify the location of the double bonds. The "alpha" carbon is the carbon closest to the carboxyl group and the "omega" is the last carbon of the chain because omega is the last letter of the Greek alphabet. Thus, linoleic acid is omega-6 fatty acid because it has a double bond six carbons away from the omega carbon. Linoleic acid plays an important role in lowering cholesterol levels. Linolenic acid is an omega-3 fatty acid because it has a double double three carbons away from the omega carbon. Therefore, oleic acid is omega-9 fatty acid.

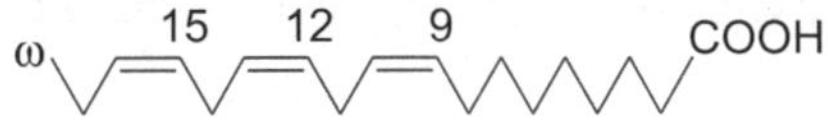

Linolenic acid (Omega-3 acid)

Linoleic acid (Omega-6 acid)

6.8 TRANS FATS AND THEIR EFFECT

A. What are Trans Fats ?

We already know that double bonds bind carbon atoms tightly and prevent rotation of the carbon atoms along the bond axis. This gives rise to configurational isomers which we call *cis* and *trans* as shown below :

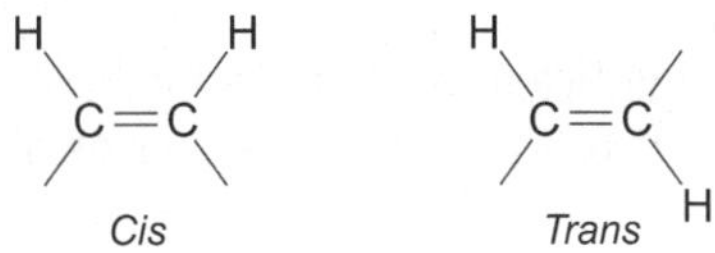

Naturally occurring fatty acids generally have the *cis* configuration. The natural form of 9-octadecenoic acid (oleic acid) found in olive oil has a "V" shape. The *trans* configuration looks like a straight line.

Low levels of trans fatty acids occur naturally in some animal food products. Polyunsaturated *trans* fatty acids cannot be used to produce useful products because the molecules have unnatural shapes that are not acted upon by enzymes. Partially hydrogenated oils contain a large proportion of diverse *trans* fatty acids. Trans fatty acids that are incorporated into the cell membranes create denser membranes that alter the normal functions of the cell.

B. Effect of Trans Fats on the Heart

Dietary *trans* fats raise the level of low density lipoproteins (LDL, or "bad cholesterol") increasing the risk of coronary heart disease. *Trans* fats also reduce high-density lipoproteins (HDL or "good cholesterol) and raise the levels of triglycerides in the blood. Both of these conditions are associated with insulin resistance which is linked to diabetes, hypertension and cardiovascular disease. Research has shown that people who take partially hydrogenated oils, which are high in *trans* fats, worsen their lipid profiles and have nearly twice the risk of heart attacks compared with those who do not consume hydrogenated oils.

C. Effect of *Trans* Fats on the Brain

Trans fats also have a detrimental effect on the brain and nervous system. Neural tissue consists mainly of lipids and fats. Studies show that trans fatty acids in the diet get incorporated into the brain cell membranes and alter the ability of neurons to communicate and may cause neural degenration and diminished mental performance.

D. Effect of Fats on Blood Cholesterol Levels

High blood serum cholesterol levels are associated with increased risk of cardiovascular diseases. Blood cholesterol levels are lowered by reducing the sources of dietary cholesterol, increasing the amount of fibre in the diet and by consuming oils high in polysaturated fatty acids while reducing the intake of saturated fats. It has been observed that myristic acid (C14 : 0) and to some extent palmitic acid (C16 : 0) increase cholesterol levels whereas polyunsaturated acids such as linoleic (C18 : 2) reduce cholesterol levels. Other saturated fatty acids (*e.g.*, stearic acid) or monounsaturated fatty acids (*e.g.*, oleic acid C18 : 1) do not affect cholesterol levels significantly.

Animal fats and tropical oils which are high in myristic acid and low in linoleic acid increase cholesterol levels. Oils high in linoleic acid such as sunflower oil and safflower oil can play a significant role in reducing blood cholesterol levels when consumed regularly as part of the diet. The effect of olive oil on cholesterol is relatively neutral.

The best quality oils are unrefined cold pressed or expeller pressed oils packed in dark glass bottles filled with an inert gas.

6.9 EXTRACTION OF OILS AND FATS

The raw materials for the fats and oils are animal by-products from the slaughter of cattle and sheep ; fatty fish and marine mammals; a few fleshy fruits like palm and olive and various oil seeds. Most oil seeds are grown specifically for processing to oils but several important vegetable oils are obtained from by-product raw materials. For example, cotton-seed is a by-product of cotton grown for fibre and corn oil is obtained from the corn germ that accumulates from the corn-milling industry whose primary products are corn grits and starch.

Fats and oils may be obtained from oil bearing tissues by three general methods with varying degrees of mechanical simplicity :

(1) rendering, (2) pressing with mechanical presses and (3) extracting with volatile solvents

A. (i) Rendering-Fruits and Seeds

This is the crudest method of rendering oil from fruits, which is still practiced in some countries. It consists of heaping fruits in piles, exposing them to the sun and collecting the oil that exudes. This process is used in the preparation of palm oil. The fresh palm fruits are boiled in water and the oil is skimmed from the surface. Such processes can be used only with seeds or fruits, such as olive and palm that contain large quantities of easily released fatty matter.

(ii) Rendering-Animal Fats

The rendering process is applied on a large scale to the production of animal fats such as tallow, bone fat and whale oil. It consists of cutting the fatty pieces into small pieces that are boiled in open vats or cooked in steam digestors. The fat, which is liberated from the cells, floats on the surface of water, where it is collected by skimming. Finally, the fat is separated from the aqueous phase by pressing in hydraulic or screw presses. Additional fat is thereby obtained. The residue is used for animal feed or as fertilizer. Several centrifugal separation processes have also been developed. Compared with conventional rendering, the centrifugal methods provide a higher yield of better-quality fat and the separated protein has potential as an edible meat product.

B. Pressing

Many different mechanical devices have been used for pressing. The screw press has been used for the production of olive oil. In India, the most important source of edible oils are seeds such as mustard, groundnut, cottonseed, sesame and sunflower. The extraction from seeds is done by a variety of methods right from the village ghani to the more sophisticated expeller. The oil bearing material is screened and crushed by passing between steel rollers.

In order to get the maximum yield the crushed material is heated at 70°–100°C in steam jacketed vessel to rupture the oil cells. The heated material is then pressed by a high pressure expeller. The modern screw press has replaced many of the hydraulic presses because it is a continuous process and has greater capacity. It also requires less labour and will generally remove more oil. The expeller consists of a perforated cylinderical tube in which a screw shaft moves. While the oil expelled from the seeds flows out of the holes in the expeller tube, the pressed oil cake is ejected from the other end of the tube.

The exhausted oil cake still contains oil. Most of the exhausted cake is used for cattle feed or subjected to solvent extraction for recovery of more oil.

C. Solvent Extraction

It is a process which involves extracting oil from oil-bearing materials by treating it with a low boiling solvents. The solvent extraction method recovers almost all the oils and leaves behind only 0.5% to 0.7% residual oil in the raw material. In case of mechanical pressing the residual oil left in the oil cake may be anywhere from 6% to 14%. The solvent extraction method can be applied directly to any low oil content materials. The method can also be used to extract pre-pressed oil cakes obtained from high oil content materials. As the percentage of recovered oil is quite high, solvent extraction has become the most popular method of extraction of oils and fats. Fat is one of the essential components of the human diet, therefore, the demand for oils and fats is increasing with the increase in population and standards of living. Today large quantities of oil cakes such as peanut, cottonseed, sunflower, linseed, neem, castor etc., are extracted. Direct extraction of rice bran and soya bean is also used.

The Process

Solvent extraction is basically a process of diffusion of a solvent into oil-bearing cells of the raw material resulting in a solution of the oil in the solvent. Various solvents can be used for extraction. Most common solvents are various grades of petroleum ether and hexane. Hexane is considered to be the best and it is exclusively used for this purpose. Thus, the extraction process consists of treating the raw material with hexane and recovering the oil by distillation of the resulting solution of the oil in hexane called miscella. Evaporation and condensation from the distillation of miscella recovers the hexane absorbed in the material. The hexane thus recovered is reused for extraction. The boiling point of hexane is very low (67°C) and oils and fats are very soluble in hexane.

A typical extraction system consists of:

1. Cleaning to remove tramp iron, foreign weed seeds and stones.
2. Removing hulls in cracking, aspirating or screening operations.
3. Cracking or rough grinding the kernels or prepressed cake.
4. Steaming (tempering or cooking of the meats.)
5. Flaking the small pieces between smooth flaking rolls.
6. Extracting the oil with solvent.
7. Separating the marc from the oil-solvent solution, miscella.
8. Removing the solvent from both the miscella and the marc.

6.10 REFINING OF OILS AND FATS

The extent of processing applied to oils and fats depends on their source, quality and ultimate use. Many oils are used for edible purposes after only a single processing step, *i.e.*, clarification by settling or filtering. Sometimes filtering is done twice. Many cold-pressed oils can be used in food products without further processing. For example, cold-pressed olive, peanut and some coconut and sunflower oils are used in this way. Large quantities of butter and lard are used without special treatment after churning or rendering. However, large demands of refined oils have led to extensive processing techniques.

A. Alkali Refining

Oils and fats may contain some free fatty acids. They can be removed by treating fats at 40° to 85°C with an aqueous solution of sodium hydroxide or sodium carbonate. The refining may be done in a tank, in which case it is called batch refining. Alkali refining may also be carried out in a continuous system. In batch refining, the aqueous emulsion of soaps formed from free fatty acids, along with other impurities settles at the bottom and is drawn off. In the continuous system the emulsion is separated with centrifuges.

After the fat has been refined, it is usually washed with water to remove traces of alkali and soapstock. Oils that have been refined with sodium carbonate or ammonia generally require a light re-refining with sodium hydroxide to improve colour. After washing with water, the oil may be dried by heating in a vacuum or by filtering through a dry filter-aid material.

The refined oil may be used for industrial purposes or may be processed further to edible oils. Usually, the refined oils are neutral, free of material that separates on heating, lighter in colour, less viscous and more susceptible to rancidity.

B. Water Refining

Water refining is usually called degumming. It is generally carried out before alkali refining. Water refining consists of treating the natural oil with a small amount of water, followed by centrifugal separation. The process is applied to many oils that contain phospholipids in significant amounts. Since the separated phospholipids are rather waxy or gummy solids, the term degumming is quite naturally applied to the separation. The separated phosopholipid emulsion layer from oils such as corn and soyabean oils may be dried and used as emulsifiers in chocolate products and emulsion paints. The degummed oil may also be used directly in industrial applications such as in paints or alkyd resins or refined with alkalis for ultimate edible consumption.

C. Bleaching

Most crude vegetable oils are deeply coloured. They must be bleached. In certain cases, some colour may go after water and alkali refining. If further colour removal is desired, the oil may be treated with various bleaching agents. Heated oils are treated with fuller's earth, activated charcoal or activated clays. Fuller's earth is a natural earthy material that will decolourise oils. Many impurities, including chlorophyll and carotenoid pigments, are absorbed onto such agents and removed by filtration.

To the dry oil dry fuller's earth is added, usually a few per cent, the quantity depending upon the character of the oil and the temperature. The mixture is then heated usually to less than 80°C and agitated for one-half to one hour. It is then pumped through a filter press which retains the fuller's earth and permits the oil to run out clear. An appreciable amount of the oil remains in the earth retained in the press at the end of the operation. A considerable proportion of this is sometimes recovered by forcing the dry steam through the press which carries out with it much of the residual oil.

Bleaching often reduces the resistance of the oils to racidity, because some material antioxidants are removed together with impurities. When many oils are heated to more than 175°C, a phenomenon known as heat bleaching takes place. The heat decomposes some pigments such as carotenoids and converts them to colourless materials.

D. Deodourisation

Many oils even after they have been refined and decolourised to transparent whiteness retain diagreeable odour and flavour. Vegetable fats especially have a relatively strong taste that is foreign to that of butter. Odourless and tasteless fats first came into high demand as ingredients for the manufacture of margarine, a product designed to duplicate the flavour and texture of butter.

The deodourisation process consists of blowing steam through heated fat or oil held under vacuum. Since the substances responsible for odours and flavours are usually volatile, they are removed leaving a neutral, virtually odourless fat or oil. Deodourisation is normally carried by blowing steam through the oil after it has been heated to a high temperature say 175°–245°C. The most effective method is to carryout this treatment in a vacuum. Originally, deodourisation was a batch process, but continuous systems are being used in which hot fat flows through an evacuated column countercurrent to the upward passage of stream.

6.11 HYDROGENATION

Unsaturated fats and oils exposed to air get oxidised to form compounds that have rancid or unpleasant odours or flavours. Hydrogenation is a commercial chemical process to add more hydrogen to natural unsaturated fats to decrease the number of double bonds and retard or eliminate the potential for rancidity. Unsaturated glycerides react with hydrogen in the presence of metal catalyst, usually nickel, to give saturated glycerides. This reaction is similar to the catalytic hydrogenation of alkenes. The hydrogenation process saturates the double bonds present in the fatty acid components of glycerides, thereby converting them to saturated acid components.

Oleic acid and linoleic acid are both converted to stearic acid when fully saturated. Thus, the liquid vegetable oil becomes a solid saturated fat. For example, glyceryltrioleate (m.p. - 5°C) upon hydrogenation yields glyceryl tristearate (m.p. 71°C)

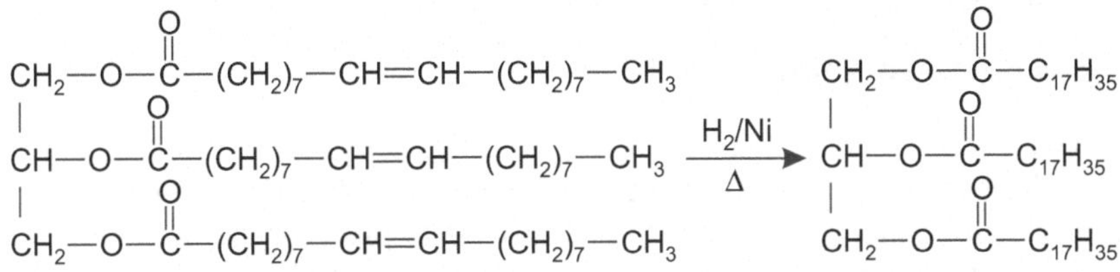

Glyceryl trioleate   Glyceryl tristreate

All the important unsaturated triglycerides *e.g.*, triolein, trilinolein and trilinolenin are liquids at ordinary temperatures. Hence, fats which contain them in considerable amounts are oils or if the amount is smaller, they are soft solids. On the other hand, the important saturated triglycerides, *e.g.*, tristearin and tripalmitin are solid and fats which contain them in considerable amounts are firm at ordinary temperatures.

Hydrogenation converts unsaturated triglycerides into saturated ones *i.e.*, it changes oils into solid fat. This process is of greatest commercial and economic importance. In practice, hydrogenation is widespread, but it is used far more for edible oils than for industrial products.

Fully saturated fats are too waxy and solid to use as food additives. So manufacturers use partially hydrogenated oils. In this, the process is stopped when the oil has the proper consistency for its application. The high temperature and catalysts used for this chemical reaction weaken the double bonds and cause a large percentage of the natural *cis* double bonds to change to *trans* double bonds. Trans fatty acids are present mainly in partially hydrogenated fats, but they are also present in hydrogenated fats because chemical reactions never achieve 100% efficiency.

6.12 IDENTIFICATION OF FATS AND OILS

A. Reichert-Meissel Value

The Reichert-Meissel value is a measure of the volatile water-soluble acid constituents of the fat. It is the number of millilitres of N/10 potassium hydroxide solution required to neutralise the volatile water-soluble fatty acids obtained by five grams of fat. The sample is heated with a standardised solution of potassium hydroxide until completely saponified. The mixture is then acidified with sulphuric acid and distilled. Acids upto C_{10} are volatile and appear in the distillate. The water-soluble acids in the distillate are neutralised with N/10 potassium hydroxide solution.

Reichert-Meissel value allows us to distinguish between butter and butter substitutes made from vegetable oils. The hydrolysis of the butter fat gives C_4, C_6, C_8

acids which are volatile in steam, while practically no acids below C_{10} are obtained from vegetable oils. Butter has a Reichert-Meissel number of 17 to 34 while that of cotton seed oil is less than 1.

B. Saponification Number

We have already seen that oils and fats on hydrolysis give glycerol and fatty acids. When this hydrolysis is carried out under alkaline conditions, the process is called *saponification* since it was first observed to take place in the manufacture of soap. Oils and fats are characterised by saponification number of an oil or a fat. It is defined as the number of milligrams of potassium hydroxide required to saponify one gram of the fat or oil *i.e.*, to convert one gram of the fat completely into glycerol and potassium soap.

Since there are three ester bonds in a molecule to hydrolyse, three equivalents of potassium hydroxide are needed to saponify one mole of any fat or oil. We can calculate the saponification number by taking the example of glycerol tripalmitate

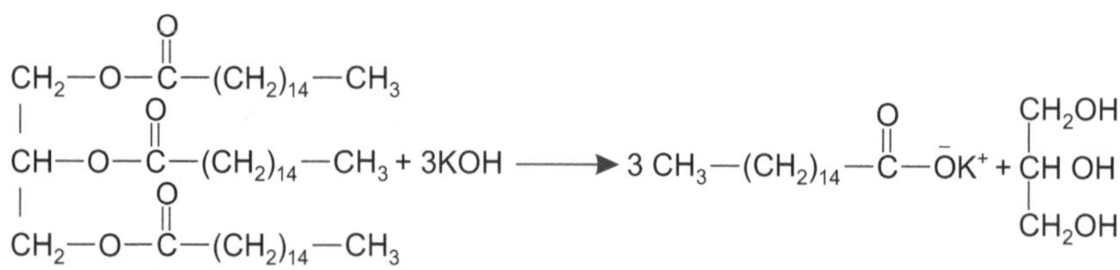

Glyceryl tripalmitate

Here, molecular weight of tripalmitate = 806

Three equivalents of potassium hydroxide are required to completely saponify

i.e., 806 g of fat = 168 g of KOH

= 168,000 milligrams of KOH

Therefore, 1 g of the fat will require $\frac{168,000}{806}$ milligrams of KOH

Saponification Number of glycerol tripalmitate = $\frac{168,000}{806}$ = 208.

By following the same procedure, we can determine the saponification number of any oil or fat. The higher the saponification number of a fat, the greater is the percentage of low-molecular glycerides it contains. In that case, the soap formed will be more soluble. This information is of special importance to soap makers. Table 6.4 gives the saponification numbers of some common commercial oils and fats.

TABLE 6.4 Saponificatioin Numbers of Common Oils and Fats

Oil or Fat	*Saponification Number*
Rapeseed oil	170 - 179
Corn oil	188 - 193
Olive oil	185 - 196
Soya bean oil	193
Linseed oil	192 - 195
Mutton tallow	192 - 195
Cottonseed oil	193 - 195
Peanut oil	190 - 196
Palm oil	196 - 205
Butter	220 - 233
Palm kernel oil	242 - 250
Coconut oil	246 - 260

Examination of the Table 6.4 shows that butter ranks with palm kernel oil and coconut oil as having a very high saponification number. This is due to the fact that its triglycerides contain appreciable quantities of myristic acid and small quantities of lauric acid. In butter, there is a high percentage of low molecular weight fatty acids, their sodium soaps are quite soluble in water. The high saponification number of coconut oil and palm kernel oil is due to large proportion of lauric acid and myristic acid that they contain. These oils, therefore, yield quite soluble soaps.

C. Iodine Number

We know oils and fats contain different saturated and unsaturated acids. When the fatty acid molecule contains the maximum of hydrogen possible, the acid is said to be a saturated fatty acid. It is saturated with respect to hydrogen. We have seen myristic, lauric, palmitic and stearic acids are such saturated acids. They are solid at ordinary temperatures. When the fatty acid molecule does not contain the maximum of hydrogen possible, the acid is said to be unsaturated fatty acid. It is unsaturated with respect to hydrogen. Such unsaturated acids are oleic, linoleic and linolenic acids. They are liquids at ordinary temperatures.

We have already discussed these unsaturated acids can be made to combine with hydrogen. This process is known as hydrogenation. However, unsaturated fatty acids can be made to combine with other substances instead of hydrogen. They may be made to take up iodine. This gives us a method to know the degree of unsaturation of a given fat or oil *i.e.*, by determining the amount of iodine that can be made to combine with the fat or oil. So, the degree of unsaturation in a fat or oil is expressed in terms of **Iodine Number** or **Iodine Value**. It is an index to the degree of unsaturation of the fat. Iodine number is defined as the number of grams of iodine that will add to 100 g of the fat or oil.

Table 6.5 gives the iodine number of some common oils and fats.

TABLE 6.5 Iodine Numbers of Common Oils and Fats

Oil or Fat	*Iodine Number*
Linseed oil	173 - 201
Tung oil	170
Whale oil	121 - 146
Soyabean oil	137 - 143
Sunflower oil	119 - 135
Corn oil	111 - 130
Cottonseed oil	108 - 110
Rapeseed oil	103 - 108
Peanut oil	83 - 100
Olive oil	79 - 88
Palm oil	51 - 57
Milk fat	26 - 50
Beef tallow	38 - 46
Palm kernel oil	13 - 17
Coconut oil	8 - 10

Examination of the above table shows that oils with the highest iodine numbers are very good drying oils *i.e.*, linseed and tung oil. As shown soya bean oil and sunflower oil are good from nutrition point of view as they contain appreciable amounts of unsaturated fatty acids.

We consider here example of glyceryl trioleate (triolein) to calculate its iodine number. Triolein adds with three molecules of iodine to give the addition product. The molecular weight of triolein is 884.

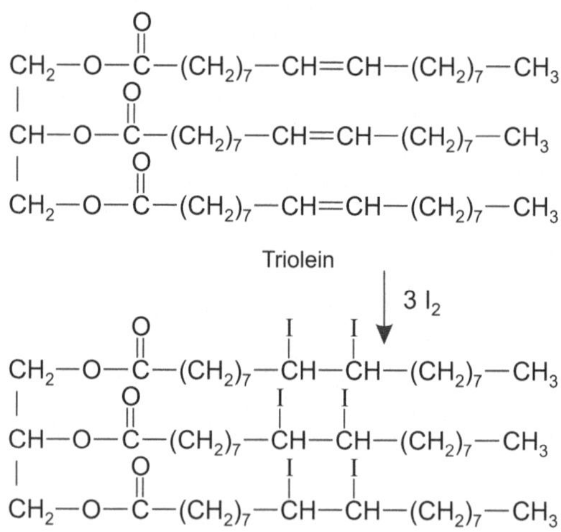

The above equation indicates that $6 \times 127 = 762$ g of iodine will add to 884 g of triolein.

The number of grams of iodine that will add to 100 g of triolein will be

$$= \frac{762 \times 100}{884}.$$

Therefore, iodine number of triolein

$$= \frac{760 \times 100}{884} = 86.$$

It is clear that the value of iodine number depends on the number of double bonds present in the acid component of the triglycerides. A high value of iodine number indicates that the glycerides contain a large number of double bonds, while a low value gives the presence of a few double bonds. The iodine number of triplamitate with no double bonds would be zero.

D. Acid Number

The acid number of an oil or fat tells us the amount of free fatty acids present in it. The acid number is expressed as the number of milligrams of potassium hydroxide required to neutralise the free fatty acids in one gram of the oil or fat. It is determined by dissolving a weighed amount of the fat in alcohol and titrating the solution against standard alkali solution using phenolphthalein as indicator. The acid number of a fat can give the extent of rancidity in a stored sample.

6.13 FLAVOUR CHANGES IN OILS AND FATS

A. Rnacidity

Fresh fats and oils are almost odourless and are neutral in reaction. However, they are quite unstable substances. When stored for any considerable length of time and the temperature is high and the air has free access to them, they deteriorate and spoil. The fats and oils acquire a peculiar disagreeable odour and flavour. They become rancid and are inedible. When the fat or oil develops a disagreeable odour, we say it is the onset of rancidity.

Rancidity is mainly caused by hydrolysis of the ester links and oxidation of the double bonds of the triglycerides. In this respect different oils and fats differ markedly. Some spoil more readily than others. Some oils and fats become rancid when kept for longer periods or exposed to high temperature. Onset of rancidity is also caused when they are kept in some metal containers.

We observe at our homes that when mustard oil is kept in brass container and the air has free access to it, the rancidity develops so fast that after a few days the colour changes from brown to green and the oil gives very offensive disagreeable odour.

The rancidity of a particular fat or oil is not necessarily the result of a long storage under unfavourtable conditions. The fat may have been spoiled and become rancid from the moment of its production. This will be true when the materials from which it was produced have undergone decomposition. Thus, the fat obtained from putrifying carcasses will be rancid. Similarly, oil obtained from fermented cottonseeds will have disagreeable odour. In other words, to obtain a good quality oil or fat, the raw material must be of good quality. It must be worked up speedily before giving it time to decompose. The fat thus obtained must be stored under favourable conditions and must be processed quickly and its consumption should not be too long delayed.

B. Different Types of Rancidity

As mentioned above there are two basic types of rancidity - hydrolytic rancidity and oxidative rancidity

1. Hydrolytic Rancidity

Fats and oils become rancid through the addition of water because water hydrolyses the bonds in the triglycerides causing it to breakdown into smaller compounds. The liberation of lower fatty acids by hydrolysis of ester links are responsible for this type of rancidity. Hydrolytic rancidity is particularly applicable to butter. Under moist and warm conditions the hydrolysis of the glycerides in butter liberates the odourous butyric acid, caproic acid, caprylic acid and capric acid. Microorganisms present in the air catalyse the hydrolytic process. The butter starts giving disagreeable odour even when hydrolysis may have proceeded to only a small extent. Thus, butter becomes unpalatable when even very small amounts of butyric acid are produced by the hydrolysis of triglycerides of butter.

In general, hydrolytic rancidity is caused by the lipase enzymes and heat. The rancidity has implications for deep-fat frying, because placing cold, wet food in heated frying oil introduces water and makes the oil prone to hydrolytic rancidity. On the other hand, fats that have not been heated are more prone to hydrolytic rancidity because the lipase enzymes have not yet been destroyed by heat. Room temperature is ideal for the activity of the lipase enzyme. If butter is left out at room temperature, it quickly decomposes. As mentioned above, butter's volatile fatty acids, such as butyric and caproic acids create a rancid odour. The long-chain fatty acids are also freed but they are not volatile and therefore, do not contribute to the odour of the rancid butter. Rancidity so caused can be prevented by keeping butter covered in a refrigerator.

2. Oxidative Rancidity

Fats and oils can also become rancid when they are exposed to the oxygen in the air. This occurs in triglycerides containing unsaturated fatty acids. The higher the degree of unsaturation, the more likely it is that the fat will be subject to oxidative rancidity. In oxidative rancidity, the double bonds of an unsaturated fatty acid react chemically with oxygen resulting in two or more smaller molecules, which in turn give rise to the formation of offensive aldehydes and acids.

Unlike hydrolytic rancidity, oxidative rancidity occurs in a series of steps. The initiation period is slow and is caused by light, high temperature, food particles in the frying oil and certain metals such as iron, copper and nickel. The initial stage is followed by quicker, irreversible and self-perpetuating chain reaction. Oxygen atoms attach to the carbon next to the double bond of the fatty acid, creating very reactive and unstable free radicals. These free radicals contribute to further breakdown of fats and oils into smaller molecules, resulting in unpleasant odours and off flavours. Once this process starts, it is difficult to stop. The free radicals generated by the reaction create more free radicals, which in turn keep producing free radicals until all the double bonds have been used in the process.

The addition of antioxidants will preserve edible oils and fats for long periods of storage. Antioxidants, found naturally in the fat or commercially added, inhibit oxidative rancidity and extend shelf life. Two antioxidants occurring in natural fats are vitamin E and ascorbic acid.

❑❑❑

INDEX

A

B

C

D

E

F

G

H

I

K

L

M

N

O

P

R

S

❑❑❑